Enzyme Handbook 12

Springer-Verlag Berlin Heidelberg GmbH

Attention all "Enzyme Handbook" users:

A file with the complete volume indexes Vols. 1 through 12 in delimited ASCII format is available for downloading at no charge from the Springer EARN mailbox. Delimited ASCII format can be imported into most databanks.

The file has been compressed using the popular shareware program "PKZIP" (Trademark of PKware inc., PKZIP is available from most BBS and shareware distributors).

This file is distributed without any expressed or implied warranty.

To receive this file send an e-mail message to:
SVSERV@DHDSPRI6.BITNET

The message must be:
GET /CHEMISTRY/ENZ_HB.ZIP

SVSERV is an automatic data distribution system. It responds to your message. The following commands are available:

HELP	returns a detailed instruction set for the use of SVSERV,
DIR *(name)*	returns a list of files available in the directory "name",
INDEX *(name)*	same as "DIR"
CD *<name>*	changes to directory "name",
SEND *<filename>*	invokes a message with the file "filename",
GET *<filename>*	same as "SEND".

D. Schomburg · D. Stephan (Eds.)
GBF– Gesellschaft für Biotechnologische Forschung

Enzyme Handbook

12

Class 2.3.2 – 2.4 Transferases

Springer

Professor Dr. Dietmar Schomburg
Dr. Dörte Stephan

GBF – Gesellschaft für Biotechnologische Forschung mbH
Mascheroder Weg 1
38124 Braunschweig
FRG

This collection of datasheets was generated from the database „BRENDA"

Library of Congress Cataloging-in-Publication Data

Enzyme handbook/ D. Schomburg, M. Salzmann (eds.). v. ⟨1–2,4–10⟩; 23 cm. Vols. 6–7 edited by D. Schomburg, M. Salzmann, D. Stephan. Vols. 9–10 edited by D. Schomburg, D. Stephan.
Loose-leaf.
Includes bibliographical references and indexes.
Contents: 1. Claas 4: Lyases – 2. Class 5: Isomerases. Class 6: Ligases – 4–5. Class 3: Hydrolases – 6. Class 1.2–1.4, Oxidoreductases – 7. Class 1.5–1.12, Oxidoreductases – 8. Class 1.13–1.97, Oxidoreductases – 9 Class 1.1, Oxidoreductases, EC 1.1.1.150 – EC 1.1.99.26 – v. 10. Class 1.1, Oxidoreductases, EC 1.1.1.150 – EC 1.1.99.26.

ISBN 978-3-642-47805-5   ISBN 978-3-642-61117-9 (eBook)
DOI 10.1007/978-3-642-61117-9

1. Enzymes-Handbooks, manuals, etc. I. Schomburg, D. (Dietmar)   II. Salzmann, M. (Margit)   III. Stephan, D. (Dörte)
QP601-E5158   1990
660'.634–dc20

© Springer-Verlag Berlin Heidelberg 1996
Originally published by Springer-Verlag Berlin Heidelberg New York Tokyo in 1996
Softcover reprint of the hardcover 1st edition 1996

Media conversion, printing and bookbinding: Brühlsche Universitätsdruckerei, Giessen
Production of the plasticfiles: Lux-Plastik oHG, Murnau
SPIN: 10076281   51/3020 - 5 4 3 2 1 0 - Printed on acid-free paper

Preface

Recent progress on enzyme immobilisation, enzyme production, coenzyme regeneration and enzyme engineering has opened up fascinating new fields for the potential application of enzymes in a large range of different areas. As more progress in research and application of enzymes has been made the lack of an up-to-date overview of enzyme molecular properties has become more apparent. Therefore, we started the development of an enzyme data information system as part of protein-design activities at GBF. The present book "Enzyme Handbook" represents the printed version of this data bank. In future a computer searchable version will be also available.

The enzymes in this Handbook are arranged according to the Enzyme Commission list of enzymes. Some 3000 "different" enzymes will be covered. Frequently enzymes with very different properties are included under the same EC number. Although we intend to give a representative overview on the characteristics and variability of each enzyme the Handbook is not a compendium. The reader will have to go to the primary literature for more detailed information. Naturally it is not possible to cover all the numerous literature references for each enzyme (for special enzymes up to 40000) if the data representation is to be concise as is intended.

It should be mentioned here that the literature data are extracted from literature and critically evaluated by qualified scientists. On the other hand the original authors' nomenclature for enzyme forms and subunits is retained as is their nomenclature for organisms and strains even if the organism is reclassified in the meantime. The cross references to the protein sequence data bank and to the Brookhaven protein 3D structure data bank are taken directly from their data files without further verification by the authors. In order to keep the tables concise redundant information is avoided as far as possible (e.g. if K_m values are measured in the presence of an obvious cosubstrate, only the name of the cosubstrate is given in parentheses as a commentary without reference to its specific role).

The authors are grateful to the following biologists and chemists for invaluable help in the compilation of data: Cornelia Munaretto, Dr. Astrid Beermann, Dr. Ida Schomburg and Dr. Astrid Haberz. In addition we would like to thank Mrs. C. Munaretto and Dr. I. Schomburg for the correction of the final manuscript.

Braunschweig, 1996

Dörte Stephan
Dietmar Schomburg

BRENDA – Compilation of Enzyme Data

To collect basic characteristics of enzymes – is that not a kind of archaic activity in the times of molecular biology and computer-aided data banks providing sequences of nucleic acids and proteins with little more delay than a few days as well as their three-dimensional structures? What should be the purpose of compiling turnover numbers, Michaelis constants, substrate specificities, sources, synonyms etc. of enzymes from sometimes remote publications? The answer sounds as simple as surprising: The aim of the compilation of data is to make use of the overwhelming abundance of structural knoweldge we owe to the new techniques of molecular biology.

Admittedly, it was not primarily enzymology which caused the explosion of knowledge in biology during the last decade. This was due to the advance of molecular biology which enabled us to isolate genes, to amplify them *ad libidum* and to elucidate their primary structure within days only. Also, the optimization and automatization of techniques for the analysis of macromolecules has provided detailed insights into a large variety of complex biomolecules nobody would have anticipated in the early seventies. Due to powerful computers it has now become feasible to propose fairly realistic models of macromolecules based solely on primary structures and homology considerations.

Nevertheless – or therefore – it appears as mandatory as rewarding to know the brave world of enzymology in which one had and often still has to come along without any detailed structural knowledge. We should not ignore that nature has not generated the multiplicity of structures, because it simply felt obliged to the principle of diversification or because it wanted to test our computing capacity to handle sequence data. It had to create new structures to cope with the steadily changing demands of a variable environment. Thus, amino acid sequences, folding of peptide chains and conformational details are only the technical tools of nature to catalyse specific biological functions. In consequence, *it is the functional profile of an enzyme which enables a biologist or physician to analyze a metabolic pathway and its disturbance; it is the substrate specificity of an enzyme which tells an analytical biochemist how to design an assay; it is the stability, specificity and efficiency of an enzyme which determines its usefulness in the biotechnical transformation of a molecule.* And the sum of all these functional data will have to be considered when the designer of artificial biocatalysts has to choose the optimum prototype to start with.

Unfortunately, it is by no means as simple to design (organize) a meaningful and systematic compilation of functional enzymological data as to enter sequences of amino acids or nucleotides into a data base. Functional data are less well defined, are never devoid of a trace of ambiguity, their selection remains inevitably subjective, and their complexity requires simplification. The present compilation of enzymological data, therefore, can and will not be a substitute for original publications but rather offer a key to the literature. But I do think that the Enzyme Handbook is indeed an excellent key to open or reopen the mysterious world of enzyme to all those who there have to find the solutions of their problems: to biologists, physicians, structural biochemists, biochemical analysts, biotechnologists and also to the molecular biologists.

Braunschweig, Spring 1993   Leopold Flohé

List of Abbreviations

A	adenosine	ER	endoplasmic reticulum
Ac	acetyl	Et	ethyl
ACP	acyl-carrier-protein	EXAFS	extended X-ray absorption fine structure
ADP	adenosine 5'-diphosphate		
Ala	alanine	FAD	flavin-adenine dinucleotide
All	allose	FMN	flavin mononucleotide (ribo-flavin 5'-monophosphate)
Alt	altrose		
AMP	adenosine 5'-monophosphate	FPLC	fast protein liquid chroma-tography
Ara	arabinose		
Arg	arginine	Fru	fructose
Asn	asparagine	Fuc	fucose
Asp	aspartic acid	G	guanosine
ATP	adenosine 5'-triphosphate	GABA	4-aminobutanoic acid
Bicine	N,N'-bis(2-hydroxyethyl) glycine	Gal	galactose
		GDP	guanosine 5'-diphosphate
C	cytidine	Glc	glucose
cal	calorie	GlcN	glucosamine
CDP	cytidine 5'-diphosphate	GlcNAc	N-acetylglucosamine
CDTA	trans-1,2-diaminocyclo-hexa-ne-N,N,N,N-tetra-aceticacid	Gln	glutamine
		Glu	glutamic acid
CHAPS	3-[(3-cholamidopropyl)-dimethylammonio]-1-propanesulfonate	Gly	glycine
		Glygly	glycylglycine
		GMP	guanosine 5'-monophosphate
CHAPSO	3-[(3-cholamidopropyl)-dimethylammonio]-2-hydroxy-1-propane-sulfonate		
		GSH	glutathione
		GSSG	oxidized glutathione
		GTP	guanosine 5'-triphosphate
CMP	cytidine 5'-monophosphate	Gul	gulose
CoA	coenzyme A	h	hour
CTP	cytidine 5'-triphosphate	H_4	tetrahydro
Cys	cysteine	HEPES	4-(2-hydroxyethyl)-1-piper-azineethane sulfonic acid
d	deoxy-		
D- and L-	prefixes indicating configuration	His	histidine
		HPLC	high performance liquid chromatography
Dap	diaminopimelic acid		
DFP	diisopropylfluorophosphate	Hyl	hydroxylysine
DNA	deoxyribonucleic acid	Hyp	hydroxyproline
DPN	diphosphopyridinium nucleotide (now NAD)	IAA	iodoacetamide
		Ig	immunoglobulin
DTNB	5,5'-dithiobis(2-nitrobenzoate)	Ile	isoleucine
DTT	dithiothreitol (i.e. Cleland's reagent)	Ido	idose
		IDP	inosine 5'-diphosphate
e	electron	IMP	inosine 5'-monophosphate
EC	number of enzyme in Enzyme Commission's system	ir	irreversible
		ITP	inosine 5'-triphosphate
E. coli	Escherichia coli	K_m	Michaelis constant
EDTA	ethylene diaminetetraacetate	L-	see D-
EGTA	ethylene glycol bis (β-amino-ethylether) tetraacetate	Leu	leucine
		Lys	lysine
EPR	electron paramagnetic resonance	Lyx	lyxose
		M	mol/l

m-	meta-	mRNA	messenger RNA
Man	mannose	rRNA	ribosomal RNA
MES	2-(N-morpholino)ethane	tRNA	transfer RNA
	sulfonate	Sar	N-methylglycine
Met	methionine		(sarcosine)
min	minute	SDS-PAGE	sodium dodecyl sulphate
MOPS	3-(N-morpholino)		polyacrylamide gel
	propane sulfonate		electrophoresis
Mur	muramic acid	Ser	serine
MW	molecular weight	SFK-525A	2-diethylaminoethyl-2,2-
NAD	nicotinamide-adenine		diphenylvalerate
	dinucleotide	sp.	species
NADH	reduced NAD	T	ribosylthymine
NADP	NAD phosphate	$t\frac{1}{2}$	time for half-completion
NADPH	reduced NADP		of reaction
NAD(P)H	indicates either NADH	Tal	talose
	or NADPH	TDP	ribosylthymine
NDP	nucleoside 5'-diphosphate		5'-diphosphate
NEM	N-ethylmaleimide	TEA	triethanolamine
Neu	neuraminic acid	TES	N-tris[hydroxymethyl]-
Nle	norleucine		methyl-2-amino-
NMN	nicotinamide		ethanesulfonic acid
	mononucleotide	THF	tetrahydrofolate
NMP	nucleoside	Thr	threonine
	5'-monophosphate	TMP	ribosylthymine
NTP	nucleoside 5'-triphosphate		5'-monophosphate
o-	ortho-	Tos-	tosyl-(p-toluenesulfonyl-)
OMP	orotidine 5-monophosphate	TPN	triphosphopyridinium
Orn	ornithine		nucleotide (now NADP)
p-	para-	Tris	tris(hydroxymethyl)-
PAPS	3'-phosphoadenylylsulfate		aminomethane
PCMB	p-chloro-mercuribenzoate	Trp	tryptophan
PEG	polyethylene glycol	TTP	ribosylthymine
PEP	phosphoenolpyruvate		5'-triphosphate
pH	$-\log_{10}[H^+]$	Tyr	tyrosine
Ph	phenyl	U	uridine
Phe	phenylalanine	U/mg	μmol/(mg·min)
PIXE	proton-induced	UDP	uridine 5'-diphosphate
	X-ray emission	UMP	uridine 5'-monophosphate
PMSF	phenylmethane-	UTP	uridine 5'-triphosphate
	sulfonylfluoride	UV	ultraviolet
Pro	proline	Val	valine
Q_{10}	factor for the change in	Xaa	symbol for an amino
	reaction rate for a 10°		acid of unknown consti-
	temperature increase		tution in peptide formula
r	reversible	XAS	X-ray absorption
Rha	rhamnose		spectroscopy
Rib	ribose	XTP	xanthosine 5'-triphosphate
RNA	ribonucleic acid	Xyl	xylose

Index

(Alphabetical order of Enzyme names)

EC-No.	Name	EC-No.	Name

2.4.1.20 Cellobiose phosphorylase
2.4.1.49 Cellodextrin phosphorylase
2.4.1.29 Cellulose synthase (GDP-forming)
2.4.1.12 Cellulose synthase (UDP-forming)
2.4.1.80 Ceramide glucosyltransferase
2.4.1.16 Chitin synthase
2.4.1.142 Chitobiosyldiphosphodolichol alpha-mannosyltransferase
2.4.1.177 Cinnamate beta-D-glucosyltransferase
2.4.1.111 Coniferyl-alcohol glucosyltransferase
2.4.1.114 2-Coumarate O-beta-glucosyltransferase
2.4.1.116 Cyanidin-3-rhamnosylglucoside 5-O-glucosyltransferase
2.4.1.85 Cyanohydrin beta-glucosyltransferase
2.4.1.19 Cyclomaltodextrin glucanotransferase
2.4.1.118 Cytokinin 7-beta-glucosyltransferase
2.4.2.23 Deoxyuridine phosphorylase
2.4.1.5   Dextransucrase
2.4.1.2   Dextrin dextranase
2.4.1.46 1,2-Diacylglycerol 3-beta-galactosyltransferase
2.4.1.157 1,2-Diacylglycerol 3-glucosyltransferase
2.4.1.104 o-Dihydroxycoumarin 7-O-glucosyltransferase
2.4.1.202 2,4-Dihydroxy-7-methoxy-2H-1,4-benzoxazin-3(4H)-one 2-D-glucosyltransferase
2.4.2.20 Dioxotetrahydropyrimidine phosphoribosyltransferase
2.4.1.26 DNA alpha-glucosyltransferase

2.4.1.27 DNA beta-glucosyltransferase
2.4.1.119 Dolichyl-diphosphooligosaccharide-protein glycotransferase
2.4.1.153 Dolichyl-phosphate alpha-N-acetylglucosaminyltransferase
2.4.1.117 Dolichyl-phosphate beta-glucosyltransferase
2.4.1.130 Dolichyl-phosphate-mannose-glycolipid alpha-mannosyltransferase
2.4.1.109 Dolichyl-phosphate-mannose-protein mannosyltransferase
2.4.1.83 Dolichyl-phosphate beta-D-mannosyltransferase
2.4.2.32 Dolichyl-phosphate D-xylosyltransferase
2.4.2.33 Dolichyl-xylosyl-phosphate-protein xylosyltransferase
2.4.2.27 dTDPdihydrostreptose-streptidine-6-phosphate dihydrostreptosyltransferase
2.4.1.185 Flavanone 7-O-beta-glucosyltransferase
2.4.2.25 Flavone apiosyltransferase
2.4.1.81 Flavone 7-O-beta-glucosyltransferase
2.4.1.159 Flavonol-3-O-glucoside L-rhamnosyltransferase
2.4.1.91 Flavonol 3-O-glucosyltransferase
2.4.2.35 Flavonol-3-O-glycoside xylosyltransferase
2.4.1.100 1,2-beta-Fructan 1^F-fructosyltransferase
2.4.1.40 Fucosylgalactose alpha-N-acetylgalactosaminyltransferase
2.4.1.37 Fucosylglycoprotein 3-alpha-galactosyltransferase

EC-No.	Name	EC-No.	Name
2.4.1.67	Galactinol-raffinose galactosyltransferase	2.4.1.134	Galactosylxylosylprotein 3-beta-galactosyltransferase
2.4.1.82	Galactinol-sucrose galactosyltransferase	2.4.1.136	Gallate 1-beta-glucosyltransferase
2.4.1.205	Galactogen 6beta-galactosyltransferase	2.4.1.62	Ganglioside galactosyltransferase
2.4.1.184	Galactolipid galactosyltransferase	2.4.1.176	Gibberellin beta-D-glucosyltransferase
2.4.1.69	Galactoside 2-alpha-L-fucosyltransferase	2.4.1.88	Globoside alpha-N-acetylgalactosaminyltransferase
2.4.1.65	Galactoside 3(4)-L-fucosyltransferase	2.4.1.154	Globotriosylceramide beta-1,6-N-acetylgalactosaminyltransferase
2.4.1.152	Galactoside 3-fucosyltransferase	2.4.1.18	1,4-alpha-Glucan branching enzyme
2.4.99.4	beta-Galactoside alpha-2,3-sialyltransferase	2.4.1.24	1,4-alpha-Glucan 6-alpha-glucosyltransferase
2.4.99.1	beta-Galactoside alpha-2,6-sialyltransferase	2.4.1.25	4-alpha-Glucanotransferase
2.4.1.163	beta-Galactosyl-N-acetylglucosaminylgalactosylglucosylceramide beta-1,3-acetylglucosaminyltransferase	2.4.1.97	1,3-beta-D-Glucan phosphorylase
		2.4.1.113	alpha-1,4-Glucan-protein synthase (ADP-forming)
2.4.1.164	Galactosyl-N-acetylglucosaminylgalactosylglucosylceramide beta-1,6-N-acetylglucosaminyltransferase	2.4.1.112	alpha-1,4-Glucan-protein synthase (UDP-forming)
		2.4.1.34	1,3-beta-Glucan synthase
		2.4.1.183	alpha-1,3-Glucan synthase
2.4.1.87	beta-D-Galactosyl-N-acetylglucosaminylglycopeptide alpha-1,3-galactosyltransferase	2.4.1.32	Glucomannan 4-beta-mannosyltransferase
		2.4.1.86	Glucosaminylgalactosylglucosylceramide beta-galactosyltransferase
2.4.99.5	Galactosyldiacylglycerol alpha-2,3-sialyltransferase	2.4.1.28	Glucosyl-DNA beta-glucosyltransferase
2.4.1.79	Galactosylgalactosylglucosylceramide beta-D-acetylgalactosaminyltransferase	2.4.1.17	Glucuronosyltransferase
		2.4.1.175	Glucuronyl-N-acetylgalactosaminylproteoglycan beta-1,4-N-acetylgalactosaminyltransferase
2.4.1.135	Galactosylgalactosylxylosylprotein 3-beta-glucuronosyltransferase		
2.4.1.146	beta-1,3-Galactosyl-O-glycosyl-glycoprotein beta-1,3-N-acetylglucosaminyltransferase	2.4.1.174	Glucuronylgalactosylproteoglycan beta-1,4-N-acetylgalactosaminyltransferase
2.4.1.102	beta-1,3-Galactosyl-O-glycosyl-glycoprotein beta-1,6-N-acetylglucosaminyltransferase	2.3.2.5	Glutaminyl-peptide cyclotransferase

EC-No.	Name	EC-No.	Name

2.3.2.4   gamma-Glutamylcyclo-
          transferase
2.3.2.1   D-Glutamyltransferase
2.3.2.2   gamma-Glutamyltrans-
          ferase
2.3.2.15 Glutathione gamma-glut-
          amylcysteinyltransferase
2.4.1.96 sn-Glycerol-3-phosphate
          1-galactosyltransferase
2.4.1.137 sn-Glycerol-3-phosphate
          2-alpha-galactosyltrans-
          ferase
2.4.1.11 Glycogen (starch) synthase
2.4.1.186 Glycogenin glucosyltrans-
          ferase
2.4.1.131 Glycolipid 2-alpha-mannosyl-
          transferase
2.4.1.132 Glycolipid 3-alpha-mannosyl-
          transferase
2.4.1.122 Glycoprotein-N-acetyl-
          galactosamine 3-beta-
          galactosyltransferase
2.4.1.68 Glycoprotein 6-alpha-L-
          fucosyltransferase
2.4.1.74 Glycosaminoglycan
          galactosyltransferase
2.4.2.15 Guanosine phosphorylase
2.4.1.48 Heteroglycan alpha-
          mannosyltransferase
2.4.1.197 High-mannose-oligosaccha-
          ride beta-1,4-N-acetyl-
          glucosaminyltransferase
2.4.1.45 2-Hydroxyacylsphingosine
          1-beta-galactosyltrans-
          ferase
2.4.1.181 Hydroxyanthraquinone
          glucosyltransferase
2.4.1.194 4-Hydroxybenzoate 4-O-
          beta-D-glucosyltransferase
2.4.1.126 Hydroxycinnamate
          4-beta-glucosyltransferase
2.4.1.158 13-Hydroxydocosanoate
          13-beta-glucosyltrans-
          ferase
2.4.1.178 Hydroxymandelonitrile
          glucosyltransferase

2.4.2.8   Hypoxanthine phospho-
          ribosyltransferase
2.4.1.121 Indole-3-acetate beta-
          glucosyltransferase
2.4.2.34 Indolylacetylinositol
          arabinosyltransferase
2.4.1.156 Indolylacetyl-myo-inositol
          galactosyltransferase
2.4.1.123 Inositol 1-alpha-galactosyl-
          transferase
2.4.1.200 Inulin fructotransferase
          (depolymerizing, difructo-
          furanose-1,2':2',1-dianhy-
          dride forming)
2.4.1.93 Inulin fructotransferase
          (depolymerizing, difructo-
          furanose-1,2':2,3'-dianhy-
          dride-forming)
2.4.1.9   Inulosucrase
2.4.1.170 Isoflavone 7-O-glucosyl-
          transferase
2.4.1.106 Isovitexin beta-glucosyl-
          transferase
2.4.1.22 Lactose synthase
2.4.1.206 Lactosylceramide 1,3-N-ace-
          tyl-beta-D-glucosaminyl-
          transferase
2.4.1.179 Lactosylceramide beta-1,3-
          galactosyltransferase
2.4.99.9 Lactosylceramide
          alpha-2,3-sialyltransferase
2.4.99.11 Lactosylceramide alpha-
          2,6-N-sialyltransferase
2.4.1.31 Laminaribiose phosphory-
          lase
2.3.2.6   Leucyltransferase
2.4.1.10 Levansucrase
2.4.1.63 Linamarin synthase
2.4.1.182 Lipid-A-disaccharide
          synthase
2.4.1.56 Lipopolysaccharide N-ace-
          tylglucosaminyltransferase
2.4.1.180 Lipopolysaccharide N-ace-
          tylmannosaminouronosyl-
          transferase

EC-No.	Name	EC-No.	Name
2.4.1.169	Xyloglucan 6-xylosyltrans-ferase	2.4.1.203	Zeatin O-beta-D-glucosyl-transferase
2.4.1.133	Xylosylprotein 4-beta-galactosyltransferase	2.4.1.204	Zeatin O-beta-D-xylosyl-transferase

1 NOMENCLATURE

EC number
    2.3.2.1

Systematic name
    Glutamine:D-glutamyl-peptide 5-glutamyltransferase

Recommended name
    D-Glutamyltransferase

Synonyms
    D-Glutamyl transpeptidase
    Glutamyltransferase, D-
    D-gamma-Glutamyl transpeptidase

CAS Reg. No.
    9030-02-8

2 REACTION AND SPECIFICITY

Catalysed reaction
    D-(or L-)Glutamine + D-glutamyl-peptide →
    → NH_3 + 5-glutamyl-D-glutamyl-peptide

Reaction type
    Aminoacyl group transfer

Natural substrates

Substrate spectrum
    1 D-Glutamine + D-glutamyl-peptide (r) [1]
    2 L-Glutamine + D-glutamic acid [1]

Product spectrum
    1 NH_3 + 5-glutamyl-D-glutamyl-peptide [1]
    2 NH_3 + gamma-glutamyl-dipeptide [1]

Inhibitor(s)

Cofactor(s)/prosthetic group(s)/activating agents

Metal compounds/salts

Turnover number (min^{-1})

Specific activity (U/mg)

K$_m$-value (mM)

pH-optimum
8.5 [1]

pH-range

Temperature optimum (°C)

Temperature range (°C)

3 ENZYME STRUCTURE

Molecular weight

Subunits

Glycoprotein/Lipoprotein
–

4 ISOLATION/PREPARATION

Source organism
Bacillus subtilis [1]

Source tissue
Cell [1]

Localization in source
Soluble (exoenzyme) [1]

Purification

Crystallization
–

Cloned
–

Renatured
–

5 STABILITY

pH

Temperature (°C)

Oxidation

Organic solvent

General stability information

Storage

6 CROSSREFERENCES TO STRUCTURE DATABANKS

PIR/MIPS code

Brookhaven code

7 LITERATURE REFERENCES

[1] Williams, W.J., Litwin, J., Thorne, C.B.: J. Biol. Chem.,212,427–438 (1955)

1 NOMENCLATURE

EC number
2.3.2.2

Systematic name
(5-L-Glutamyl)-peptide:amino-acid 5-glutamyltransferase

Recommended name
gamma-Glutamyltransferase

Synonyms
Glutamyl transpeptidase
Glutamyltransferase, gamma-
alpha-Glutamyl transpeptidase
gamma-Glutamyl peptidyltransferase
gamma-Glutamyl transpeptidase
gamma-GPT
gamma-GT
gamma-GTP
L-gamma-Glutamyl transpeptidase
L-gamma-Glutamyltransferase
L-Glutamyltransferase
More (cf. EC 2.3.2.14)

CAS Reg. No.
9046-27-9

---

2 REACTION AND SPECIFICITY

Catalysed reaction
(5-L-Glutamyl)-peptide + an amino acid →
→ peptide + 5-L-glutamyl amino acid (mechanism [11, 25, 27, 29, 33–35, 58], structure [20, 35, 43])

Reaction type
Aminoacyl group transfer

Natural substrates
Glutathione + an amino acid (involved in gamma-L-glutamyl cycle of glutathione metabolism [35, 36], and polyglutamic acid synthesis [47]) [35, 36, 47]
Leukotriene D_4 + glutamate [32]

Substrate spectrum

1 (5-L-Glutamyl)-peptide + acceptor (Glu-donors are gamma-L-Glu-p-nitro-anilide [36, 42, 48, 58], GSH [33, 34, 36, 42, 48, 51, 55, 58] and its S-sub-stituted derivatives [36], S-acetophenoneglutathione [33, 42] and S-acetyl-glutathione [42], glutathione disulfide [33–35, 55], gamma-L-Glu-anilide [36, 58], ophthalmic acid, poly-gamma-Glu-derivatives, L-gamma-Glu-7-amino-4-methylcoumarin [33], gamma-ethyl-Glu and beta-benzyl-Glu (hydrolyzed by pig enzyme) [33], leukotriene C [32, 33], gamma-L-Glu-naphthylamide [31, 36], gamma-L-Glu-3-sulfonic-4-nitroani-lide [36], donor specificity [47, 48, 55, 56, 58], strict L-stereospecificity of gamma-Glu-acceptor site [35, 38], donor binding site reacts with L- and to a lesser extent D-enantiomer [35, 38], acceptor specificity (rat kidney en-zyme [34]) [27, 28, 34–36, 40, 42, 46, 51, 54], acceptors are amino acids or dipeptides [27, 30, 36, 42], dipeptides are better acceptors than free amino acids [47], concurrent reactions: autotranspeptidation (with an-other donor molecule as acceptor) and hydrolase reaction (with H_2O as acceptor) [34, 47–58], poor substrates are beta-substituted amino acids [40], free amino acids [57], no donors are Glu, alpha-L-Glu-L-Ala [47, 55], L-Asn, L-homoglutamine, beta-aminoglutaryl-L-alpha-aminobutyrate, L-gamma-(beta-methyl)Glu-L-alpha-aminobutyrate, leucyl-beta-naphthyl-amide, N-(alpha-L-Glu)-beta-naphthylamide, L-gamma-(N-methyl)Glu-L-alpha-aminobutyrate [33]) [27, 28, 30–36, 38, 40, 42, 46–59]

2 gamma-L-Glutamyl-p-nitroanilide + L-glutamine (best acceptor (high mole-cular weight enzyme variant [7]) [3, 7], glutamyl transfer at 43% the rate of the reaction with glycylglycine [24]) [3, 6, 7, 11, 15, 18, 24, 27, 28, 30, 40, 42, 51, 55]

3 gamma-L-Glutamyl-p-nitroanilide + L-methionine (best acceptor [27], glutamyl transfer at 39% [24], 44% (viral enzyme) [26] the rate of the re-action with glycylglycine, L-Met or D-Met [30]) [3, 6, 11, 15, 18, 24, 26–28, 30, 40, 42, 51, 55]

4 gamma-L-Glutamyl-p-nitroanilide + L-cystine (best acceptor (low molecu-lar weight enzyme variant [7]) [6, 7, 33]) [6, 7, 18, 33, 36, 40, 54]

5 gamma-L-Glutamyl-p-nitroanilide + glycylglycine (best acceptor [18, 24, 27, 36]) [6, 11, 12, 15, 18, 20, 22–24, 26–28, 30, 35, 36, 39, 42, 45–51, 53–55, 57]

6 gamma-L-Glutamyl-p-nitroanilide + glycine (glutamyl transfer at 17.2% [24], 19% (viral enzyme) [26] the rate of the reaction with glycylglycine, glycyl-tri- to hexapeptides can act as acceptors [27, 30]) [24, 26, 27, 30, 51, 54, 55]

7 gamma-L-Glutamyl-p-nitroanilide + citrulline [11, 18]

8 gamma-L-Glutamyl-p-nitroanilide + L-alanine (glutamyl transfer at 32% the rate of the reaction with glycylglycine [24], L-Ala or D-Ala [30]) [6, 18, 24, 27, 28, 30, 42, 51, 54, 55]

9 gamma-L-Glutamyl-p-nitroanilide + L-aspartate (glutamyl transfer at 8.6%
the rate of the reaction with glycylglycine [24]) [24, 30, 40, 51, 54, 55]

10 gamma-L-Glutamyl-p-nitroanilide + L-asparagine [30, 42, 54]

11 gamma-L-Glutamyl-p-nitroanilide + L-cysteine [28, 40, 51, 54, 55]

12 gamma-L-Glutamyl-p-nitroanilide + L-lysine (glutamyl transfer at 26% the
rate of the reaction with glycylglycine [24]) [18, 24, 28, 30, 51, 54, 55]

13 gamma-L-Glutamyl-p-nitroanilide + L-arginine (glutamyl transfer at 30%
the rate of the reaction with glycylglycine [24]) [18, 24, 28, 30, 51, 54]

14 gamma-L-Glutamyl-p-nitroanilide + L-glutamate [18, 27, 28, 30, 42, 51,
54, 55]

15 gamma-L-Glutamyl-p-nitroanilide + L-proline [27, 51, 55]

16 gamma-L-Glutamyl-p-nitroanilide + L-histidine (glutamyl transfer at 8.6%
the rate of the reaction with glycylglycine [24]) [24, 30, 51, 54]

17 gamma-L-Glutamyl-p-nitroanilide + L-hydroxyproline (glutamyl transfer at
17% the rate of the reaction with glycylglycine [24]) [24, 51]

18 gamma-L-Glutamyl-p-nitroanilide + L-isoleucine (glutamyl transfer at 17%
the rate of the reaction with glycylglycine [24]) [24, 30, 54, 55]

19 gamma-L-Glutamyl-p-nitroanilide + L-phenylalanine (glutamyl transfer at
26% the rate of the reaction with glycylglycine [24]) [24, 28, 30, 51, 54,
55]

20 gamma-L-Glutamyl-p-nitroanilide + L-serine (glutamyl transfer at 28% the
rate of the reaction with glycylglycine [24]) [24, 27, 42, 51, 54, 55]

21 gamma-L-Glutamyl-p-nitroanilide + L-threonine (glutamyl transfer at 20%
the rate of the reaction with glycylglycine [24]) [24, 51, 55]

22 gamma-L-Glutamyl-p-nitroanilide + L-valine (glutamyl transfer at 20% the
rate of the reaction with glycylglycine [24]) [24, 30, 51, 54, 55]

23 gamma-L-Glutamyl-p-nitroanilide + L-tryptophan [51, 54, 55]

24 gamma-L-Glutamyl-p-nitroanilide + aminobutyrate (alpha- [27, 51], beta-
or gamma-enantiomer of aminobutyrate [27]) [27, 51]

25 gamma-L-Glutamyl-p-nitroanilide + L-leucine (glutamyl transfer at 20%
the rate of the reaction with glycylglycine [24]) [24, 30, 51, 54]

26 gamma-L-Glutamyl-p-nitroanilide + gamma-L-glutamyl-p-nitroanilide (autotranspeptidation, L- or (to a lesser extent) D-enantiomer [30]) [7, 23,
27, 30, 32]

27 gamma-L-Glutamine + gamma-L-glutamine [47]

28 Glutathione + acceptor (best donor [33], acceptor: hydroxylamine [51])
[11, 17, 26, 30, 31, 33, 34, 36, 42, 51]

29 Glutathione + H_2O (hydrolase reaction, concurrent to (auto-)transpepti-
dation) [29, 30, 35, 41]

30 Glutathione + glutathione (autotranspeptidation) [29, 30, 32, 34, 35]

31 Leukotriene C_4 + acceptor (i.e. glutathione containing derivative of ara-
chidonic acid, acceptors are H_2O, amino acids or dipeptides [33]) [32,
33]

32 Lentinic acid + amino acid (best acceptor is S-alkylcysteine sulfoxide) [51, 52]
33 gamma-L-Glutamyl-7-amino-4-methylcoumarin + glycylglycine [19, 21, 33, 35, 40]
34 gamma-L-Glutamyl-3-carboxy-4-nitroanilide + glycylglycine (i.e. glucana) [25, 36–38]
35 gamma-L-Glutamyl-3-carboxy-4-nitroanilide + gamma-L-glutamyl-3-carboxy-4-nitroanilide [37, 38]
36 L-alpha-Methyl-gamma-glutamyl-L-alpha-aminobutyrate + acceptor (hydrolysis in the absence of acceptor, methyl group at alpha-C prevents autotranspeptidation) [29]
37 gamma-L-Glutamyl-cis-3-amino-L-proline + acceptor [55]

Product spectrum
1 Peptide + 5-L-glutamyl amino acid [47, 48, 55, 56, 58]
2 ?
3 ?
4 ?
5 gamma-Glutamylglycylglycine + p-nitroaniline [27, 35, 36]
6 ?
7 ?
8 ?
9 ?
10 ?
11 ?
12 ?
13 ?
14 ?
15 ?
16 ?
17 ?
18 ?
19 ?
20 ?
21 ?
22 ?
23 ?
24 ?
25 ?
26 gamma-L-Glutamyl-gamma-L-glutamyl-p-nitroanilide + p-nitroaniline [7, 23, 27, 29, 30, 32]
27 gamma-Poly-glutamic acid + glutamate [47]
28 Cysteinylglycine + gamma-L-glutamyl-acceptor [17]
29 Cysteinylglycine + glutamate [29, 30, 35]
30 Cysteinylglycine + gamma-glutamyl-glutathione [29, 30, 35]

31 Leukotriene D_4 + glutamyl-acceptor [32, 33]
32 Desglutamyl lentinic acid + L-gamma-glutamylamino acid [51, 52]
33 ?
34 L-gamma-Glutamyl-glycylglycine + 3-carboxy-4-nitroaniline [37, 38]
35 L-gamma-Glutamyl-L-gamma-glutamyl-3-carboxy-4-nitroanilide + 3-carb-oxy-4-nitroaniline [37, 38]
36 L-alpha-Methyl-gamma-glutamyl acceptor + L-alpha-aminobutyrate [29]
37 L-gamma-Glutamyl-acceptor + cis-3-amino-L-proline [55]

Inhibitor(s)

Phenobarbital (inactivation, serine/borate or GSH protects, maleate protects slightly, kinetics [22]) [22, 33]; Thiobarbituric acid (inactivation, serine/bo-rate or GSH protects, maleate protects slightly, kinetics [22]) [22, 33]; 6-Diazo-5-oxo-L-norleucine (i.e. DON, strong, irreversible modification of glutamyl-binding site of light subunit [39, 42], maleate accelerates inactivation [39], S-methylglutathione [39], gamma-glutamyl-donor [42] protect transpeptidase and hydrolase activity, not 5-carbon derivative [42]) [20, 23, 34, 35, 39, 42, 48, 50]; L-(alphaS,5S)-alpha-Amino-3-chloro-4,5-dihydro-5-iso-xazole acetic acid (i.e. AT-125, strong, irreversible, 1 mM [48]) [34, 35, 48]; 5-Iodoacetamidofluorescein [23, 39]; L-Azaserine (inactivation [35], trans-peptidase and hydrolase reaction, glutamyl-donor protects [42], not 5-car-bon derivative [42]) [23, 34, 35, 39, 42, 48]; Antiserum to rat kidney glutamyl-transferase [23]; IAA (irreversible [34], not [17]) [7, 26, 34, 50]; PCMB (weak [19, 21], not [17, 34, 48, 55]) [19, 21, 51]; Iodoacetate (weak [41, 48, 50], not [17]) [41, 48, 50]; p-Hydroxymercuribenzoate (weak) [41]; DTNB (weak [17], not [34]) [17, 24]; NEM (weak [24, 41], not [17, 34, 48, 50]) [24, 41]; PMSF (weak [48]) [35, 48]; 1-Chloro-3-tosylamido-7-amino-2-heptanone (weak) [48]; L-1-Tosylamido-2-phenylethylchloromethylketone [48]; L-gamma-Glucana (enzyme from human tissues, not serum) [36, 38]; L-alpha-Glucana (weak) [38]; Glycine (human [3, 36], weak [50]) [3, 36, 50]; Glutathione (L-Glu-p-nitroanilide as substrate [18], not [50]) [18, 26, 28, 41]; L-Alanine [48, 50]; L-Glutamine (not D-) [48, 50]; beta-Chloro-L-alanine [50]; Aminooxyacetate (hydrolase reaction) [50]; Glycylglycine (high concentra-tion, with L-Glu-p-nitroanilide, not glutathione as donor [11], human [36]) [11, 36]; L- or D-Serine/borate (1:1 mixture, strong [51], reversible [23]) [23, 24, 26, 32, 34, 35, 48, 50–52, 55]; Amino acids (overview, hydrolase reaction [58]) [34, 50, 58]; Tris(hydroxymethyl)aminomethane [38]; 1-gamma-L-Glutamyl-2-(2-carboxyphenyl)hydrazine (i.e. anthglutin, strong, in vivo and in vitro, p-derivative less effective [35], transpeptidation, kinetics [53]) [35, 53]; alpha-Ketoglutarate-gamma-glutamylhydrazone (i.e. GSH-analogue) [26]; Phenylhydrazines [34]; gamma-Glutamylhydrazones [34]; Sulfophthalein derivatives [34]; Sulfobromophthalein derivatives [34]; Bromocresol green [34]; Free bile acids and their glycine and taurine conju-gates [35]; Acetazolamide [34] Maleate (transpeptidase, not hydrolase reaction [27, 35]) [7, 15, 27, 34, 35]; Hippurate (transpeptidase, not hydrolase reaction) [35]; Zn^{2+} (strong [51], reversible by EDTA [21]) [17, 19, 21, 24, 51]; Hg^{2+} (strong [51], hydrolase, not transferase reaction [48]) [24, 48, 50, 51]; Co^{2+} (weak) [41]; Cu^{2+} (strong) [51]; Pb^{2+} [54]; Mg^{2+} (above 0.1 M [11], 0.15 M [15], not [19, 21]) [11, 15]; Ca^{2+} (0.15 M [15], not [3]) [15]; Na^+ (above 0.1 M [11], 0.1 M [15]) [11, 15]; NH_4^+ (weak) [41]; Phosphate [51, 57]; EDTA (weak [24], not [17, 19, 21]) [24]; More (product inhibiton [56], no inhibiton by phorbol ester, polyamines, cAMP [3], 5-amino-2-nitrobenzoic acid [38], L-Glu, L-Ser, L-Asp, L-alpha-aminobutyrate [48], DTT, 2-mer-captoethanol [48, 50], glutathione disulfide, cysteine, cystine [50]) [3, 38, 48, 50, 56]

Cofactor(s)/prosthetic group(s)/activating agents

Metal compounds/salts

Mg^{2+} (activation, transpeptidase (not hydrolase [17, 30, 55]), slight [17], not [21, 41, 46]) [17, 30]; Ca^{2+} (activation, transpeptidase (not hydrolase [17, 30, 55]), slight [17], not [41, 46]) [17, 30]; K^+ (activation, transpeptidase (not hydrolase [30, 55]), not [41, 46]) [30]; Na^+ (activation, transpeptidase (not hydrolase [30, 55]), not [41, 46]) [30]; Li^+ (activation, not [41, 46]) [30]; Carb-oxylic acids (activation, e.g. sodium citrate [59]) [58, 59]; Hippurate (stimu-lates hydrolase reaction, inhibits transpeptidation) [35]; Maleate (stimulates hydrolase reaction, inhibits transpeptidase reaction, with L-Glu as substrate [42], not [55]) [33, 35, 42]; Free bile acids (and their glycine and taurine conjugates [35], activation) [33, 35]; More (no activation by Zn^{2+} and Mn^{2+} [41, 46], phosphate, arsenate, carbonate [55]) [41, 46, 55]

Turnover number (min^{-1})

69000 (p-nitroanilide, human); 75000 (p-nitroanilide, rat) [42]

Specific activity (U/mg)

More [3, 19, 21, 23, 25, 48, 51, 54, 55, 57, 58]; 6.0–7.7 (viral enzyme) [26]; 16 [39, 50]; 21.2 (milk fat-globule membrane enzyme) [26]; 62.9 [47]; 75.7 [44]; 81 [45]; 104.5 [24]; 120 (hepatoma [43]) [43, 56]; 165 [40]; 250 (liver) [43]; 407 [27]; 423 [45]; 510 (without acceptor) [30]; 630 [23]; 792 (rat) [33]; 810 [35]; 956 [28]; 1320 [18]

K_m-value (mM)

More (kinetic study of renatured large subunit [20], kinetic study [25, 29, 33, 56], kinetic data [32], hydrolysis data [55]) [20, 25, 29, 32, 33, 55, 56]; 0.005–0.0068 (L-Glu-p-nitroanilide, hydrolase reaction) [35, 54]; 0.0056 (leu-kotriene C) [32]; 0.0057 (glutathione) [32]; 0.029–0.035 (L-cystine, rat kid-ney enzyme [34], glutathione [48], D-Glu-p-nitroanilide [35], hydrolase reac-tion [35, 48]) [34, 35, 48]; 0.068 (L-Glu-p-nitroanilide, hydrolase reaction) [48]; 0.13–0.18 (L-Glu-p-nitroanilide [54], GSH [50], L-Cys-bis-glycine, rat kidney enzyme [34]) [34, 50, 54]; 0.21–0.32 (L-Glu-p-nitroanilide [3],

L-Glu-Gly, L-Met-Gly, rat kidney enzyme [34]) [3, 34]; 0.39–0.49
(gamma-Glu-donor [19], gamma-Glu-7-amino-4-methylcoumarin [21],
L-Glu-p-nitroanilide [50]) [19, 21, 50]; 0.65 (L-Glu-3-carboxy-4-nitroanilide)
[38]; 0.67 (L-Glu-p-nitroanilide or GSH) [26]; 0.76 (L-Gln, rat kidney enzyme)
[34]; 0.80–1.33 (L-Glu-p-nitroanilide (+ glycylglycine [15], without any acceptor [28])) [12, 15, 17, 18, 23, 24, 28, 39, 41, 45]; 1.09
(L-Glu-3-carboxy-4-nitroanilide [1], L-Cys-Gly, rat kidney enzyme [34]) [1,
34]; 2.9–2.96 (glycylglycine, rat kidney enzyme [34]) [34, 50]; 3.61
(D-Glu-3-carboxy-4-nitroanilide) [38]; 4.0–4.7 (glycylglycine [3], L-Met, rat
kidney enzyme [34]) [3, 34]; 7.6 (glycylglycine) [24]; 10–12.4 (glycylglycine)
[12, 45]; 32 (L-Phe) [50]; 48 (glycylglycine) [54]; 210 (L-Arg) [48]; 590 (glycylglycine) [48]

pH-optimum
    More (several isozymic forms with different content of sialic acid [42], 12
    isozymes with pI-values ranging from 5 to 8 [35], pI: 3.0 [17], 3.4–3.45 [5],
    3.63 [12], 3.85 [27], 4.5 [41], 6.0 [1]) [1, 5, 12, 17, 27, 35, 41, 42]; 5.0 (hy-
    drolase reaction) [4]; 6.0 (hydrolase reaction) [50]; 6.5 (hydrolase reaction)
    [52]; 7.5–8.0 (transpeptidation) [50]; 7.6 [51]; 7.9 [38]; 8–9 [54]; 8.0 [23, 39];
    8.1 [46]; 8.2 (transpeptidation, isoform II and III [55]) [24, 41, 55]; 8.2–8.5
    [1, 2]; 8.5 (with acceptor [27]) [27, 47]; 8.5–9.0 [57]; 8.6 (L-Glu-p-nitroanilide
    + glycylglycine [7], transpeptidation, isoform I [55]) [7, 55]; 8.6–9.0 (hydro-
    lase reaction, isoform III) [55]; 8.8 (transpeptidation) [30, 48]; 8.8–9.0 [17];
    9.0–9.2 (isoform I) [55]; 9.0 (broad [27], transpeptidation [52], without ac-
    ceptor [27], hydrolase reaction, isoform II [55]) [25, 27, 52, 55]; 9.0–9.2 (hy-
    drolase reaction, isoform I) [55]; 9.2 [4]; 9.4 (L-Glu-p-nitroanilide +
    L-Glu-p-nitroanilide) [7]; 9.5 (hydrolase reaction [30, 48]) [30, 48, 58]; 10.0
    (hydrolase reaction) [25]

pH-range
    6.5–9.5 (about half-maximal activity at pH 6.5 and 9.5) [51]; 7.3–8.5 (about
    half-maximal activity at pH 7.3, maximal activity at pH 8.5) [27]; 7.5–9.0
    (about half-maximal activity at pH 7.5, about 90% of maximal activity at pH
    9.0) [47]; 8.1–9.6 (about 70% of maximal activity at pH 8.1, about 75% of
    maximal activity at pH 9.6) [17]

Temperature optimum (°C)
    25 (assay at) [28]; 30 (assay at) [25, 38, 57]; 37 (assay at [17, 21–24, 26,
    27, 29, 30, 32, 33, 39–42, 44, 46–48, 52, 53, 58]) [50]; 45 [54]; 60 [47]

Temperature range (°C)
    45–80 (about half-maximal activity at 45°C and about 70% of maximal activi-
    ty at 80°C) [47]

3 ENZYME STRUCTURE

Molecular weight

More (multiple forms of different molecular weights [10], in the presence of Triton X-100: molecular aggregation [21, 39, 41], protease solubilized enzymes have a reduced MW [39], viral enzyme of at least MW 400000 [26], amino acid composition [17, 23, 28, 41, 44], immunological characterization [44]) [2, 10, 17, 18, 21, 23, 26, 28, 39, 41, 44]

54000 (human, gel filtration) [1]

57000 (E. coli isoform B, gel filtration) [48]

58000 (E. coli isoform A, gel filtration [48], Bacillus subtilis, HPLC-gel filtration [47]) [47, 48]

68000 (rat, gel filtration) [23]

70000 (rat, gel filtration) [24]

71000 (human, papain-solubilized, gel filtration) [39]

78000 (human seminal plasma, prostate, testis) [2]

80000 (bovine [27], Proteus mirabilis [50], gel filtration [27, 50], pig, sucrose gradient centrifugation [31]) [27, 31, 50]

86000 (human, gel filtration) [42]

90000 (human [8, 13, 45], Saccharomyces cerevisiae [56], gel filtration [8, 56], PAGE [13, 45], sheep, PAGE in 8 M urea, after cross-linkage of subunits with dimethylsuberimidate [30]) [8, 13, 30, 45, 56]

98000 (human, Triton X-100-solubilized, PAGE) [13]

102000 (Morchella esculenta isoform III, gel filtration) [55]

110000 (bovine, PAGE [28], human, gel filtration [45]) [28, 45]

113000 (rat, analytical ultracentrifugation) [17]

120000 (Penicillium roquefortii, gel filtration) [54]

155000 (Morchella esculenta isoform I, gel filtration) [55]

160000 (human, Triton X-100 solubilized, gel filtration) [39]

180000 (kidney bean, gel filtration) [58]

219000 (Morchella esculenta isoform II, gel filtration) [55]

250000 (human [44], rat [21], gel filtration in the presence of Triton X-100) [21, 44]

500000 (human, gel filtration without Triton X-100) [44]

Subunits

Dimer (1 × 38000 + 1 × 14000, human, SDS-PAGE [1], 1 × 38600 + 1 × 22000, E. coli isoform B, SDS-PAGE [48], 1 × 39200 + 1 × 22000, E. coli isoform A, SDS-PAGE [48], 1 × 43000 + 1 × 25000, rat, SDS-PAGE [23], 1 × 45000 + 1 × 22000, Bacillus subtilis, SDS-PAGE [47], 1 × 45000 + 1 × 23000, rat, SDS-PAGE [24], 1 × 46000 + 1 × 22000, rat, papain solubilized, SDS-PAGE [35], 1 × 47000 + 1 × 22000, human, SDS-PAGE [45], 1 × 47000 + 1 × 28000, Proteus mirabilis, SDS-PAGE [50], 1 × 51000 + 1 × 22000, rat, Triton X-100 solubilized, SDS-PAGE [35], 1 × 53000 + 1 × 20000, human, papain-solubilized, SDS-PAGE [39], 1 × 55000 + 1 × 25000, bovine, SDS-PAGE [27],

1 × 57000 + 1 × 21000, human, SDS-PAGE, presence of urea, 2-mercaptoethanol [41], 1 × 57000 + 1 × 25500, bovine, SDS-PAGE [14], 1 × 61000 + 1 × 27000, human, SDS-PAGE [8], 1 × 62000 + 1 × 20000, human, SDS-PAGE [39], 1 × 62000 + 1 × 22000, human, SDS-PAGE [42], 1 × 64000 + 1 × 23000, Saccharomyces cerevisiae, SDS/2-mercaptoethanol-PAGE [56], 1 × 64000 + 1 × 29000, human, SDS-PAGE [43], 1 × 65000 + 1 × 27000, sheep, SDS-PAGE, 8 M urea-PAGE [30], 1 × 66000 + 1 × 55000, Penicillium roquefortii, SDS-PAGE [54], 1 × 68000 + 1 × 27000, bovine, SDS-PAGE [28], 1 × 71000 + 1 × 28000, bovine, SDS-PAGE [12], 1 × 150000 + 1 × 95000, human, SDS-PAGE [44]) [1, 8, 12, 14, 23, 24, 27, 28, 30, 35, 39, 41–45, 47, 48, 50, 54, 56]
More (the enzyme is composed of two non-identical catalytically active subunits [20], protease solubilization leads to a reduced MW of the larger subunit [39]) [20, 39]

Glycoprotein/Lipoprotein

Glycoprotein (sialoglycoprotein [8, 15, 16, 27], carbohydrate rich [36], 20% carbohydrate [27], contains neutral and amino sugars and sialic acid [28], both subunits are glycoproteins, structure [35, 43]) [1, 5, 8, 12, 14–16, 27, 28, 35, 36, 39, 42, 43, 55, 56]

---

4 ISOLATION/PREPARATION

Source organism

Human [1, 2, 5–8, 10, 13, 36–39, 41–46, 53]; Rat (male, Donryu [17], Wistar (female [24]) [19, 20, 23, 24], Sprague-Dawley strain [21, 22]) [3, 9, 16–24, 33–36, 40]; Mouse (Swiss albino and RIII strain, both infected with mammary tumour virus, uninfected: C57BL strain [26]) [26, 35]; Rabbit (male, Fauve de Bourgogne strain [25]) [25, 36]; Pig [11, 31–33]; Sheep [29, 30, 34]; Bovine (Holstein breed [27]) [12, 14, 15, 27, 28]; Mammals [34, 35]; Marthasterias glacialis [4]; Kidney bean [58, 59]; Lentinus edodes (shiitake mushroom) [51, 52]; Penicillium roquefortii [54]; Morchella esculenta Fr. [55]; Saccharomyces cerevisiae 1278b [56]; Aspergillus oryzae MA-27-IM [57]; Bacillus subtilis (natto strain NR-1) [47]; E. coli K-12 (strain MG1655) [48, 49]; Proteus mirabilis [50]; Oncornavirinae of various origins [26]

Source tissue

Liver (human [53], tissue, serum and bile contain multiple forms of different molecular weights [10], 80–90% of activity in biliary tract [21], hyperplastic hepatic nodules, primary and Yoshida ascites (AH 13) hepatomas [16], hepatocellular carcinoma [43]) [10, 13, 16, 19, 21, 25, 28, 35, 36, 43, 45, 53]; Bile [46, 53]; WI-38 Lung fibroblasts (fetal) [39]; Pancreas (normal, pancreatic carcinoma and pancreatic cancer cell line HPC-Y1 [5]) [5, 8, 23, 35, 36]; Kidney (cortex [30], brush border of convoluted tubules [30], luminal membrane of proximal tubules [36]) [2, 3, 9, 15, 16, 22, 30–33, 35, 36, 40–42,

53]; Urine (high- and low-molecular weight variants [7]) [7, 53]; Intestine (small intestine [11], jejunum [35], duodenum [36], ileum and rectum [53]) [11, 35, 36, 53]; Parotid gland [12]; Blood (serum) [13, 37, 38, 53]; Mammary gland (lactating, milk-membranes [14], mammary adenocarcinoma MT 13762 [18], mammary cell products formed by exocytosis: milk fat-globule membrane, mammary tumour virus [26]) [14, 18, 26]; Colostrum [27]; Deciduoma [24]; Testis [1, 2]; Seminal plasma (and reproductive tissue [2]) [2, 35]; Epididymis [35]; Prostate [2]; Brain (cortex microvessels) [6]; Spleen [35]; Ciliary body [15]; Pyloric cacae [4]; Fruit (green beans) [58, 59]; Mycelium [54, 55]; Fruit body [51, 52]; Koji-culture (solid culture on wheat bran) [54, 57]; Cell [47–50, 56]; More (distribution) [35]

Localization in source
Membrane-bound (extracellular (outer surface of plasma membrane) [32], intrinsic membrane protein [35]) [5, 6, 11, 14, 15, 17–19, 24–28, 30, 32, 33, 35, 36, 39, 40]; Microsomes [15, 17, 51]; Periplasmic space [48, 49]; Extracellular [47]; Cytoplasma [56]

Purification
Human (partial (soluble in the absence of detergents [45]) [10, 13, 45], solubilized with Triton X-100 [6, 13, 39, 41], Lubrol W [42, 45], deoxycholate [41, 45], bromelain [5, 42], papain [13] or with various detergents and proteases [6]) [1, 2, 5, 6, 8, 10, 13, 39, 41–46]; Sheep [30]; Pig (solubilized with ficin [31]) [31, 33]; Bovine (partial, solubilized with Emulphogene BC720 [15], papain [27] or Lubrol WX [14]) [12, 14, 15, 27, 28]; Rabbit (solubilized with Lubrol/deoxycholate) [25]; Rat (partial [3, 21], post-translational multiple forms differing in sugar portion [16], solubilized with either detergents or proteinases [35], e.g. deoxycholate [17], Triton X-100 [18, 24, 33], Lubrol WX/deoxycholate [19] or papain [24, 33], affinity chromatography [40]) [3, 16–19, 21, 23, 24, 33, 35, 40]; Mouse (partial) [26]; Kidney bean [58]; Marthasterias glacialis [4]; Penicillium roquefortii [54]; Morchella esculenta (partial, 3 isoforms) [55]; Bacillus subtilis [47]; E. coli [48]; Proteus mirabilis [50]

Crystallization
(E. coli [48], Proteus mirabilis [50]) [48, 50]

Cloned
–

Renatured
(rat, circular dichroic spectra [20]) [20, 35]

5 STABILITY

pH
4 (50% loss of activity after 60 min, 37°C) [50]; 5–9 (at least 60 min stable, 37°C) [50]; 6–9 (stable) [54]; 7–8 (stable) [47]; 10 (50% loss of activity after 60 min, 37°C) [50]

Temperature (°C)
4–30 (at least 10 min stable) [54]; 40 (at least 15 min stable [47], 10% loss of activity within 10 min [50], stable below [51]) [47, 50, 51]; 45 (12% loss of activity within 15 min [48], 37% loss of activity within 10 min [50]) [48, 50]; 48 ($t_{1/2}$: 15 min) [48]; 50 (95% loss of activity within 10 min [50], 92% loss of activity within 15 min [48]) [48, 50]; 55 (inactivation within 15 min at pH 8.0) [47]; 58 ($t_{1/2}$: 5 min [24], 12 min [17], 20% of initial activity retained after 60 min [17], in the presence of glutathione, 0.02 M, at least 60 min stable [17]) [17, 24]; 65 (inactivation, GSH and serine/borate protect) [51]

Oxidation

Organic solvent

General stability information
PMSF stabilizes [7]; Dilution inactivates [55]; Glycerol, 5% v/v, stabilizes [55]; Freezing or lyophilization inactivates purified enzyme [58]

Storage
–80°C, 9 months [38]; –70°C, at least 30 days [24]; –30°C, freeze-dried, at least 3 months [44]; –22°C, at least 3 months [17]; –20°C, crude, at least 1 year [55]; –19°C, at least 2 months [56]; 0°C, at least 3 weeks [58]; 2–8°C, 7 days [37]; 4°C, crystalline, at least 3 months [50]; 4°C, at least 5 days [38]; 4°C, $t_{1/2}$: 2 months [41]

---

6 CROSSREFERENCES TO STRUCTURE DATABANKS

PIR/MIPS code
PIR3:S58286 (Arabidopsis thaliana); PIR1:EKECEX (Escherichia coli); PIR3:JC4570 (mouse); PIR2:S05532 (pig); PIR2:B47739 (light chain II lung human); PIR2:A05225 (precursor rat); PIR2:A41125 (related protein human); PIR1:EKHUEX (type 1 precursor human); PIR2:A35074 (type 1 precursor short splice form human); PIR2:A36742 (type II human (fragment)); PIR2:B45946 (hepatic bovine (fragments)); PIR2:A45946 (hepatic dog (fragments))

Brookhaven code

7 LITERATURE REFERENCES

[1] Yoshida, K.I., Arai, K., Kobayashi, N., Saitoh, H.: Andrologia,22,239–246 (1990)
[2] Arai, K., Yoshida, K., Komoda, T., Sakagishi, Y.: Clin. Biochem.,23,105–112 (1990)
[3] Kimm, S.W., Kim, E.G., Park, S.C.: Korean J. Biochem.,18,135–143 (1986)
[4] Glynn, B.P., Johnson, D.B.: Comp. Biochem. Physiol.B Comp. Biochem., 80B,941–948 (1985)
[5] Sugimoto, M., Yamaguchi, N., Keiichi, K.: Gastroenterol. Jpn.,19,227–231 (1984)
[6] Vesely, J., Cernoch, M.: Neurochem. Res.,9,927–934 (1984)
[7] Rambabu, K., Pattabiraman, T.N.: J. Biosci.,4,287–294 (1982)
[8] Masuike, M., Ogawa, M., Kosaki, G., Minamiura, N., Yamamoto, T.: Enzyme,27, 163–170 (1982)
[9] Frielle, T., Curthoys, N.P.: Biophys. J.,37,193–195 (1982)
[10] Echetebu, Z.O., Moss, D.W.: Enzyme,27,1–8 (1982)
[11] Nakamura, Y., Kato, H., Suzuki, F., Nagata, Y.: Biomed. Res.,2,509–516 (1981)
[12] Hata, K., Hayakawa, M., Abiko, Y., Takiguchi, H.: Int. J. Biochem.,13,681–692 (1981)
[13] Tsuji, A., Matsuda, Y., Katunuma, N.: Clin. Chim. Acta,104,361–364 (1980)
[14] Baumrucker, C.R.: J. Dairy Sci.,63,49–54 (1980)
[15] Das, N.D., Shichi, H.: Exp. Eye Res.,29,109–121 (1979)
[16] Tsuchida, S., Hoshino, K., Sato, T., Ito, N.: Cancer Res.,39,4200–4205 (1979)
[17] Taniguchi, N.: J. Biochem.,75,473–480 (1974)
[18] Jaken, S., Mason, M.: Biochim. Biophys. Acta,568,331–338 (1979)
[19] Ding, J.L., Smith, G.D., Peters, T.J.: Biochem. Soc. Trans.,8,77 (1980)
[20] Horiuchi, S., Inoue, M., Morino, Y.: Eur. J. Biochem.,105,93–102 (1980)
[21] Ding, J.L., Smith, G.D., Peters, T.J.: Biochim. Biophys. Acta,657,334–343 (1981)
[22] Sachdev, G.P., Leahy, D.S., Chace, K.V.: Biochim. Biophys. Acta,749,125–129 (1983)
[23] Takahashi, S., Steinman, H.M., Ball, D.: Biochim. Biophys. Acta,707,66–73 (1982)
[24] Tarachand, U.: J. Appl. Biochem.,6,278–288 (1984)
[25] Bagrel, D., Petitclerc, C., Schiele, F., Siest, G.: Biochim. Biophys. Acta,658,220–231 (1981)
[26] François, C., Calberg-Bacq, C.M., Gosselin, L., Kozma, S., Osterrieth, P.M.: Biochim. Biophys. Acta,567,106–115 (1979)
[27] Yasumoto, K., Iwami, K., Fushiki, T., Mitsuda, H.: J. Biochem.,84,1227–1236 (1978)
[28] Furukawa, M., Higashi, T., Tateishi, N., Ochi, K., Sakamoto, Y.: J. Biochem.,93, 839–846 (1983)
[29] Karkowsky, A.M., Bergamini, M.V.W., Orlowski, M.: J. Biol. Chem.,251,4736–4743 (1976)
[30] Zelazo, P., Orlowski, M.: Eur. J. Biochem.,61,147–155 (1976)
[31] Leibach, F.H., Binkley, F.: Arch. Biochem. Biophys.,127,292–301 (1968)
[32] Örning, L., Hammarström, S.: Biochem. Biophys. Res. Commun.,106,1304–1309 (1982)
[33] Bernström, K., Örning, L., Hammarström, S.: Methods Enzymol.,86,38–45 (1982) (Review)
[34] Allison, D.: Methods Enzymol.,113,419–437 (1985) (Review)
[35] Tate, S.S., Meister, A.: Methods Enzymol.,113,400–419 (1985) (Review)
[36] Shaw, L.M. in "Methods Enzymol. Anal.",3rd Ed. (Bergmeyer, H.U., Ed.) ,3,349–352 (1983) , Verlag Chemie, Weinheim (Review)

[37] Wahlefeld, A.W., Bergmeyer, H.U. in "Methods Enzymol. Anal.",3rd Ed. (Bergmeyer, H.U., Ed.) ,3,352–356 (1983) , Verlag Chemie, Weinheim (Review)

[38] Shaw, L.M., Stromme, J.H. in "Methods Enzymol. Anal.",3rd Ed. (Bergmeyer, H.U., Ed.) ,3,357–364 (1983) , Verlag Chemie, Weinheim (Review)

[39] Takahashi, S., Zukin, R.S., Steinman, H.M.: Arch. Biochem. Biophys.,207,87–95 (1981)

[40] Cook, N.D., Peters, T.J.: Biochim. Biophys. Acta,828,205–212 (1985)

[41] Miller, S.P., Awasthi, Y.C., Srivastava, S.K.: J. Biol. Chem.,251,2271–2278 (1976)

[42] Tate, S.S., Ross, M.E.: J. Biol. Chem.,252,6042–6045 (1977)

[43] Yamashita, K., Totani, K., Iwaki, Y., Takamisawa, I., Tateishi, N., Higashi, T., Sakamoto, Y., Kobata, A.: J. Biochem.,105,728–735 (1989)

[44] Abe, S., Kochi, H., Hiraiwa, K.: Biochim. Biophys. Acta,1077,259–264 (1991)

[45] Huseby, N.: Biochim. Biophys. Acta,483,46–56 (1977)

[46] Indirani, N., Hill, P.G.: Biochim. Biophys. Acta,483,57–62 (1977)

[47] Ogawa, Y., Hosoyama, H., Hamano, M., Motai, H.: Agric. Biol. Chem.,55,2971–2977 (1991)

[48] Suzuki, H., Kumagai, H., Tochikura, T.: J. Bacteriol.,168,1325–1331 (1986)

[49] Suzuki, H., Kumagai, H., Tochikura, T.: J. Bacteriol.,168,1332–1335 (1986)

[50] Nakayama, R., Kumagai, H., Tochikura, T.: J. Bacteriol.,160,341–346 (1984)

[51] Iwami, K., Yasumoto, K., Nakamura, K., Mitsuda, H.: Agric. Biol. Chem.,39,1933–1940 (1975)

[52] Iwami, K., Yasumoto, K., Nakamura, K., Mitsuda, H.: Agric. Biol. Chem.,39,1941–1946 (1975)

[53] Minato, S.: Arch. Biochem. Biophys.,192,235–240 (1979)

[54] Tomita, K., Yano, T., Tsuchida, T., Kumagai, H., Tochikura, T.: J. Ferment. Bioeng.,70,128–130 (1990)

[55] Moriguchi, M., Yamada, M., Suenaga, S., Tanaka, H., Wakasugi, A., Hatanaka, S.-I.: Arch. Microbiol.,144,15–19 (1986)

[56] Penninckx, M.J., Jaspers, C.J.: Phytochemistry,24,1913–1918 (1985)

[57] Tomita, K., Ito, M., Yano, T., Kumagai, H., Tochikura, T.: Agric. Biol. Chem.,52,1159–1163 (1988)

[58] Goore, M.Y., Thompson, J.F.: Biochim. Biophys. Acta,132,15–26 (1967)

[59] Goore, M.Y., Thompson, J.F.: Biochim. Biophys. Acta,132,27–32 (1967)

1 NOMENCLATURE

EC number
2.3.2.3

Systematic name
L-Lysyl-tRNA:phosphatidylglycerol 3-O-lysyltransferase

Recommended name
Lysyltransferase

Synonyms

CAS Reg. No.
37357-20-8

2 REACTION AND SPECIFICITY

Catalysed reaction
L-Lysyl-tRNA + phosphatidylglycerol →
→ tRNA + 3-phosphatidyl-1'-(3'-O-L-lysyl)glycerol

Reaction type
Aminoacyl group transfer

Natural substrates

Substrate spectrum
1 L-Lysyl-tRNA + phosphatidylglycerol (2'- (not 3'-) deoxy-analogue of phosphatidylglycerol can also act as L-lysyl-acceptor) [1]

Product spectrum
1 tRNA + 3-phosphatidyl-1'-(3'-O-L-lysyl)glycerol [1]

Inhibitor(s)

Cofactor(s)/prosthetic group(s)/activating agents

Metal compounds/salts
Anionic surfactant (e.g. sodium-salt of a fatty acid, activation) [1]; High ionic
strength (activation) [1]

Turnover number (min^{-1})

Specific activity (U/mg)

K_m-value (mM)

pH-optimum

pH-range

Temperature optimum (°C)

Temperature range (°C)

3 ENZYME STRUCTURE

Molecular weight

Subunits

Glycoprotein/Lipoprotein

–

4 ISOLATION/PREPARATION

Source organism
  Staphylococcus aureus [1]

Source tissue
  Cell [1]

Localization in source
  Membrane-bound [1]

Purification
  Staphylococcus aureus (partial) [1]

Crystallization
  –

Cloned
  –

Renatured
  –

5 STABILITY

pH

Temperature (°C)

Oxidation

Organic solvent

General stability information

Storage
    −20°C, native membrane-bound enzyme, several weeks [1]; −20°C, mem-
    brane-bound enzyme extracted with organic solvents, $t_{1/2}$: 3–5 days [1]

6 CROSSREFERENCES TO STRUCTURE DATABANKS

PIR/MIPS code

Brookhaven code

7 LITERATURE REFERENCES

[1] Lennarz, W.J., Bonsen, P.P.M., Van Deenen, L.L.M.: Biochemistry,6,2307–2312
    (1967)

1 NOMENCLATURE

EC number
2.3.2.4

Systematic name
(5-L-Glutamyl)-L-amino-acid 5-glutamyltransferase (cyclizing)

Recommended name
gamma-Glutamylcyclotransferase

Synonyms
Cyclotransferase, gamma-glutamyl
gamma-Glutamyl-amino acid cyclotransferase
gamma-L-Glutamylcyclotransferase
L-Glutamic cyclase

CAS Reg. No.
9045-44-7

---

2 REACTION AND SPECIFICITY

Catalysed reaction
(5-L-Glutamyl)-L-amino acid →
→ 5-oxoproline + L-amino acid (mechanism [12])

Reaction type
Aminoacyl group transfer

Natural substrates
gamma-L-Glutamyl-L-amino acid (enzyme of L-glutamyl-cycle [4], involved in sclerotization process of white pupae [13]) [4, 13]

Substrate spectrum
1 gamma-L-Glutamyl-L-amino acid (stereospecific [10], specificity (sheep and human tissue [9], overview) [9, 10], preferred substrates: gamma-L-Glu-gamma-L-Glu-NH-CHR-COOH [7, 10]) [1, 2, 5–10, 12]
2 gamma-L-Glutamyl-gamma-L-glutamyl-L-amino acid (substrates are the derivatives of L-alanine, L-glutamine, L-glutamate, glycine, L-lysine [6], L-valine, L-leucine, L-tyrosine, L-phenylalanine, L-proline [6, 7]) [6, 7]
3 gamma-L-Glutamyl-L-methionine (specific, not D-Glu-L-methionine [1], best substrate [5], 70% as effective as L-Glu-L-phenylalanine [13]) [1, 5–7, 13]

4 gamma-L-Glutamyl-L-cysteine [1, 6]
5 gamma-L-Glutamyl-L-phenylalanine (best substrate [13], poor substrate
  [6, 7, 9, 10]) [6, 7, 9, 10, 13]
6 gamma-L-(threo-beta-Methyl)glutamyl-L-alpha-aminobutyrate (model
  substrate, in vivo and in vitro) [6, 8]
7 gamma-(beta-Methyl)glutamyl-L-alpha-aminobutyrate (substrate analog,
  90% as effective as gamma-Glu-L-alpha-aminobutyrate) [10]
8 gamma-L-Glutamyl-L-glutamine (best substrate [5–7], not [13]) [2, 5–7, 9]
9 gamma-L-Glutamyl-L-gamma-L-glutamyl-L-p-nitroanilide (best substrate
  [2, 9], 25% as effective as gamma-L-Glu-L-phenylalanine [13], not
  D-Glu-isomer [13]) [2, 5, 6, 9, 13]
10 gamma-L-Glutamyl-L-alanine (not [13]) [2–4, 6, 7, 9, 12]
11 gamma-L-Glutamyl-L-alpha-aminobutyrate (not [13]) [2, 5–11]
12 gamma-L-Glutamyl-alpha-naphthylamide (best substrate, human) [10]
13 gamma-L-Glutamyl-L-valine (10% as effective as
   gamma-L-Glu-L-phenylalanine [13], poor substrate [6]) [6, 13]
14 gamma-L-Glutamyl-glycine (poor substrate [5–7], not [13]) [2, 5–7, 9]
15 gamma-L-Glutamyl-L-S-methyl-L-cysteine [2, 9]
16 gamma-L-Glutamyl-L-epsilon-N-benzyloxycarbonyl-L-lysine [5]
17 gamma-L-Glutamyl-L-glutamate (poor substrate [5], not [13]) [5–7, 10]
18 gamma-L-Glutamyl-L-leucine (poor substrate [5], not [13]) [5, 9]
19 gamma-L-Glutamyl-cystine [6]
20 gamma-L-Glutamyl-L-aspartate [6, 7]
21 alpha-N-(gamma-L-Glutamyl)L-lysine [5]
22 gamma-Glutamyl-glutathione [10]
23 More (poor substrates are L-Glu-L-leucine [6, 10], L-Glu-L-tyrosine [6, 7,
   9], L-Glu-L-isoleucine [6], L-Glu-L-proline [6, 7], no substrates are
   alpha-L-Glu-glycine, alpha-L-Glu-L-alanine, alpha-L-Glu-L-tyrosine,
   alpha-L-Glu-beta-naphthylamide, alpha-L-Glu-4-methoxy-2-beta-naphthyl-
   amide [13], glutathione (i.e. gamma-L-Glu-L-cysteinylglycine) [2, 13],
   gamma-Glu-p-nitroanilide [2], LD-, DD- and DL-isomers of
   gamma-L-Glu-gamma-L-Glu-p-nitroanilide [2], L-glutamine (pig [10]) [9,
   10, 13], L-Glu-beta-alanine, L-Glu-beta-aminoisobutyrate [9],
   D-Glu-alpha-aminobutyrate, L-(alpha-methyl)Glu-alpha-aminobutyrate,
   beta-aminoglutaryl-alpha-aminobutyrate,
   L-gamma-(N-methyl)Glu-alpha-aminobutyrate,
   gamma-(gamma-methyl)Glu-alpha-aminobutyrate [11],
   L-Glu-L-Glu-L-phenylalanine and L-Glu-L-Glu-alpha-naphthylamide [13])
   [2, 6, 7, 9–11, 13]

Product spectrum

1 5-L-Oxoproline + L-amino acid (i.e. 2-pyrrolidone-5-carboxylic acid) [1, 2, 5–8, 12]
2 ?
3 ?
4 ?
5 5-L-Oxoproline + L-phenylalanine [13]
6 3-Methyl-5-oxoproline + L-alpha-aminobutyrate [6]
7 ?
8 ?
9 5-L-Oxoproline + gamma-L-glutamyl-L-p-nitroanilide [2]
10 5-L-Oxoproline + L-alanine [3]
11 ?
12 ?
13 ?
14 ?
15 ?
16 ?
17 ?
18 ?
19 ?
20 ?
21 ?
22 5-Oxoproline + glutathione [10]
23 ?

Inhibitor(s)

Cofactor(s)/prosthetic group(s)/activating agents

Metal compounds/salts

Turnover number (min^{-1})

Specific activity (U/mg)

8.9–40.0 [6]; 33.8 [3]; 70.7 [13]; 100 (human) [2]

K_m-value (mM)

More (kinetics) [12]; 0.6 (gamma-L-Glu-gamma-L-Glu-p-nitroanilide) [5]; 2.8–3.0 (gamma-L-Glu-L-methionine [5], gamma-L-Glu-L-phenylalanine [13]) [5, 13]; 5.0 (gamma-L-(threo-beta-methyl)Glu-L-alpha-aminobutyrate) [8]; 6.25–6.6 (gamma-L-Glu-L-alpha-aminobutyrate) [6–8]; 8–12 (gamma-L-Glu-L-alpha-aminobutyrate) [5]; 10–18 (gamma-L-Glu-L-glutamine) [5]; 100 (alpha-N-(gamma-L-Glu)L-lysine) [5]

pH-optimum

More (two isozymic forms of different pI-values: pI 4.6 and 5.1) [6, 7]; 7.1–7.3 [13]; 7.5–8.0 [5–7]; 7.8–8.2 (borate buffer [9]) [2, 9]; 9.0 [3]

Enzyme Handbook © Springer-Verlag Berlin Heidelberg 1996
Duplication, reproduction and storage in data banks are only
allowed with the prior permission of the publishers

pH-range
6.0–9.5 (about half-maximal activity at pH 6 and 9.5) [7]; 6.5–8.1 (about half-maximal activity at pH 6.5 and 8.1) [13]; 6.5–11.5 (about half-maximal activity at pH 6.5 and 11.5) [3]

Temperature optimum (°C)
35 [1]

Temperature range (°C)

---

3 ENZYME STRUCTURE

Molecular weight
25250 (human, gel filtration) [3]
27500 (rat, gel filtration) [5]
30000 (Musca domestica, gel filtration) [13]
More (amino acid analysis) [7]

Subunits
Monomer (1 × 27000, rat, SDS-PAGE) [6]

Glycoprotein/Lipoprotein
–

---

4 ISOLATION/PREPARATION

Source organism
Nicotiana tabacum (tobacco, cv. Samsun) [1]; Musca domestica (house fly) [13]; Human [2–4, 9, 10, 12]; Sheep [2, 9, 11]; Pig [10]; Rat (Sprague-Dawley [5], Holtzman [7], female Wistar [10] strain) [5–7, 10]; Mouse [8]; More (distribution) [2]

Source tissue
Cell (cell suspension culture) [1]; Pupae [13]; Brain [2, 5, 8, 11]; Erythrocytes [3, 4, 10, 12]; Kidney [5–8, 10]; Liver (pig [10]) [5, 8, 10]; More (distribution, in rat and human tissues [10]) [5, 8–10]

Localization in source
Cytoplasm (soluble) [1–7, 9, 10, 13]

Purification
Musca domestica [13]; Human [2, 3, 9]; Sheep [2, 9]; Rat (2 isozymic, immunologically identical forms [5]) [5–7]; Pig [10]

Crystallization
–

Cloned
–

Renatured

–

---

5 STABILITY

pH

Temperature (°C)
57 (purified: inactivation within 5 min, crude: 5 min stable) [2]

Oxidation

Organic solvent

General stability information
Freezing, stable to [2]; Lyophilization, stable to [2]; Dialysis against 0.05 M
Tris-HCl buffer, pH 8.0 containing 5 mM EDTA, stable to [5]

Storage
–20°C, stable [2]; 0°C, at least 1 month (human [9]) [2, 9]; 0°C, 30% loss of
activity within 3 weeks, pI 5.1-isoform [6]; 0°C, at least 2 months, pI 4.6-iso-
form [6]

---

6 CROSSREFERENCES TO STRUCTURE DATABANKS

PIR/MIPS code

Brookhaven code

---

7 LITERATURE REFERENCES

[1] Steinkamp, R., Schweihofen, B., Rennenberg, H.: Physiol. Plant.,69,499–503 (1987)
[2] Orlowski, M., Richman, P.G., Meister, A.: Biochemistry,8,1048–1055 (1969)
[3] Board, P.G., Moore, K.A., Smith, J.E.: Biochem. J.,173,427–431 (1978)
[4] York, M.J., Kuchel, P.W., Chapman, B.E.: J. Biol. Chem.,259,15085–15088 (1984)
[5] Orlowski, M., Meister, A.: J. Biol. Chem.,248,2836–2844 (1973)
[6] Meister, A.: Methods Enzymol.,113,428–445 (1981) (Review)
[7] Taniguchi, N., Meister, A.: J. Biol. Chem.,253,1799–1806 (1978)
[8] Bridges, R.J., Griffith, O.W., Meister, A.: J. Biol. Chem.,255,10787–10792 (1980)
[9] Orlowski, M., Meister, A. in "The Enzymes",3rd Ed. (Boyer, P.D., Ed.) ,4,123–151
    (1971) , Academic Press, New York (Review)
[10] Szweczuk, A., Connell, G.E.: Can. J. Biochem.,53,706–712 (1975)
[11] Griffith, O.W., Meister, A.: Proc. Natl. Acad. Sci. USA,74,3330–3334 (1977)
[12] York, M.J., Crossley, M.J., Hyslop, S.J., Fisher, M. L., Kuchel, P.W.: Eur. J. Bio-
    chem.,184,97–101 (1989)
[13] Bodnaryk, R.P., McGirr, L.: Biochim. Biophys. Acta,315,352–362 (1973)

---

1 NOMENCLATURE

EC number
2.3.2.5

Systematic name
L-Glutaminyl-peptide gamma-glutamyltransferase (cyclizing)

Recommended name
Glutaminyl-peptide cyclotransferase

Synonyms
Glutaminyl cyclase
Cyclotransferase, glutaminyl-transfer ribonucleate
Glutaminyl-tRNA cyclotransferase

CAS Reg. No.
37257-21-9

2 REACTION AND SPECIFICITY

Catalysed reaction
L-Glutaminyl-peptide →
→ 5-oxoprolyl-peptide + NH_3

Reaction type
Aminoacyl group transfer

Natural substrates
L-Glutaminyl-peptide (involved in posttranslational modification of the N-terminal glutamine of peptide hormones or neurotransmitters, such as thyrotropin releasing hormone, luteinizing hormone releasing hormone, gastrin, heavy chain of gamma-globulin) [1]

Substrate spectrum
1 L-Glutaminyl-peptide (e.g. Gln-His-Pro (i.e. thyrotropin-releasing hormone) [1], Gln-His-Pro-Gly (i.e. Gly[4]-thyrotropin-releasing hormone), gonadotropin-releasing hormone and L-Gln-Tyr-Ala [2]) [1–3]
2 More (no substrates are D-Gln-Tyr-Ala, Lys-Arg-Gln-His-Pro-Gly-Lys-Arg (thyrotropin releasing hormone precursor)) [2]

Product spectrum
1 5-Oxoprolyl-peptide + NH_3 (i.e. pyroglutamyl-peptide) [1–3]
2 ?

Inhibitor(s)

Cofactor(s)/prosthetic group(s)/activating agents
More (no activation by ATP or glucose/NADP$^+$/hexokinase/glucose-6-phosphate-system) [2]

Metal compounds/salts
EDTA (increase of activity [1], not [2]) [1]; NaCl (increase of activity [1], not [2]) [1]

Turnover number (min^{-1})

Specific activity (U/mg)
0.000217 (human) [1]; 0.00025 (rat) [1]; 0.0065–0.0115 [2]; 0.023 (pig, activity 1) [1]; 0.038 (pig, activity 2) [1]

K$_m$-value (mM)
0.063 (gonadotropin releasing hormone) [2]; 0.088 (Gly^4-thyrotropin releasing hormone) [2]; 0.132 (Gln-Thr-Ala) [2]

pH-optimum
7.2–7.5 (pig) [1]; 8.0 [2]

pH-range
5–8 (detectable activity) [1]

Temperature optimum (°C)
37 (assay at) [2]

Temperature range (°C)

---

3 ENZYME STRUCTURE

Molecular weight
25000 (papaya, gel filtration) [3]
43000–50000 (bovine, gel filtration) [2]
55000 (pig pituitary, gel filtration) [1]

Subunits
Monomer (1 × 43000–50000, bovine, gel filtration in the presence of 6 M urea) [2]

Glycoprotein/Lipoprotein
Glycoprotein [1]

---

4 ISOLATION/PREPARATION

Source organism
Rat (neonatal [1]) [1, 2]; Human [1]; Pig [1]; Bovine [1, 2]; Papaya [3]

Source tissue
Brain (rat, hypothalamus [2]) [1, 2]; Pituitary gland (pig [1]) [1, 2]; B-Lymphocytes (secretory and non-secretory cell line P3X63Ag8) [1]; Adrenal medulla (bovine) [1]

Localization in source
Soluble (chromaffin vesicle fraction [1], secretory granules [2]) [1, 2]

Purification
Pig [1]; Rat [1]; Human [1]

Crystallization
–

Cloned
–

Renatured
–

---

5 STABILITY

pH

Temperature (°C)

Oxidation

Organic solvent

General stability information
Dialysis against 0.05 M Tris-buffer, pH 7.0, leads to 75% loss of activity [1]

Storage

---

6 CROSSREFERENCES TO STRUCTURE DATABANKS

PIR/MIPS code
PIR2:A41535 (precursor bovine)

Brookhaven code

---

7 LITERATURE REFERENCES

[1] Busby, W.H., Quackenbush, G.E., Humm, J., Youngblood, W.W., Kizer, J.S.: J. Biol. Chem.,262,8532–8536 (1987)
[2] Fischer, W.H., Spiess, J.: Proc. Natl. Acad. Sci. USA,84,3628–3632 (1987)
[3] Messer, M., Ottesen, M.: Biochim. Biophys. Acta,92,409–411 (1964)

---

1 NOMENCLATURE

EC number
  2.3.2.6

Systematic name
  L-Leucyl-tRNA:protein leucyltransferase

Recommended name
  Leucyltransferase

Synonyms
  Leucyl, phenylalanine-tRNA-protein transferase
  Leucyl-phenylalanine-transfer ribonucleate-protein aminoacyltransferase
  Leucyl-phenylalanine-transfer ribonucleate-protein transferase

CAS Reg. No.
  37257-22-0

2 REACTION AND SPECIFICITY

Catalysed reaction
  L-Leucyl-tRNA + protein →
  → tRNA + L-leucyl-protein

Reaction type
  Aminoacyl group transfer

Natural substrates

Substrate spectrum
  1 L-Leucyl-tRNA + acceptor protein (acceptor protein: bovine serum albumin [1, 2], alpha-casein [3, 5], acceptor peptide specificity [3], dipeptide specificity [5], acceptors with arginine or lysine as initial NH_2-terminal residue [4], basic NH_2-terminal is absolute determinant of specificity [3, 5]) [1–5]
  2 L-Phenylalanyl-tRNA + acceptor protein [1–5]
  3 L-Methionyl-tRNA + acceptor protein (methionyl-$tRNA_m^{Met}$ preferred to methionyl-$tRNA_f^{Met}$ [5]) [4, 5]

Product spectrum
  1 tRNA + L-leucyl-protein [1–4]
  2 ?
  3 ?

Inhibitor(s)
Mg^{2+} [2]; Ca^{2+} [2]; Mn^{2+} [2]; Puromycin (not the aminonucleoside derivative) [2]; Ribonuclease [1]; More (no inhibition by free lysine [3], chloramphenicol, streptomycin [2]) [2, 3]

Cofactor(s)/prosthetic group(s)/activating agents

Metal compounds/salts
Na^+ (requirement) [1]; K^+ (requirement, 0.15 M [2]) [1, 2]; NH_4^+ (requirement) [1]; More (no Mg^{2+}-requirement [1], divalent cations inhibit [2]) [1, 2]

Turnover number (min^{-1})

Specific activity (U/mg)
0.00003 (phenylalanine) [1]; 0.000146 (leucine) [1]; 0.0011–0.0012 [2]; 0.01 (methionine) [5]; 0.025 (leucine) [5]; 0.03 (phenylalanine) [5]; 0.061–0.063 [3, 4]

K_m-value (mM)

pH-optimum
7.6–8.2 (leucine) [2]; 8.2–8.6 (phenylalanine) [2]

pH-range
7.0–9.0 (about 80% of maximal activity at pH 7.0 and 9.0) [2]

Temperature optimum (°C)
37 (assay at) [1–4]

Temperature range (°C)

3 ENZYME STRUCTURE

Molecular weight

Subunits

Glycoprotein/Lipoprotein
–

4 ISOLATION/PREPARATION

Source organism
E. coli (B [1–3, 5], K12 strain W4977 and revertant strain R18 of mutant strain MS845 [5]) [1–5]

Source tissue
Cell [1–5]

Localization in source
    Soluble [1–5]

Purification
    E. coli (partial [1]) [1–5]

Crystallization
    –

Cloned
    –

Renatured
    –

---

5 STABILITY

pH

Temperature (°C)
    60 ($t_{1/2}$: 1.5 min, complete inactivation after 8.5 min) [2]

Oxidation

Organic solvent

General stability information
    Desalting procedures inactivate [1, 2]; Repeated freeze-thawing, stable to
    [3]; Glycerol, 50%, stabilizes during storage [3]

Storage
    –20°C, concentrated enzyme solution, at least 2 months [2]; –20°C, concen-
    trated enzyme solution, at least 6 months [3]; 0°C, 1 month [1, 2]

---

6 CROSSREFERENCES TO STRUCTURE DATABANKS

PIR/MIPS code
    PIR2:A36888 (Escherichia coli)

Brookhaven code

---

7 LITERATURE REFERENCES

[1] Leibowitz, M.J., Soffer, R.L.: Biochem. Biophys. Res. Commun.,36,47–53 (1969)
[2] Leibowitz, M.J., Soffer, R.L.: J. Biol. Chem.,245,2066–2073 (1970)
[3] Soffer, R.L.: J. Biol. Chem.,248,8424–8428 (1973)
[4] Deutch, C.E.: Methods Enzymol.,106,198–205 (1984) (Review)
[5] Scarpulla, R.C., Deutch, C.E., Soffer, R.L.: Biochem. Biophys. Res. Commun.,71,
    584–589 (1976)

---

1 NOMENCLATURE

EC number
2.3.2.7

Systematic name
L-Asparagine:hydroxylamine gamma-aspartyltransferase

Recommended name
Aspartyltransferase

Synonyms
beta-Aspartyl transferase
Aspartotransferase [1]

CAS Reg. No.
37257-23-1

2 REACTION AND SPECIFICITY

Catalysed reaction
L-Asparagine + hydroxylamine →
→ NH_3 + L-aspartylhydroxamate

Reaction type
Aminoacyl group transfer

Natural substrates
L-Asparagine + hydroxylamine (involved in initial step of asparagine metabolism) [1]

Substrate spectrum
1 Asparagine + hydroxylamine (strict specificity, L- and D-enantiomer
equally effective, no substrate: L-glutamine) [1]

Product spectrum
1 NH_3 + beta-L-aspartohydroxamic acid [1]

Inhibitor(s)
Co^{2+} [1]; Cu^{2+} [1]; L-Aspartate (20 mM) [1]; L-Cysteine (5 mM) [1]; Streptomycin [1]; PCMB (reversible by GSH, 2-mercaptoethanol or 2,3-dimer-
captopropanol (i.e. British Anti-Lewisite or BAL), not L-cysteine) [1]; More
(no inhibition by NaN_3, isonicotinic acid hydrazide, L-homoserine, L-glutamic
acid, terminal amino acids of aspartic acid pathway, i.e. lysine, threonine,
methionine and isoleucine (up to 20 mM)) [1]

Cofactor(s)/prosthetic group(s)/activating agents
Isonicotinic acid hydrazide (activation) [1]

Metal compounds/salts
L-Cystine (slight activation) [1]; More (no metal ion requirement) [1]

Turnover number (min^{-1})

Specific activity (U/mg)
0.0245 [1]

K_m-value (mM)
7.14 (L-asparagine) [1]; 19.3 (hydroxylamine) [1]

pH-optimum
9.0 [1]

pH-range
8.5–9.4 (about half-maximal activity at pH 8.5 and 9.4) [1]

Temperature optimum (°C)
50 [1]

Temperature range (°C)
40–60 (about half-maximal activity at 40°C and about 80% of activity at 60°C) [1]

3 ENZYME STRUCTURE

Molecular weight

Subunits

Glycoprotein/Lipoprotein
–

4 ISOLATION/PREPARATION

Source organism
Mycobacterium tuberculosis H37Ra (strain number 7417) [1]

Source tissue

Localization in source

Purification
Mycobacterium tuberculosis [1]

Crystallization
–

Cloned

–

Renatured

–

5 STABILITY

pH

Temperature (°C)
   28 (6 h stable) [1]; 37 (about 8% loss of activity after 6 h) [1]; 60 (about 45% loss of activity after 10 min) [1]; 65 (about 66% loss of activity after 10 min) [1]; 70 ($t_{1/2}$: 1 min, about 55% loss of activity after 2 min) [1]

Oxidation

Organic solvent

General stability information

Storage

6 CROSSREFERENCES TO STRUCTURE DATABANKS

PIR/MIPS code

Brookhaven code

7 LITERATURE REFERENCES

[1] Jayaram, H.N., Ramakrishnan, T., Vaidyanathan, C.S.: Indian J. Biochem.,6,106–110 (1969)

1 NOMENCLATURE

EC number
2.3.2.8

Systematic name
L-Arginyl-tRNA:protein arginyltransferase

Recommended name
Arginyltransferase

Synonyms
Arginine transferase
Arginyl-transfer ribonucleate-protein aminoacyltransferase
Arginyl-transfer ribonucleate-protein transferase
Arginyl-tRNA protein transferase

CAS Reg. No.
37257-24-2

---

2 REACTION AND SPECIFICITY

Catalysed reaction
L-Arginyl-tRNA + protein →
→ tRNA + L-arginyl-protein

Reaction type
Aminoacyl group transfer

Natural substrates
L-Arginyl-tRNA + acceptor protein (catalyzes post-translational ribosome-in-dependent modification of certain acceptor proteins [1, 7], possibly involved in degradation of proteins with acidic NH_2-termini [7]) [1, 7]

Substrate spectrum
1 L-Arginyl-tRNA + acceptor protein (specific for arginine [2], addition to amino terminal [1], acceptor proteins require an acidic amino terminal, an Asp- or Glu-residue at the acceptor site [6], in vitro acceptors are bovine serum albumin [1–3, 7], bovine insulin (less effective [2]) [1, 2], bovine thyroglobulin [1, 2], bovine alpha-lactalbumin [4, 7], rabbit liver fructose diphosphatase (less effective [2]) [1, 2], kappa-light chain of immunoglobulin, soybean trypsin inhibitor [7], di- and tripeptides [3], non-peptide-derivatives of dicarboxylic amino acids with blocked alpha-carboxyl group and unsubstituted beta- or gamma-carboxyl group, such as isoasparagine and isoglutamine [3], bovine serum albumin ac-

cepts 1 mol arginine/mol, thyroglobulin 2 mol/mol [2], no acceptors are rabbit muscle aldolase [1], heme proteins [1, 2], Candida transaldolases I and III [2], lysozyme, ox RNase, human alpha-lactalbumin, beta-casein, beta-lactoglobulin [7], heparin [4]) [1–7]

2 L-Arginyl-tRNA + L-Glu-L-Ala (other dipeptides: overview, not dipeptides with D-Glu) [3]

3 L-Arginyl-tRNA + L-Glu-L-Ala-L-Ala (other tripeptides: overview) [3]

4 L-Arginyl-tRNA + L-cystinyl-bis-L-Ala (poor substrate) [3]

5 L-Arginyl-tRNA + L-Asp-L-Ala (other dipeptides: overview) [3]

6 L-Arginyl-tRNA + L-aspartic acid (poor substrate) [3]

7 L-Arginyl-tRNA + L-glutamic acid (poor substrate) [3]

Product spectrum

1 tRNA + L-arginyl-acceptor protein [1–7]

2 tRNA + L-Arg-L-Glu-L-Ala [3]

3 ?

4 ?

5 ?

6 tRNA + L-Arg-L-Asp [3]

7 tRNA + L-Arg-L-Glu [3]

Inhibitor(s)

Di- and tripeptides (with glutamyl-, asparagyl- and to a lesser extent cystinyl-NH_2-terminal residues, competitive to bovine serum albumin acceptor) [3]; L-Glu-L-Val-L-Phe [7]; Isoasparagine [3]; Isoglutamine [3]; Heparin (kinetics, competitive to L-arginyl-tRNA) [4]; Spermine (competitive to L-arginyl-tRNA) [4]; Spermidine (competitive to L-arginyl-tRNA) [4]; Antibodies to hog enzyme (prepared in rabbit) [5]; More (no inhibition by puromycin [1], chondroitinsulfate A, B or C, hyaluronic acid, D-glucosamine N-sulfate, D-glucose 6-sulfate, D-glucosamine, D-galactosamine, N-acetyl-D-glucosamine, N-acetyl-D-galactosamine, D-xylose, D-glucuronic acid [4]) [1, 4]

Cofactor(s)/prosthetic group(s)/activating agents

2-Mercaptoethanol (requirement, 0.1 M [2]) [1, 2]; Dithiothreitol (requirement, 0.01 M) [2]

Metal compounds/salts

K^+ (0.2 M [2], requirement for monovalent cations) [1, 2]; Na^+ (requirement for monovalent cations) [1]; NH_4^+ (requirement for monovalent cations) [1]; More (no requirement for Mg^{2+} or GTP) [1]

Turnover number (min^{-1})

Specific activity (U/mg)

0.00084 [1]; 0.0047 [7]; 0.114 [2, 6]; 2.3 [4]; 7.6 [5]

K_m-value (mM)

pH-optimum
   9.0 (about) [2]

pH-range
   8.0–9.8 (about half-maximal activity at pH 8.0, about 80% of maximal activity
   at pH 9.8) [2]

Temperature optimum (°C)
   37 (assay at) [1, 2]

Temperature range (°C)

---

3 ENZYME STRUCTURE

Molecular weight
   50000 (rabbit, gel filtration) [7]

Subunits
   ? (x × 35000, pig, SDS-PAGE) [5]

Glycoprotein/Lipoprotein
   –

---

4 ISOLATION/PREPARATION

Source organism
   Rabbit [1–3, 6, 7]; Sheep [8]; Pig (hog) [4, 5]; Rat [4]; Baker's yeast [4]

Source tissue
   Liver [1–3, 6]; Reticulocytes [7]; Thyroid [8]; Kidney [4, 5]; Spleen (rat) [4];
   Cell [4]

Localization in source
   Cytoplasm (soluble) [1, 2, 4–6, 8]

Purification
   Rabbit (partial [1, 7]) [1, 2, 6, 7]; Pig (affinity chromatography on heparin-Se-
   pharose [4, 5] and angiotensin II-Sepharose combined with affinity elution
   with lactalbumin [5]) [4, 5]

Crystallization
   –

Cloned
   –

Renatured

---

5 STABILITY

pH

Temperature (°C)

Oxidation

Organic solvent

General stability information
Freezing inactivates [1]; Polyethylene glycol, 30%, stabilizes concentrated solutions during storage [2]; Glycerol, 20%, stabilizes concentrated enzyme solutions during storage [2]; 2-Mercaptoethanol or dithiothreitol stabilize during storage [2]; Phosphate, 0.04 M, stabilizes [5]

Storage
−20°C, crude, at least 6 months [1]; 0°C, partially purified, 75% loss of activity within 1 week [1]; 0°C, concentrated enzyme solution, 20% glycerol, $t_{1/2}$: about 1 week [2, 6]; 4°C, concentrated enzyme solution, 0.5 M phosphate-KOH buffer, pH 7.8, 0.1 M 2-mercaptoethanol, up to 6 months [5]; 4°C, purified, 0.05 M Tris-HCl buffer, pH 8.0, 0.1 M 2-mercaptoethanol, 0.09 M KCl, 0.02 M spermine, 1 week [5]

---

6 CROSSREFERENCES TO STRUCTURE DATABANKS

PIR/MIPS code
PRR2:S31566 (yeast (Saccaromyces cerevisiae))

Brookhaven code

---

7 LITERATURE REFERENCES

[1] Soffer, R.L., Horinishi, H.: J. Mol. Biol.,43,163–175 (1969)
[2] Soffer, R.L.: J. Biol. Chem.,245,731–737 (1970)
[3] Soffer, R.L.: J. Biol. Chem.,248,2918–2921 (1973)
[4] Kato, M.: J. Biochem.,94,2015–2022 (1983)
[5] Kato, M., Nozawa, Y.: Anal. Biochem.,143,361–367 (1984)
[6] Deutch, C.E.: Methods Enzymol.,106,198–205 (1984) (Review)
[7] Ciechanover, A., Ferber, S., Ganoth, D., Elias, S., Avram, H., Arfin, S.: J. Biol. Chem.,263,11155–11167 (1988)
[8] Soffer, R.L.: Biochim. Biophys. Acta,155,228–240 (1968)

1 NOMENCLATURE

EC number
2.3.2.9

Systematic name
(gamma-L-Glutamyl)-N^1-(4-hydroxymethylphenyl)hydrazine:(acceptor) gamma-glutamyltransferase

Recommended name
Agaritine gamma-glutamyltransferase

Synonyms
Glutamyl transferase, agaritine gamma-

CAS Reg. No.
37257-25-3

---

2 REACTION AND SPECIFICITY

Catalysed reaction
Agaritine + acceptor →
→ 4-hydroxymethylphenylhydrazine + gamma-L-glutamyl-acceptor

Reaction type
Aminoacyl group transfer

Natural substrates
beta-N-(gamma-L(+)-Glutamyl)-4-hydroxymethylphenylhydrazine + acceptor (i.e. agaritine, unique enzyme for gamma-L-glutamyl transfer in the basidiomycete family Agaricaceae) [1]

Substrate spectrum
1 beta-N-(gamma-L(+)-Glutamyl)-4-hydroxymethylphenylhydrazine + acceptor (i.e. agaritine, can be replaced by gamma-L-glutamyl-1-naphthylhydrazine, donor-specificity: gamma-L-glutamyl-residue is essential and not replaceable by acetyl-, beta-aspartyl-, delta-homoglutaryl-, glutaryl-, gamma-D-glutamyl-, gamma-L-(alpha-N-acetyl)glutamyl and gamma-L-(alpha-glutamyl)glycine-residues, acceptors may be hydroxylamine, phenylhydrazine, p-hydroxyaniline (r), NH_4^+, not glycine, phenylalanine and aspartic acid, in the absence of acceptor the gamma-glutamyl-moiety is irreversibly transferred to H_2O, hydrolyzes gamma-glutamyl-amides and esters as well as hydrazides, no donors are gamma-glutamyl-derivatives of ammonia or hydrazine and their corresponding N-alkylated analogues, glutathione, gamma-glutamyl-phenylalanine, gamma-glutamyl-beta-aminoisobutyric acid) [1]

   2 Glutamine + phenylhydrazine [1]
   3 N-gamma-L-Glutamylcyclohexylamine + phenylhydrazine (most effective
     donor) [1]
   4 N-gamma-L-Glutamylcyclohexylamine + NH_4^+ (at high NH_4^+-concentra-
     tions) [1]
   5 gamma-L-Glutamyl-p-hydroxyaniline + phenylhydrazine [1]
   6 gamma-Benzyl-L-glutamate + phenylhydrazine [1]

Product spectrum
   1 4-Hydroxymethylphenylhydrazine + gamma-L-glutamyl-acceptor [1]
   2 ?
   3 Cyclohexylamine + gamma-L-glutamylphenylhydrazine [1]
   4 Cyclohexylamine + glutamine [1]
   5 p-Hydroxyaniline + gamma-L-glutamylphenylhydrazine [1]
   6 ? + gamma-L-glutamylphenylhydrazine (L-glutamate in the absence of
     phenylhydrazine) [1]

Inhibitor(s)
beta-N-(gamma-L-Glutamyl)-1-naphthylhydrazine (competitive inhibition of
agaritine hydrolysis) [1]; gamma-L-Glutamylhydrazine (competitive inhibition
of agaritine hydrolysis) [1]; N-(gamma-L-Glutamyl)-4-hydroxyaniline (competitive inhibition of agaritine hydrolysis) [1]; N-(gamma-L-Glutamyl)-cyclo-
hexylamine (competitive inhibition of agaritine hydrolysis) [1];
gamma-(O-Benzyl)-L-glutamic acid (competitive inhibition of agaritine hydrolysis) [1]; L-Glutamine (competitive inhibition of agaritine hydrolysis) [1];
4-Hydroxyaniline (agaritine hydrolysis) [1]; Iodoacetate [1]; p-Hydroxymercuribenzoate [1]; Hg^{2+} [1]; Cu^{2+} [1]; Zn^{2+} [1]; Diisopropylfluorophosphate
[1]; NaN_3 [1]; Glycerol (40%) [1]; Phenylalanine (hydroxylamine as substrate) [1]; Glycine (hydroxylamine as substrate) [1]; More (no inhibition by
NH_4^+ (high concentrations), leucine, aspartic acid) [1]

Cofactor(s)/prosthetic group(s)/activating agents
More (no activation by ADP or ATP) [1]

Metal compounds/salts
More (no activation by Mg^{2+} or Mn^{2+}) [1]

Turnover number (min^{-1})

Specific activity (U/mg)
   1.4 [1]

K_m-value (mM)

pH-optimum
   7.0 [1]

pH-range
6.6–7.8 (about half-maximal activity at pH 6.6 and 7.8) [1]

Temperature optimum (°C)
30 (assay at) [1]

Temperature range (°C)

---

3 ENZYME STRUCTURE

Molecular weight

Subunits

Glycoprotein/Lipoprotein
–

---

4 ISOLATION/PREPARATION

Source organism
Agaricus bisporus (basidiomycete) [1]; Agaricus edulis [1]; Agaricus patter-
sonii [1]; Agaricus perrarus [1]; Agaricus xanthodermus [1]; More (no activity in Agaricus sterlingii and subrutilescens, no activity in spinach leaves,
carrot roots, avocado fruit, rat and pigeon liver, E. coli, Eremothecium ash-
byii, Neurospora crassa, sporophores of Caprinus sp., Boletus sp., Colosy-
phia sp., Discina sp., Cortinarius sp., Xeromphalina sp., Pleurotus sp. and
Helvella sp.) [1]

Source tissue
Sporophores [1]

Localization in source
Cytoplasm [1]

Purification
Agaricus bisporus (partial) [1]

Crystallization
–

Cloned
–

Renatured
–

5 STABILITY

pH

Temperature (°C)

Oxidation

Organic solvent
Glycerol, 40%, inactivates [1]

General stability information
2-Mercaptoethanol stabilizes during purification [1]; Freeze-thawing inactivates [1]

Storage

---

6 CROSSREFERENCES TO STRUCTURE DATABANKS

PIR/MIPS code

Brookhaven code

---

7 LITERATURE REFERENCES

[1] Gigliotti, H.J., Levenberg, B.: J. Biol. Chem.,239,2274–2284 (1964)

1 NOMENCLATURE

EC number
    2.3.2.10

Systematic name
    L-Alanyl-tRNA:UDP-N-acetylmuramoyl-L-alanyl-D-glutamyl-L-lysyl-D-alanyl-D-alanine N^6-alanyltransferase

Recommended name
    UDP-N-acetylmuramoylpentapeptide-lysine N^6-alanyltransferase

Synonyms
    Alanyltransferase, uridine diphosphoacetylmuramoylpentapeptide lysine N^6-
    Alanyl-transfer ribonucleate-uridine diphosphoacetylmuramoylpentapeptide transferase
    UDP-N-acetylmuramoylpentapeptide lysine N^6-alanyltransferase
    Uridine diphosphoacetylmuramoylpentapeptide lysine N^6-alanyltransferase

CAS Reg. No.
    37257-26-4

---

2 REACTION AND SPECIFICITY

Catalysed reaction
    L-Alanyl-tRNA +
    UDP-N-acetylmuramoyl-L-alanyl-D-glutamyl-L-lysyl-D-alanyl-D-alanine →
    → tRNA +
    UDP-N-acetylmuramoyl-L-alanyl-D-glutamyl-N^6-(L-alanyl)-L-lysyl-D-alanyl-D-alanine

Reaction type
    Aminoacyl group transfer

Natural substrates
    L-Alanyl-tRNA +
    UDP-N-acetylmuramoyl-L-alanyl-D-glutamyl-L-lysyl-D-alanyl-D-alanine (involved in peptidoglucan metabolism, responsible for interpeptide bridges)
    [1]

Substrate spectrum

1 L-Alanyl-tRNA^{Ala} +
UDP-N-acetylmuramoyl-L-alanyl-D-glutamyl-L-lysyl-D-alanyl-D-alanine (ir
[1], no strict specificity for tRNA-carrier: tRNA^{Cys} can replace tRNA^{Ala},
tRNA from other species can substitute to some extent for the homologous tRNA from Lactobacillus viridescens, higher specificity for amino
acid transferred: amino acids with larger substituents are no substrates,
UDP-acetylmuramyl-L-Ala-D-Glu-meso-diaminopimelic acid-D-Ala-D-Ala
can replace UDP-acetylmuramyl-L-Ala-D-Glu-L-Lys-D-Ala-D-Ala at 15% the
transfer rate, poor or no acceptors are acetylmuramyl- and phospho-
acetylmuramyl-L-Ala-L-Glu-L-Lys-D-Ala-D-Ala or
UDP-acetylmuramyl-L-Ala-L-Glu-L-Lys) [1]

2 L-Seryl-tRNA +
UDP-N-acetylmuramoyl-L-alanyl-D-glutamyl-L-lysyl-D-alanyl-D-alanine
(about half as effective as L-alanyl-tRNA) [1]

3 L-Cysteine-tRNA +
UDP-N-acetylmuramoyl-L-alanyl-D-glutamyl-L-lysyl-D-alanyl-D-alanine [1]

4 L-Glycine-tRNA +
UDP-N-acetylmuramoyl-L-alanyl-D-glutamyl-L-lysyl-D-alanyl-D-alanine (poor
substrate) [1]

Product spectrum

1 tRNA +
UDP-N-acetylmuramoyl-L-alanyl-D-glutamyl-N^6-(L-alanyl)-L-lysyl-D-alanyl-D-
alanine [1]

2 tRNA +
UDP-N-acetylmuramoyl-L-alanyl-D-glutamyl-N^6-(L-seryl)-L-lysyl-D-alanyl-D-
alanine [1]

3 ?

4 ?

Inhibitor(s)
More (no inhibition by EDTA or metal ions) [1]

Cofactor(s)/prosthetic group(s)/activating agents

Metal compounds/salts
More (no activation by EDTA or metal ion requirement) [1]

Turnover number (min^{-1})

Specific activity (U/mg)
0.28 [1]

K$_m$-value (mM)
0.0002 (UDP-acetylmuramoylpentapeptide) [1]

pH-optimum
   6.8–7.2 (L-alanyl-tRNA) [1]; 7.2–7.6 (L-seryl-tRNA) [1]

pH-range
   5.2–8.6 (L-alanyl-tRNA, about 80% of maximal activity at pH 5.2 and 8.6) [1];
   6.1–8.6 (L-seryl-tRNA, about half-maximal activity at pH 6.1 and about 90%
   of maximal activity at pH 8.6) [1]

Temperature optimum (°C)
   30 (assay at) [1]

Temperature range (°C)

---

3 ENZYME STRUCTURE

Molecular weight
   40000 (Lactobacillus viridescens, PAGE) [1]

Subunits

Glycoprotein/Lipoprotein
   –

---

4 ISOLATION/PREPARATION

Source organism
   Lactobacillus viridescens [1]

Source tissue
   Cell [1]

Localization in source
   Soluble [1]

Purification
   Lactobacillus viridescens [1]

Crystallization
   –

Cloned
   –

Renatured
   –

5 STABILITY

pH

Temperature (°C)

Oxidation

Organic solvent

General stability information

Storage

6 CROSSREFERENCES TO STRUCTURE DATABANKS

PIR/MIPS code

Brookhaven code

7 LITERATURE REFERENCES

[1] Plapp, R., Strominger, J.L.: J. Biol. Chem.,245,3673–3682 (1970)

1 NOMENCLATURE

EC number
2.3.2.11

Systematic name
L-Alanyl-tRNA:phosphatidylglycerol alanyltransferase

Recommended name
Alanylphosphatidylglycerol synthase

Synonyms
Synthase, O-alanylphosphatidylglycerol
Alanyl phosphatidylglycerol synthetase [1]

CAS Reg. No.
37257-27-5

2 REACTION AND SPECIFICITY

Catalysed reaction
L-Alanyl-tRNA + phosphatidylglycerol →
→ tRNA + 3-O-L-alanyl-1-O-phosphatidylglycerol

Reaction type
Aminoacyl group transfer

Natural substrates

Substrate spectrum
1 L-Alanyl-tRNA^{Ala} + phosphatidylglycerol [1]
2 More (no substrates are N-acetylalanyl-tRNA, lactyl-tRNA (alanyl-tRNA treated with HNO_3), alanyl-tRNA^{Cys} and phenylalanyl-tRNA^{Ala}) [1]

Product spectrum
1 tRNA^{Ala} + 3-O-L-alanyl-1-O-phosphatidylglycerol [1]
2 ?

Inhibitor(s)

Cofactor(s)/prosthetic group(s)/activating agents

Metal compounds/salts

Turnover number (min^{-1})

Specific activity (U/mg)

K$_m$-value (mM)

pH-optimum
5.7 [1]

pH-range

Temperature optimum (°C)
30 (assay at) [1]

Temperature range (°C)

---

3 ENZYME STRUCTURE

Molecular weight

Subunits

Glycoprotein/Lipoprotein
–   .

---

4 ISOLATION/PREPARATION

Source organism
Clostridium welchii [1]

Source tissue
Cell [1]

Localization in source
Particulate [1]

Purification

Crystallization
–

Cloned
–

Renatured
–

5 STABILITY

pH

Temperature (°C)

Oxidation

Organic solvent

General stability information
Glycerol stabilizes during storage [1]

Storage
−60°C, several months [1]; −20°C, 3 weeks, 10 mM Tris/HCl, pH 8.0, 10 mM
2-mercaptoethanol, 30% glycerol [1]

6 CROSSREFERENCES TO STRUCTURE DATABANKS

PIR/MIPS code

Brookhaven code

7 LITERATURE REFERENCES

[1] Gould, R.M., Thornton, M.P., Liepkalns, V., Lennarz, W.J.: J. Biol. Chem.,243, 3096–3104 (1968)

1 NOMENCLATURE

EC number
  2.3.2.12

Systematic name
  Peptidyl-tRNA:aminoacyl-tRNA N-peptidyltransferase

Recommended name
  Peptidyltransferase

Synonyms
  Transpeptidase
  Ribosomal peptidyltransferase [4, 6]

CAS Reg. No.
  9059-29-4

2 REACTION AND SPECIFICITY

Catalysed reaction
  Peptidyl-tRNA$_1$ + aminoacyl-tRNA$_2$ →
  → tRNA$_1$ + peptidyl-aminoacyl-tRNA$_2$

Reaction type
  Aminoacyl group transfer

Natural substrates
  Peptidyl-tRNA$_1$ + alpha-aminoacyl-tRNA$_2$ (involved in protein biosynthesis, peptide chain elongation) [2]

Substrate spectrum
  1 Peptidyl-tRNA$_1$ + alpha-aminoacyl-tRNA$_2$ (transfer of polypeptide from 2'-OH or 3'-OH of tRNA to alpha-NH$_2$ of incoming aminoacyl-tRNA with elimination of first tRNA [2], (Phe)$_n$-tRNA [2], or (Lys)$_n$-tRNA [3, 4] can be substrates in vitro [2–4], puromycin substitutes for aminoacyl-tRNA [2–4, 7, 9–11, 13–15]) [2–4, 6–11, 13–15]

Product spectrum
  1 tRNA$_1$ + peptidyl-aminoacyl-tRNA$_2$ [2, 6, 9]

Inhibitor(s)

Spermine (mechanism, kinetics) [1]; EDTA [2]; Puromycin (splits bond between poly-Phe and tRNA) [2, 9]; Chlortetracycline [2]; Chloramphenicol (elongation factor G plus GTP protect [8], mechanism, kinetics [14]) [2, 8, 14]; Thiamphenicol (mechanism) [14]; Tenevel (mechanism) [14]; Gougerotin (elongation factor G plus GTP protect [8]) [4, 8]; Blasticidin S (most effective [4], kinetics [13]) [4, 13]; Amicetin (elongation factor G plus GTP protect [8]) [4, 7, 8]; Plicacetin (weak [7]) [4, 7]; Bamicetin (weak [7]) [4, 7]; 40S-Ribosomal subunits (methanol reduces inhibition in vitro) [5]; Poly-(U) [5]; Poly-(A) [5]; Oxamicetin [7]; Amicetin-analogs [7]; Lincomycin (antagonized by elongation factor G plus GTP) [8]; Griseoviridin (antagonized by elongation factor G plus GTP) [8]; Cytidylyl-(3'-5')-2'(3')-O-L-phenylalanyl-adenosine (reversible by puromycin) [9]; 2'(3')-O-L-Phenylalanyl-L-adenosine (weak) [9]; Cytidylyl-3'-5'-L-adenosine (weak) [9]; Phenylboric acid and derivatives (kinetics) [11]; 2'(3')-O-Cycloleucyl- and alpha-aminobutyryl-derivatives of cytidylyl-(3'-5')-L-adenosine [12]

Cofactor(s)/prosthetic group(s)/activating agents

Poly-(U) (activation, only with acetyl-phenylalanine-tRNA as substrate) [5]; Poly-(A) (slight increase of activity, with acetyl-lysine-tRNA as substrate) [5]; Alcohol (requirement, in vitro, ethanol or methanol, 45–50% [10]) [5, 10]

Metal compounds/salts

Mg^{2+} (requirement, 0.01 M [3], 0.004 M [5], for $(Lys)_n$-tRNA-binding to ribosomes [5]) [1, 3, 5, 7, 8]; K^+ (requirement, for aminoacyl-tRNA-binding to ribosomes [3], 0.3 M [5]) [3, 5]; NH_4^+ (requirement) [2, 3]; NaCl (activation, less efficient than KCl) [5]

Turnover number (min^{-1})

Specific activity (U/mg)

K_m-value (mM)

More (kinetics with different model substrates [10], kinetics [13, 15]) [10, 13, 15]

pH-optimum

7.5–8.3 (puromycin-reaction) [5]

pH-range

Temperature optimum (°C)

20 (puromycin-reaction) [5]

Temperature range (°C)

15–30 (about 80% of maximal activity at 15°C and 30°C, puromycin-reaction) [5]

3 ENZYME STRUCTURE

Molecular weight

Subunits

Glycoprotein/Lipoprotein

—

4 ISOLATION/PREPARATION

Source organism
E. coli (strain B [1, 3, 4, 7, 13–15], strain MRE-600 [8, 10]) [1–4, 7–15]; Rat [5]

Source tissue
Liver [5]; Cell [1–4, 7–15]

Localization in source
Ribosomes (60S-subunit [1–15], not 40S-subunit [5]) [1–15]

Purification

Crystallization
—

Cloned
—

Renatured
—

5 STABILITY

pH

Temperature (°C)
12 ($t_{1/2}$: 30 min, puromycin-reaction) [5]; 20 ($t_{1/2}$: about 20 min, puromycin-reaction) [5]

Oxidation

Organic solvent

General stability information

Storage

Enzyme Handbook © Springer-Verlag Berlin Heidelberg 1996
Duplication, reproduction and storage in data banks are only
allowed with the prior permission of the publishers

6 CROSSREFERENCES TO STRUCTURE DATABANKS

PIR/MIPS code

Brookhaven code

---

7 LITERATURE REFERENCES

[1] Kalpaxis, D.L., Drainas D.: Arch. Biochem. Biophys.,300,629–634 (1993)
[2] Traut, R.R., Monro, R.E.: J. Mol. Biol.,10,63–72 (1964)
[3] Rychlík, I.: Biochim. Biophys. Acta,114,425–427 (1966)
[4] Cerná, J., Rychlík, I., Lichtenthaler, F.W.: FEBS Lett.,30,147–150 (1973)
[5] Thompson, H.A., Moldave, K.: Biochemistry,13,1348–1353 (1974)
[6] Fahnestock, S., Neumann, H., Rich, A.: Methods Enzymol.,30 Pt.F,489–497 (1974) (Review)
[7] Lichtenthaler, F.W., Cerná, J., Rychlík, I.: FEBS Lett.,53,184–187 (1975)
[8] Spirin, A.S., Asatryan, L.S.: FEBS Lett.,70,101–104 (1976)
[9] Bhuta, P., Zemlicka, J.: Biochem. Biophys. Res. Commun.,83,414–420 (1978)
[10] Streltsov, S., Kosenjuk, A., Kukhanova, M., Krayevsky, A., Gottikh, B.: FEBS Lett.,104,279–283 (1979)
[11] Cerná, J., Rychlík, I.: FEBS Lett.,119,342–348 (1980)
[12] Chladek, S., Bhuta, P.: Biochim. Biophys. Acta,696,212–217 (1982)
[13] Kalpaxis, D.L., Theocharis, D.A., Coutsogeorgopoulos, C.: Eur. J. Biochem.,154, 267–271 (1986)
[14] Drainas D., Kalpaxis, D.L., Coutsogeorgopoulos, C.: Eur. J. Biochem.,164,53–58 (1987)
[15] Synetos, D., Coutsogeorgopoulos, C.: Biochim. Biophys. Acta,923,275–285 (1987)

1 NOMENCLATURE

EC number
2.3.2.13

Systematic name
Protein-glutamine:amine gamma-glutamyltransferase

Recommended name
Protein-glutamine gamma-glutamyltransferase

Synonyms
Transglutaminase
Factor XIIIa
Fibrinoligase
Fibrin stabilizing factor [2]
Glutamyltransferase, glutaminylpeptide gamma-
Glutaminylpeptide gamma-glutamyltransferase
Polyamine transglutaminase
Tissue transglutaminase
R-Glutaminyl-peptide:amine gamma-glutamyl transferase [4]

CAS Reg. No.
80146-85-6

2 REACTION AND SPECIFICITY

Catalysed reaction
Protein glutamine + alkylamine →
→ protein N^5-alkylglutamine + NH_3 (mechanism and structure [2, 16], mechanism [23, 40], structure, function, evolution [12, 17, 40])

Reaction type
Aminoacyl group transfer

Natural substrates
Protein-bound gamma-glutamine + alkylamine (involved in a wide variety of cellular processes, including growth, differentiation, stabilization of cytoskeleton [13], last step in blood coagulation [2, 8, 13, 15], production of vaginal plug by postejaculatory clotting of rodent seminal plasma, formation of chemically resistant envelope of the stratum [15, 17], mediates membrane-structural changes [32], epidermal enzyme involved in formation of cornified envelope [33]) [2, 8, 13, 15, 17, 32, 33, 39]

Substrate spectrum

1 Protein-bound gamma-glutamine + alkylamine (acyl-transfer reaction [5, 40], hydrolysis and aminolysis of certain aliphatic amides and active esters (p-nitrophenyl esters [15, 16] and thiolesters [16]) [15, 16, 40], catalyzes post-translational protein modifications by transamidation of glutamine residues [12], forms intramolecular isopeptide bonds between fibrin molecules [2, 8, 12], donors: gamma-carboxamide groups of protein-bound glutamine, acceptors: epsilon-amino groups of protein-bound lysine [8, 40], transglutaminase B: simultaneously gamma-polymer and alpha-polymer formation [27], amine donors are primary amines [15], diamines and polyamines [36], overview [40], no amine donors are tyrosinamide, glycine, Gly-Leu, gamma-aminobutyric acid [40], broad specificity towards amine acceptor [15, 40], in the absence of amine acceptors H_2O acts as substrate [15], synthetic protein acceptors [9, 40], identification of natural protein substrates [35], substrates are membrane-associated erythrocyte proteins [7], coagulation factor V, alpha$_2$-macroglobulin, platelet myosin, actin [2], fibronectin [2, 12], plasminogen-activator inhibitor type-2 (factor XIIIa, tissue transglutaminase) [10], acetylated B-chains of oxidized insulin [15], pepsin, thrombin, cellulase, creatine kinase [36], fibrinogen, beta-lactoglobulin, casein, insulin [36, 40], carbobenzoxy-L-Gln-Gly [15]) [2, 4, 5, 7–10, 12, 15, 16, 27, 35, 36, 38, 40]

2 Putrescine + casein (in vitro acceptor [3], alpha- or beta-casein [15]) [3–5, 15, 31, 33, 39]

3 Putrescine + fibronectin (in vivo acceptor [3]) [3, 9]

4 Putrescine + bovine muscle actin (preferred substrate) [39]

5 Putrescine + N,N'-dimethylcasein (spermidine [32, 36, 39], spermine [36, 39], diaminopropane and cadaverine can replace putrescine [32, 36, 39]) [7, 13, 14, 22, 28–30, 34, 36, 39]

6 Dansylcadaverine + casein (casein can be replaced by various synthetic peptide acceptors [9]) [6, 8, 9]

7 Dansylcadaverine + F-actin [8]

8 Histamine + acetyl-alpha$_{S1}$-casein [37]

9 Histamine + maleyl-bovine serum albumin [37]

10 Methylamine + succinyl-beta-casein (transglutaminase B (TGB)) [27]

11 Hydroxylamine + carbobenzoxy-Gln-Gly (other substrates are carbobenzoxy-Gln-Gln-Gly, Gly-Gln-Gln-Gly, Gly-Gly-Gln-Gly with 38%, 13% and 28% efficiency, respectively) [38]

12 GTP + H_2O (intrinsic GTPase activity) [5]

13 p-Nitrophenyl acetate + alanine ethylester [40]

14 More (no substrates are catalase [36], native bovine serum albumin [36, 38], bovine myosin, histone mixture, human serum fibronectin, spinach ribulose 1,5-diphosphate carboxylase-oxygenase, carbobenzoxyglutamine, carbobenzoxy-Asn-Gly [38], Z-L-glutaminylglycine, Z-alpha-L-glutamyl(gamma-p-nitrophenyl ester) glycine (guinea pig hair follicle enzyme [40])) [36, 38, 40]

Product spectrum

1 Protein N^5-alkylglutamine + NH_3 (resulting bonds are covalent and stable to proteolysis [12], with H_2O as acceptor the product is peptide bound glutamic acid [15]) [12, 15, 40]
2 ?
3 ?
4 ?
5 ?
6 ?
7 ?
8 ?
9 ?
10 ?
11 Carbobenzoxy-Gln-Gly-hydroxamate + ? [38]
12 GDP + phosphate [5]
13 N-Acetylalanine ethylester + ? [40]
14 ?

Inhibitor(s)

IAA (factor XIIIa [8], irreversible, pH 6.8 in the presence of Ca^{2+} [1, 40], incorporation of 1 mol carbamidomethyl/mol enzyme, substrate protects [1, 40], mechanism [40], no inhibition without Ca^{2+}, not calmodulin regulated transglutaminases [8]) [1, 8, 27, 40];
1,5-Iodoacetyl-5'-(sulfonyl-1-naphthyl)-ethylenediamine (0.01 mM) [27]; Monoiodoacetate (1 mM, weak [38]) [35, 38]; PCMB (0.1 mM [39], reversible by GSH [1]) [1, 27, 38, 39]; Hydroxymercuribenzoate [19]; p-Substituted mercuribenzoate [40]; Cu^{2+} (strong, 1 mM [35], trace amounts, 0.45 mM diethyldithiocarbamate stimulates crude preparation [36], weak [38], KCN or DTT restores activity [40], mechanism [40]) [35, 36, 38, 40]; Zn^{2+} [38, 40]; Pb-acetate [38]; Hg^{2+} [40]; Fe^{2+} [40]; DTNB (irreversible, carbobenzoxy-Phe protects [4], not reversible by GSH [1], reversible by DTT [40]) [1, 4, 40]; o-Phenanthroline (not reversible by Ca^{2+}) [36]; Tetrathionate (inactivation, not reversible by DTT) [40]; N-Ethylmaleimide (strong, 1 mM [38], 0.1 mM [39], substrate protects [1], not without Ca^{2+} [7]) [1, 7, 19, 38, 39]; DTT (1.5–16.5 mM) [36]; alpha-Difluoromethylornithine (suicide substrate, in the presence of casein, Ca^{2+}, putrescine, irreversible, competitive to putrescine or fibrinonectin) [3]; Ornithine (weak, suicide substrate in the presence of casein) [3]; Cadaverine (strong, putrescine as substrate) [36]; Monodansylcadaverine [32]; Hydroxylamine (0.1 M) [35]; Putrescine (0.1–0.2 mM, substrate inhibition) [39]; Spermidine (0.1–0.2 mM, substrate inhibition) [39]; Spermine (0.1–0.2 mM, substrate inhibition) [39]; GTP-gamma-S (GTP-hydrolysis) [5]; GDP (GTP-hydrolysis) [5]; GTP (50% as effective as ATP [29], tissue enzyme [12, 13], strong at suboptimal Ca^{2+}-level [14], not (0.1–0.5 mM [30], up to 1 mM at 5 mM Ca^{2+} [35]) [30, 35]) [12–14, 18, 29]; GDP (at

low Ca^{2+}-levels) [14]; GMP (at low Ca^{2+}-levels [14], not [5]) [14]; ATP (concentration-dependent [29], 3 mM [18], reversible, non-competitive to putrescine, not (up to 1 mM at 5 mM Ca^{2+} [35]) [14, 35]) [18, 29]; ADP (reversible) [18, 29]; AMP (weak) [18, 29]; CTP (3 mM, reversible, non-competitive to putrescine [18], not (up to 1 mM at 5 mM Ca^{2+} [35]) [14, 35]) [18, 29]; UTP (50% as effective as ATP, not (up to 1 mM at 5 mM Ca^{2+} [35]) [14, 35]) [18, 29]; EDTA (strong [35], complete inactivation, above 10 mM [30], 5 mM [35, 39], irreversible [39], not (1 mM) [38]) [11, 30, 35, 39, 40]; EGTA (2 mM, irreversible [39], weak, reversible by Ca^{2+} [36]) [36, 39]; Sodium citrate (above 10 mM, complete inactivation) [30]; Diethyl dicarbonate (not without Ca^{2+}) [7]; Chlorpromazine (reverses calmodulin enzyme stimulation) [8]; NH_4^+ [30]; NaCl (above 0.5 M) [35]; KCl (above 0.5 M) [35]; Tb^{3+} (strong, factor XIIIa, at high Ca^{2+}-levels, not reversible by Ca^{2+}) [11]; La(III)-ions (not reversible by Ca^{2+}) [11]; Ce(III)-ions (not reversible by Ca^{2+}) [11]; Gd(III)-ions (not reversible by Ca^{2+}) [11]; More (no inhibition by adenosine or adenine [18, 29], PMSF [27], diisopropylfluorophosphate [27, 35], Ba^{2+}, Co^{2+}, Fe^{3+}, K^+, Mg^{2+}, Mn^{2+}, Na^+, Ni^{2+}, Sr^{2+} [38], in the absence of Ca^{2+} no inhibition by sulfhydryl-reagents [27]) [18, 27, 29, 35, 38]

Cofactor(s)/prosthetic group(s)/activating agents

Thrombin (activation, of proenzyme [40], slight [8], human (not chicken) [8], Ca^{2+}-dependent proteolytical activation, in the presence of fibrin [15], removes blocked NH_2-terminal peptide of factor XIII and unmasks reactive thiol-group at Cys-314 to yield catalytically active factor XIIIa (i.e. fibrinoligase) from inactive zymogen XIII [2, 15], particulate enzyme [27], guinea pig epidermal enzyme [27], not (tissue-type) transglutaminase C (TGC) from rat, rabbit or guinea pig [27]) [2, 8, 12, 15, 27, 40]; DTT (requirement [20], activation (weak [35])) [20, 24, 26, 35]; Trypsin (activation, in the presence of Ca^{2+} [15], blocked by trypsin-inhibitors [20]) [15, 20]; Calmodulin/Ca^{2+} (activation, 10–200 nM (no activation of factor XIIIa), inhibition above 300 nM, kinetics) [8]

Metal compounds/salts

Ca^{2+} (requirement, 0.5 mM [30], 1 mM [40], 8 mM [35], Cys-thiol active binding-site identified [2], catalytically active monomeric metal-enzyme complex [15], mechanism [40], no activity below 1.25 mM, maximal activity at 2.5 mM and above (K_m: 1.5 mM), bovine enzyme [22], Ca^{2+} leads to plasma factor XIIIa dissociation into a'- and b-dimers [15], activation of GTPase activity [5], only with N,N'-dimethylcasein as substrate, no activation with plant proteins as substrate [36], not [38]) [1–3, 5, 7–9, 11–15, 19–22, 26–37, 39, 40]; Mn^{2+} (requirement [7], activation, 3.5% as effective as Ca^{2+} [19], slight [27]) [7, 19, 27, 40]; Sr^{2+} (activation, 27% as effective as Ca^{2+} [19], slight [27], not [7]) [19, 27, 40]; Tb^{3+} (requirement, not [10]) [7]; Ga^{3+} (requirement, not [10]) [7]; La^{3+} (requirement, 0.01–0.1 mM) [11]; Mg^{2+} (activation, slight [19, 40], stimulation of GTPase activity, 5 mM [5], not [7, 27]) [5, 9, 19, 40]; Ba^{2+} (activation, can replace Ca^{2+} to a lesser extent) [40]; Zn^{2+} (activation, 24.6% as effective as Ca^{2+} [19], not [27, 40]) [19]; Chaotropic salts (activation, in the presence of Ca^{2+}) [15]; More (no activation by trivalenic lanthanide ions, 0.01–0.1 mM [11], Cu^{2+} [27]) [11, 27]

Turnover number (min^{-1})

Specific activity (U/mg)
More (938 amine incorporation units/min [35]) [5, 6, 13, 15, 30, 35]; 0.005 [39]; 0.526 [31]; 1.0 [14]; 1.2 [37]; 2.5–3.0 [7]; 12.5 [1]; 14 [15]; 22.6 [38]; 546 [4]

K$_m$-value (mM)
More (hair follicle and liver enzyme with identical kinetic features [15], kinetic mechanism [16, 40], kinetic studies of membrane-associated and soluble enzyme [28], kinetic study with various transglutaminases: human factor XIIIa, guinea pig liver TGC, rat chondrosarcoma TGB [27], K$_m$ of N,N'-dimethylcasein: 0.44–0.52 mg/ml [28]) [15, 16, 27, 28, 40]; 0.0029 (acetyl-alpha$_{S1}$-casein, recombinant enzyme) [37]; 0.0032 (acetyl-alpha$_{S1}$-casein, native enzyme) [37]; 0.0044 (GTP, GTPase activity) [5]; 0.006 (casein, plasma factor XII, actin, chicken) [8]; 0.007 (actin, human) [8]; 0.011–0.012 (casein, chicken or human) [8]; 0.0214 (spermidine) [39]; 0.0317 (spermine) [39]; 0.04–0.05 (putrescine [28, 37], N-dimethylcasein [19]) [19, 28, 37]; 0.051 (putrescine, retinoic-acid-induced enzyme) [33]; 0.098–0.106 (putrescine, enzyme induced by 12-O-tetradecanoyl-phorbol-13-acetate or Ca^{2+}) [33]; 0.17 (putrescine) [3]; 0.203 (putrescine, bovine) [22]; 0.38 (histamine, recombinant enzyme) [37]; 0.52 (histamine, native enzyme) [37]; 0.6 (putrescine) [7]; 2.1 (alpha-difluoromethylornithine) [3]; 66 (carbobenzoxy-L-glutaminylglycine) [1]

pH-optimum
More (optima depend on amine donor substrate [40], pI: 5.6 [7], pI: 6.1 [39], pI: 7.6 (crude) [30], pI: 8.5–8.7 (purified enzyme) [30], pI: 8.9 [38]) [7, 30, 38–40]; 6 (about, hydrolysis reaction) [40]; 6–7 [38]; 9.0 [19]; 9.5 [33]; 10.0 [20]

pH-range

Temperature optimum (°C)
25 (assay at) [24]; 30 (assay at) [6, 7]; 35 (assay at) [36, 39]; 37 (assay at) [3–5, 22, 29, 32, 34, 35, 37, 38]; 50 (pH 6.0) [38]

Temperature range (°C)

3 ENZYME STRUCTURE

Molecular weight

More (MW of anionic isozyme is reduced by protease treatment without loss of catalytic activity [34], MW of zymogen factor XIII: 280000–330000 [40], amino acid composition [1, 40]) [1, 34, 40]

40000 (Streptoverticillium sp., gel filtration) [38]

50000 (mouse epidermal and hair follicle cationic isozyme, gel filtration) [34]

54000 (guinea pig (hair follicle [40]), gel filtration) [15, 40]

55000 (human epidermal enzyme [15], mouse [20], gel filtration [15, 20], guinea pig hair follicle, PAGE [40]) [15, 20, 40]

65000 (human, sucrose gradient centrifugation [7], rat transglutaminase C (TGC), gel filtration [30]) [7, 30]

76900 (guinea pig, sedimentation and diffusion) [1, 40]

77000 (Physarum polycephalum, FPLC gel filtration) [39]

80000 (human [6], rat (transglutaminase C (TGC) [27]) [22, 27], gel filtration) [6, 22, 27]

83005 (human, primary structure) [2]

86000–94000 (guinea pig, sedimentation equilibrium) [1, 40]

88000 (bovine, gel filtration) [22]

90000 (guinea pig, sedimentation equilibrium, meniscus depletion method, IAA-incorporation studies [1], mouse epidermis anionic isozyme, gel filtration [34]) [1, 34]

100000 (rat lung transglutaminase B (TGB), gel filtration) [26, 27]

Subunits

Monomer (1 × 40000, Streptoverticillium sp., SDS-PAGE [38], 1 × 50000, mouse hair follicle, SDS-PAGE [34], 1 × 55000, human epidermal enzyme, SDS-PAGE [15], 1 × 75000, guinea pig, SDS-PAGE [15], 1 × 80000, guinea pig [5], human [6], SDS-PAGE [5, 6], 1 × 80000–90000, guinea pig, gel filtration in 6 M guanidine [40], 1 × 85000, guinea pig, SDS-PAGE [40], 1 × 86000, Tachypleus tridentatus, SDS-PAGE [35], 1 × 92000, human, SDS-PAGE with or without 2-mercaptoethanol [7]) [5–7, 15, 34, 35, 38, 40]

Dimer (2 × 27000, guinea pig (hair follicle [40]), SDS-PAGE [15, 40], 2 × 39600, Physarum polycephalum, SDS-PAGE [39]) [15, 39, 40]

Tetramer (4 × 71000, human plasma factor XIIIa (a$'_2$b$_2$), SDS-PAGE) [15]

More (zymogens of plasma and platelet-coagulation factor XIIIa with different subunit structure: a$_2$b$_2$ and a$_2$, respectively [12, 15, 17], a-subunits (MW 75000) being the catalytically active, that in case of plasma enzyme are stabilized by the b-subunits (MW 80000) [17], tissue-type and epidermal TG are monomers, hair follicle TG and coagulation factor XIIIa (i.e. fibrinoligase) are dimers [17]) [12, 15, 17]

Glycoprotein/Lipoprotein

Glycoprotein (rat, chondrosarcoma [27], glycosylated with mannosyl-residues (not terminal position), substituted with saturated acyl residues and phosphoinositol [30]) [27, 30]; Lipoprotein (membrane-bound enzyme of keratocytes is anchored via palmitate and myristate) [12]; More (no carbohydrate [15, 40], not glycosylated [2, 12], guinea pig TGC [12]) [2, 12, 15, 40]

4 ISOLATION/PREPARATION

Source organism

Rat (tissue-type transglutaminase (TGC) [18], distribution [26], mature male Wistar [30], male Sasco/King (SD)BR strain [28]) [12, 18, 19, 22, 26–30]; Human (coagulation factor XIIIa [15, 40] and zymogen factor XIII (proenzyme of XIIIa) [40]) [2, 5–8, 10–15, 17, 18, 21, 29, 40]; Mouse (BALB/c strain [20, 33], newborn, CF57 strain [34]) [12, 20, 33, 34]; Bovine (calf [25]) [22, 25]; Guinea pig (Hartley strain [1], cloned in E. coli [37]) [1, 3–5, 12, 15–17, 23, 24, 31, 37, 40]; Rabbit [9]; Chinese hamster [26]; Pigeon [32]; Chicken [8]; Tachypleus tridentatus (Japanese horseshoe crab) [35]; Pisum sativum (pea, var. Kelvedon Wonder) [36]; Streptoverticillium sp. (tentatively classified, strain S-8112) [38]; Physarum polycephalum (slime mould, strain M3cV) [39]

Source tissue

Liver (guinea pig [12], rat [18, 22]) [1, 3–5, 12, 15, 16, 18, 22–24, 27–29, 31, 37, 38, 40]; Lung [26]; Brain (human) [18]; Macrophages (mouse [12]) [12]; Epidermis (stratum corneum [13, 21], callus, keratinocytes [12], epidermal carcinoma cell line A341 [13], primary and MCA3A1-cell line, retinoic acid induced enzyme differs from normal epidermal enzyme [33], subcellular distribution [33], 2 isoforms: anionic and cationic [34]) [12, 13, 20, 21, 33, 34]; Hair follicle (outer root shear cells [12], 2 cationic isoforms [34], guinea pig (not related to liver enzyme [40]) [15, 40]) [12, 15, 34, 40]; Chondrocytes (malignant, rat swarm chondrosarcoma) [27]; Lens (cortex [25]) [9, 25]; Blood platelets (human, identical with enzyme from plasma [2], placenta [2, 17], uterus, macrophages [17], coagulation factor XIIIa [15]) [2, 8, 15, 17, 40]; Blood plasma (human, identical with enzyme from placenta or platelets [2], coagulation factor XIIIa [15]) [2, 15, 40]; Erythrocytes (human [10], ghosts [32]) [5–7, 10, 14, 32]; Hemocytes [35]; Endothelial cells (human [12], aortic (cell suspension culture) [22]) [12, 22]; Gizzard (smooth muscle) [8]; Ovary [26]; Placenta (identical with enzyme from plasma [2], platelets [2, 17], macrophages [17]) [2, 12, 17, 40]; Uterus [40]; Cytotrophoblasts (3rd trimester of pregnancy) [10]; Syncytiotrophoblasts [10]; Coagulating gland secretion (not immunologically related to tissue-type enzyme or blood factor XIIIa) [30]; Hepatocytes [12]; Seedlings (apical meristematic tissue) [36]; Spherules [39]; Cell [37]; Cell suspension culture filtrate [38]; More (tissue distribution [12]) [12, 26]

Localization in source

Extracellular [38]; Soluble (depending on state of cell-proliferation, intracellular distribution [22], retinoic acid induced enzyme [33]) [1, 6, 13, 22, 26, 32, 33, 35, 36, 38, 40]; Cytoplasm (tissue transglutaminase, factor XIIIa [12]) [12, 13, 17]; Particulate (tissue transglutaminase of rat hepatocytes [12], on the surface of monocytes and tissue macrophages [12], rat [27]) [12, 26, 27]; Membrane-bound (keratocytes [12], plasma-membrane associated (lateral domain) [28]) [12, 28]; More (soluble in cells and organs devoid of significant association with extensive filamentous structure or extracellular matrix, particulate in organs with extensive filamentous structure or extracellular matrix) [26]

Purification

Guinea pig (one-step purification by monoclonal antibody immunoadsorbent [31], affinity chromatography on phenylalanine-Sepharose [4] or GTP-Agarose [5], hair-follicle, liver [15], recombinant enzyme expressed in E. coli [37]) [1, 4, 5, 15, 31, 37, 40]; Human (tissue [13], epidermal [15] transglutaminase TGC [15], plasma and platelet factor XIII [15, 40], partial [18, 29], 2 forms of epidermal enzyme [21], erythrocytes [14]) [13–15, 18, 21, 29, 40]; Mouse (partial [20], cationic isozyme) [20, 34]; Bovine (partial) [22]; Rat (partial [26–28], lung, matrix-bound enzyme solubilized, 3 isoforms: A, B and C [26], selectively solubilized with glycerol [28], Triton X-100, deoxycholate or n-octylglucoside only 20–30% effective [28], tissue-type transglutaminase [29]) [18, 19, 26–30]; Chicken (co-purified with alpha-actinin) [8]; Tachypleus tridentatus [35]; Physarum polycephalum [39]; Streptoverticillium sp. [38]

Crystallization

–

Cloned

(human placenta FXIIIa-gene [2, 12], TGC- and TGK-transglutaminases [12], guinea pig liver enzyme [17], cloned and expressed in E. coli (expression plasmid pKTG1) [37]) [2, 12, 17, 37]

Renatured

–

---

5 STABILITY

pH

5–9 (10 min stable, 37°C) [38]; 6–9 (stable, 25°C) [19]; 9 ($t_{1/2}$: 1–2 min stable at 37°C, at least 20 min stable at 4°C [13], thermolabile: $t_{1/2}$: 3–4 min, 37°C, (retinoic acid induced enzyme), 20% loss of activity within 20 min (epidermal enzyme) [33]) [13, 33]

Temperature (°C)
  4 (at least 20 min stable, pH 9.0) [13]; 25 (at least 30 min stable, pH 9.0 (retinoic acid induced enzyme)) [33]; 37 ($t_{1/2}$: 1–2 min, pH 9.0 [13], 3–4 min,
  pH 9 (retinoic acid induced enzyme), 25% loss of activity after 20 min, pH
  9.0 (epidermal enzyme) [33], 10 min stable, pH 5–9 [38]) [13, 33, 38]; 40
  (10 min stable, pH 7) [38]; 44 (inactivation within 20 min) [19]; 50 (26% loss
  of activity within 10 min, pH 7.0) [38]; 52 (inactivation within 4 min) [19]; 56
  (heating in the presence of Ca^{2+} increases activity 25-fold, human epidermal
  enzyme [15], stable for 45 min with Ca^{2+} [20]) [15, 20]; 60 (inactivation
  within 1 min) [19]

Oxidation

Organic solvent

General stability information
  EDTA, 1–2 mM, stabilizes [1]; DTT stabilizes [28, 37]; EDTA stabilizes in
  combination with DTT [28]; Ca^{2+} stabilizes in combination with DTT [28];
  Glycerol, 50% v/v, solubilizes and stabilizes [28]; Dialysis against Ca^{2+}-free
  buffers inactivates [28]; Gel electrophoresis inactivates [6]; Alkaline conditions destabilize [31]; Repeated freeze-thawing results in some loss of activity, DTT restores [15]; Ion-exchange chromatography on DEAE-cellulose,
  with 0.4 M NaCl containing buffer with or without Ca^{2+}, stable to [28]; Chromatography on DEAE-cellulose, anionic isozyme, rapid decrease of activity
  [34]; Repeated freeze-thawing cycles, cationic isozyme, stable to [34]; Lyophilization, cationic isozyme, stable to [34]

Storage
  –70°C, concentrated enzyme solution, at least 3 months [31]; –30°C, human
  zymogen factor XIII, over 1 year [15]; –30°C, lyophilized human zymogen
  factor XIII, several years [15]; –20°C, concentrated guinea pig enzyme solutions, several months [15]; –20°C, partially purified, 3 years [39]; 4°C, at
  least 3 months [35]; 4°C, at least 4 months [6]; 4°C, epidermal transglutaminase, at least 1 month [15]; 4°C, 5 mM Tris-HCl/2 mM EDTA buffer, pH 7.5,
  up to 3 months [1]; 4°C, 50 mM Tris, pH 7.5, 1 mM EDTA/0.5 mM DTT, up to
  2 weeks [13]; 4°C, membrane preparation, rapid inactivation, DTT protects
  in combination with Ca^{2+} or EDTA [28]; 20°C, pH 6–8, cationic isozyme, long
  periods [34]; Dilute enzyme solutions are unstable to storage at 4°C [13];
  Human factor XIIIa is more stable to storage, when Ca^{2+} is omitted [15]

6 CROSSREFERENCES TO STRUCTURE DATABANKS

PIR/MIPS code

PIR2:S19680 (bovine); PIR2:A47203 (chicken); PIR2:A29996 (guinea pig); PIR2:A26209 (guinea pig (fragment)); PIR2:A45321 (horseshoe crab (Tachypleus tridentatus)); PIR2:A39045 (human); PIR2:JC2501 (human); PIR2:B39045 (mouse); PIR2:A33477 (rabbit (fragment)); PIR2:B38423 (rat); PIR2:JC2090 (Streptoverticillium sp.); PIR2:JC2089 (precursor) Streptoverticillium sp.); PIR1:TGHUM1 (epidermal human); PIR3:PD0001 (liver guinea pig (fragment)); PIR1:EKHUX (plasma human); PIR2:PC2237 (tissue-type red sea bream (fragments))

Brookhaven code

1GGT (Human (Homo Sapiens) factor xiii recombinant in yeast (Saccharomyces cerevisiae))

7 LITERATURE REFERENCES

[1] Folk, J.E., Cole, P.W.: J. Biol. Chem.,241,5518–5525 (1966)
[2] Takahashi, N., Takahashi, Y., Putnam, F.W.: Proc. Natl. Acad. Sci. USA,83,8019–8023 (1986)
[3] Delcros, J.-G., Roch, A.-M., Quash, G.: FEBS Lett.,171,221–226 (1984)
[4] Brookhart, P.P., McMahon, P.L., Takahashi, M.: Anal. Biochem.,128,202–205 (1983)
[5] Lee, K.N., Birckbichler, P.J., Patterson, M.K.: Biochem. Biophys. Res. Commun., 162,1370–1375 (1989)
[6] Ando, Y., Imamura, S., Yamagata, Y., Kikuchi, T., Murachi, T., Kannagi, R.: J. Biochem.,101,1331–1337 (1987)
[7] Signorini, M., Bortolotti, F., Poltronieri, L., Bergamini, C.M.: Biol. Chem. Hoppe-Seyler,369,275–281 (1988)
[8] Puzkin, E.G., Raghuraman, V.: J. Biol.Chem.,260,16012–16020 (1985)
[9] Parameswaran, K.N., Velasco, P.T., Wilson, J., Lorand, L.: Proc. Natl. Acad. Sci. USA,87,8472–8475 (1990)
[10] Jensen, P.H., Lorand, L., Ebbesen, P., Gliemann, J.: Eur. J. Biochem.,214,141–146 (1993)
[11] Achyuthan, K.E., Mary, A., Greenberg, C.S.: Biochem. J.,257,331–338 (1989)
[12] Greenberg, C.S., Birckbichler, P.J., Rice, R.H.: FASEB J.,5,3071–3077 (1991) (Review)
[13] Dadabay, C.Y., Pike, L.J.: Biochem. J.,264,679–685 (1989)
[14] Bergamini, C.M., Signorini, M., Poltronieri, L.: Biochim. Biophys. Acta,916,149–151 (1987)
[15] Folk, J.E., Chung, S.I.: Methods Enzymol.,113,358–375 (1985) (Review)
[16] Folk, J.E.: Methods Enzymol.,87,36–42 (1982) (Review)
[17] Ichinose, A., Bottenus, R.E., Davie, E.W.: J. Biol.Chem.,265,13411–13414 (1990)
[18] Kawashima, S.: Experientia,47,709–712 (1991)
[19] Wong, W.S.D., Batt, C., Kinsella, J.E.: Int. J. Biochem.,22,53–59 (1990)
[20] Nakayama, J., Osaki, M., Nagae, S., Asahi, M., Urabe, H.: J. Dermatol.,13,448–455 (1986)

[21] Negi, M., Colbert, M.C., Goldsmith, L.A.: J. Invest. Dermatol.,85,75–78 (1985)
[22] Korner, G., Schneider, D.E., Purdon, M.A., Bjornsson, T.D.: Biochem. J.,
     262,633–641 (1989)
[23] Coussons, P.J., Price, N.C., Kelly, S.M., Fothergill-Gilmore, L.A.: Biochem. Soc.
     Trans.,20,48S (1991)
[24] Coussons, P.J., Price, N.C., Kelly, S.M., Smith, B., Sawyer, L.: Biochem. J.,
     283,803–806 (1992)
[25] Berbers, G.A.M., Bentlage, H.C.M., Brans, A.M.M., Bloemendal, H., de Jong, W.W.:
     Eur. J. Biochem.,135,315–320 (1983)
[26] Cocuzzi, E.T., Chung, S.I.: J. Biol. Chem.,261,8122–8127 (1986)
[27] Chang, S.K., Chung, S.I.: J. Biol.Chem.,261,8112–8121 (1986)
[28] Slife, C.W., Morris, G.S., Snedeker, S.W.: Arch. Biochem. Biophys.,257,39–47
     (1987)
[29] Kawashima, S.: Experientia,47,709–712 (1991)
[30] Seitz, J., Keppler, C., Hüntemann, S., Rausch, U., Aumüller, G.: Biochim. Biophys.
     Acta,1078,139–146 (1991)
[31] Ikura, K., Sakurai, H., Okumura, K., Sasaki, R., Chiba, H.: Agric. Biol. Chem.,
     49,3527–3531 (1985)
[32] Porta, R., De Santis, A., Esposito, C., Draetta, G.F., Di Donato, A., Illiano, G.: Bio-
     chem. Biophys. Res. Commun.,138,596–603 (1986)
[33] Lichti, U., Ben, T., Yuspan, S.H.: J. Biol. Chem.,260,1422–1426 (1985)
[34] Martinet, N., Kim, H.C., Girard, J.E., Nigra, D.H., Strong, D.H., Chung, S.I., Folk,
     J.E.: J. Biol. Chem.,263,4236–4241 (1988)
[35] Tokunaga, F., Yamada, M., Miyata, T., Ding, Y.-L., Hiranaga-Kawabata, M., Muta, T.,
     Iwanaga, S.: J. Biol. Chem.,268,252–261 (1993)
[36] Icekson, I., Apelbaum, A.: Plant Physiol.,84,972–974 (1987)
[37] Ikura, K., Tsuchiya, Y., Sasaki, R., Chiba, H.: Eur. J. Biochem.,187,705–711 (1990)
[38] Ando, H., Adachi, M., Umeda, K., Matsuura, A., Nonaka, M., Uchio, K., Tanaka, H.,
     Motoki, M.: Agric. Biol. Chem.,53,2613–2617 (1989)
[39] Klein, J.D., Guzman, E., Kuehn, G.D.: J. Bacteriol.,174,2599–2605 (1992)
[40] Folk, J.E., Chung, S.I.: Adv. Enzymol. Relat. Areas Mol. Biol.,38,109–191 (1973)
     (Review)

1 NOMENCLATURE

EC number
2.3.2.14

Systematic name
L-Glutamine:D-alanine gamma-glutamyltransferase

Recommended name
D-Alanine gamma-glutamyltransferase

Synonyms
More (cf. EC 2.3.2.2)

CAS Reg. No.

2 REACTION AND SPECIFICITY

Catalysed reaction
L-Glutamine + D-alanine →
→ NH_3 + gamma-L-glutamyl-D-alanine

Reaction type
Aminoacyl group transfer

Natural substrates

Substrate spectrum
1 L-Glutamine + D-alanine (best substrate) [1]
2 L-Glutamine + D-phenylalanine [1]
3 L-Glutamine + D-alpha-amino-n-butyric acid [1]
4 gamma-L-Glutamylethylester + D-alanine (equally effective as L-gluta-mine) [1]
5 GSH + D-alanine (50% as effective as L-glutamine) [1]
6 More (no substrates are D-valine, D-leucine, D-aspartic acid, D-glutamic acid, D-serine, D-proline) [1]

Product spectrum
1 NH_3 + gamma-L-glutamyl-D-alanine [1]
2 NH_3 + gamma-L-glutamyl-L-phenylalanine
3 ?
4 ?
5 ?
6 ?

Inhibitor(s)
   L-Leucine [1]; 6-Diazo-5-oxo-L-norleucine [1]; L-Serine borate [1]; Citrate (to
   some extent) [1]; More (no inhibition by Ca^{2+}, Mg^{2+}, NH_4^+, maleate) [1]

Cofactor(s)/prosthetic group(s)/activating agents

Metal compounds/salts

Turnover number (min^{-1})

Specific activity (U/mg)
   0.245 [1]

K_m-value (mM)
   2.0 (L-glutamine) [1]; 2.9 (D-alanine) [1]

pH-optimum
   9.5 [1]

pH-range

Temperature optimum (°C)
   37 (assay at) [1]

Temperature range (°C)

3 ENZYME STRUCTURE

Molecular weight

Subunits

Glycoprotein/Lipoprotein
   –

4 ISOLATION/PREPARATION

Source organism
   Pisum sativum (pea, cv. Alaska) [1]

Source tissue
   Decotyledonized seedlings [1]

Localization in source

Purification
   Pisum sativum (partial) [1]

Crystallization
   –

Cloned

–

Renatured

–

5 STABILITY

pH

Temperature (°C)

Oxidation

Organic solvent

General stability information

Storage

6 CROSSREFERENCES TO STRUCTURE DATABANKS

PIR/MIPS code

Brookhaven code

7 LITERATURE REFERENCES

[1] Kawasaki, Y., Ogawa, T., Sasaoka, K.: Biochim. Biophys. Acta,716,194–200 (1982)

1 NOMENCLATURE

EC number
2.3.2.15

Systematic name
Glutathione:poly(4-glutamyl-cysteinyl)glycine 4-glutamylcysteinyltransferase

Recommended name
Glutathione gamma-glutamylcysteinyltransferase

Synonyms
Synthase, phytochelatin
gamma-Glutamylcysteine dipeptidyl transpeptidase
Phytochelatin synthase

CAS Reg. No.
125390-02-5

2 REACTION AND SPECIFICITY

Catalysed reaction
Glutathione + [Glu(-Cys)]$_n$-Gly $\rightarrow$
$\rightarrow$ Glycine + [Glu(-Cys)]$_{n+1}$-Gly

Reaction type
Aminoacyl group transfer

Natural substrates
Glutathione + [Glu(-Cys)]$_n$-Gly (involved in phytochelatin biosynthesis, constitutive enzyme of higher plants, not induced by heavy metal ions) [1]

Substrate spectrum
1 Glutathione + [Glu(-Cys)]$_n$-Gly (acts on phytochelatins of various chain length) [1]
2 S-Monobromobimane-glutathione + [Glu(-Cys)]$_n$-Gly [1]

Product spectrum
1 Glycine + [Glu(-Cys)]$_{n+1}$-Gly
2 ?

Inhibitor(s)
Phytochelatin (chelates activating heavy metal ions, regulation) [1]

Cofactor(s)/prosthetic group(s)/activating agents

Metal compounds/salts

Cd^{2+} (requirement, 0.1 mM) [1]; Ag^{2+} (requirement, 0.1 mM, can replace Cd^{2+} with 58% efficiency) [1]; Bi^{3+} (requirement, 0.1 mM, can replace Cd^{2+} with 56% efficiency) [1]; Pb^{2+} (requirement, 0.1 mM, can replace Cd^{2+} with 43% efficiency) [1]; Zn^{2+} (requirement, 0.1 mM, can replace Cd^{2+} with 33% efficiency) [1]; Cu^{2+} (requirement, 0.01 mM, can replace Cd^{2+} with 27% efficiency) [1]; Hg^{2+} (requirement, 0.1 mM, can replace Cd^{2+} with 26% efficiency) [1]; Au^+ (requirement, 0.1 mM, can replace Cd^{2+} with 12% efficiency) [1]; More (no activation by Al^{3+}, Ca^{2+}, Fe^{2+}, Mg^{2+}, Mn^{2+}, Na^+ or K^+) [1]

Turnover number (min^{-1})

12 (glutathione) [1]

Specific activity (U/mg)

0.0278 [1]

K_m-value (mM)

1.5 (S-monobromobimane-glutathione) [1]; 6.7 (glutathione) [1]

pH-optimum

More (pl: 4.8) [1]; 7.9 [1]

pH-range

7.5–8.8 (about half-maximal activity at pH 7.5 and 8.8) [1]

Temperature optimum (°C)

35 [1]

Temperature range (°C)

20–47 (about half-maximal activity at 20°C and 47°C) [1]

3 ENZYME STRUCTURE

Molecular weight

95000 (Silene cucubalus, HPLC gel filtration, dissociates during purification into catalytically active dimer of MW 50000) [1]

Subunits

Tetramer (4 × 25000, Silene cucubalus, SDS-PAGE) [1]

Glycoprotein/Lipoprotein

—

4 ISOLATION/PREPARATION

Source organism
   Silene cucubalus (Caryophyllaceae) [1]; Podophyllum peltatum (Berberida-
   ceae) [1]; Eschscholtzia californica (Papaveraceae) [1]; Beta vulgaris (Che-
   nopodiaceae) [1]; Equisetum giganteum (Equisetaceae) [1]

Source tissue
   Cell suspension culture [1]

Localization in source

Purification
   Silene cucubalus [1]

Crystallization
   –

Cloned
   –

Renatured
   –

5 STABILITY

pH

Temperature (°C)
   4 ($t_{1/2}$: 140 h) [1]; 22 ($t_{1/2}$: 34 h) [1]; 35 ($t_{1/2}$: 6 h) [1]

Oxidation

Organic solvent

General stability information

Storage

6 CROSSREFERENCES TO STRUCTURE DATABANKS

PIR/MIPS code

Brookhaven code

7 LITERATURE REFERENCES

[1] Grill, E., Löffler, S., Winnacker, E.-L., Zenk, M.H.: Proc. Natl. Acad. Sci.
    USA,86,6838–6842 (1989)

1 NOMENCLATURE

EC number
2.4.1.1

Systematic name
1,4-alpha-D-Glucan:orthophosphate alpha-D-glucosyltransferase

Recommended name
Phosphorylase

Synonyms
Amylopectin phosphorylase
Amylophosphorylase
Glucan phosphorylase
alpha-Glucan phosphorylase
1,4-alpha-Glucan phosphorylase
Glucosan phosphorylase
Glycogen phosphorylase
Granulose phosphorylase
Maltodextrin phosphorylase
Muscle phosphorylase
Muscle phosphorylase a and b
Myophosphorylase
Phosphorylase, alpha-glucan
Polyphosphorylase
Potato phosphorylase
Starch phosphorylase
More (the recommended name should be qualified in each instance by
adding the name of the natural substrate, e.g. maltodextrin phosphorylase,
starch phosphorylase, glycogen phosphorylase)

CAS Reg. No.
9035-74-9

---

2 REACTION AND SPECIFICITY

Catalysed reaction
$(1,4\text{-alpha-D-Glucosyl})_n$ + phosphate $\rightarrow$
$\rightarrow (1,4\text{-alpha-D-glucosyl})_{n-1}$ + alpha-D-glucose 1-phosphate

Reaction type
Hexosyl group transfer

Natural substrates

$(1,4$-alpha-D-Glucosyl$)_{n-1}$ + alpha-D-glucose 1-phosphate (pathway in starch [1, 4] or glycogen metabolism [19, 56, 63], key step in cellular differentiation of Dictyostelium discoideum [23]) [1, 4, 19, 23, 56, 63]

Substrate spectrum

1 $(1,4$-alpha-D-Glucosyl$)_{n-1}$ + alpha-D-glucose 1-phosphate (r [1–60, 62, 63], favoured reaction [13], catalyzes incorporation of glucose into alpha-1,4-glucosidic linkage on exterior chains of primer [13], polyglucose primer required (not [4, 50, 51]) [1, 13, 27, 31, 33, 34, 37, 39], unprimed reaction: lag-phase, product presumably protein-bound glucan [51], strict specificity for glucose 1-phosphate [34, 45], no substrates: glucose 6-phosphate or fructose 6-phosphate, fructose 1,6-diphosphate [34, 45], ribose 5-phosphate [34], ineffective primers are glucose [31], Schardinger dextrin, cellulose [8, 34], sucrose [34], pullulan [14, 41], dextran [14]) [1–60, 62, 63]

2 Granulose + glucose 1-phosphate [16]

3 Amylose + glucose 1-phosphate (r [36, 37], potato amylose [37], effective primer [37], best substrate [34], glucosylation at 7% the rate of maltoheptaose glucosylation [14], not [15]) [8, 14, 34, 36, 37, 49]

4 Starch + glucose 1-phosphate (r [21, 29, 30, 33, 34, 36, 40, 45, 46, 48, 49], best substrate [34, 49], glucosylation at 32% the rate of dextrin glucosylation [15], poor substrate [21], soluble starch [21, 29, 30, 33, 34, 36, 40, 45, 46, 48, 49]. Glucosylation of native starch granules occurs at 82% (isozyme P-1) or 11.5% (isozyme P-2) the rate of soluble starch, phosphorolysis at the same rate as soluble starch [36]) [8, 11, 15, 21, 29, 30, 33, 34, 36, 40, 45, 46, 48, 49, 52, 53]

5 Amylopectin + glucose 1-phosphate (r [30], best substrate [31], effective primer [37], waxy maize amylopectin [14], potato amylopectin [30, 44], glucosylation at 23% the rate of maltoheptaose glucosylation [14], less effective than starch, amylose or glycogen [34]) [8, 14, 16, 30, 31, 34, 37, 40, 41, 44]

6 Debranched amylopectin + glucose 1-phosphate (r, most effective primer glucan for chloroplastic enzyme) [37]

7 Glycogen + glucose 1-phosphate (r, favoured reaction [26], best substrate [34], glucosylation at 14.4% the rate of maltoheptaose glucosylation [14], about half as effective as dextrin [15], yeast or rabbit enzyme [29], not phosphorylase B [33], glycogen from: rabbit (not bovine [31]) liver (not muscle [31]) [14, 21, 31, 37], Streptococcus salivarius [17], shell fish [19], oyster [15, 17, 21, 27, 29, 31, 33], or endogenous [21]) [8, 11, 13–15, 17, 19, 21, 22, 24, 26–29, 31, 33, 34, 37, 40, 50, 53, 54, 56, 58–60, 62, 63]

8 Maltose + glucose 1-phosphate (less effective than starch, amylose or glycogen [34], not phosphorylase B [33], not [31, 37, 39, 40]) [11, 33, 34]

9 Maltotriose + glucose 1-phosphate (r [34], ir [39], less effective than starch, amylose or glycogen [34], not [31, 37, 40]) [34, 39]
10 Maltotetraose + glucose 1-phosphate (r [14, 37], ir [29, 31, 39], best
   substrate [31], glucosylation at 22% the rate of maltoheptaose
   glucosylation [14], cytoplasmic enzyme: poor substrate [37], not
   (chloroplastic enzyme [37]) [37, 40]) [14, 29, 31, 37, 39]
11 Maltopentaose + glucose 1-phosphate (r [29, 31, 36, 37, 39, 40], isozyme I [36], glucosylation at the same (cytoplasmic enzyme) or 50%
   (chloroplastic enzyme) the rate of glucosylation of debranched amylopectin [37]) [29, 31, 36, 37, 39, 40]
12 Maltohexaose + glucose 1-phosphate (r) [29, 31, 39]
13 Maltoheptaose + glucose 1-phosphate (r [14, 29, 31, 36, 39], best substrate [29], isozyme I [36]) [14, 29, 31, 36, 39]
14 Maltooctaose + glucose 1-phosphate (r) [31, 39]
15 Maltodecaheptaose + glucose 1-phosphate (r) [31]
16 Maltodextrin + glucose 1-phosphate (r [14, 29], glucosylation at 96% the
   rate of maltoheptaose glucosylation [14], minimum chain length requirement for efficient activity: 4 glucose units per maltodextrin molecule
   [29]) [14, 29]
17 Dextrin + glucose 1-phosphate (r [15, 21, 27, 33, 34], corn dextrin [15],
   achrodextrin [33], best substrate [15], less effective than starch, amylose or glycogen [34], not phosphorylase B [33]) [15, 21, 27, 33, 34]
18 Glycogen + phosphate (r [18, 21, 56, 63], best substrate [18], poor substrate [39], glycogen from: oyster [20, 27], rabbit (not bovine [31]) liver
   [23, 31, 43] or phytoglycogen [43]) [18, 20, 21, 23, 26, 27, 31, 33, 34,
   37, 39, 41–43, 56, 63]
19 Dextrin + phosphate (r [18, 41], corn dextrin, phosphorolysis at 85% the
   rate of glycogen phosphorolysis [18]) [18, 41]
20 Maltodextrin + phosphate (maltodextrin with up to 11 glucose units [42],
   minimum chain length requirement for efficient activity: 5 glucose units
   per maltodextrin molecule [29]. Phosphorolysis activity ceases when
   maltodextrins are degraded to maltotetraose [31]) [29, 31, 39, 42]
21 Starch + phosphate (r [18, 29, 30, 33, 34, 36, 40–42], best substrate
   [42], soluble starch [29, 30, 33, 34, 36, 40–42] or native chloroplast
   grains [41]. Phosphorolysis at 31% the rate of glycogen phosphorolysis
   [18], potato enzyme [29]) [18, 29, 30, 33, 34, 36, 40–42]
22 Amylopectin + phosphate (best substrate [39], potato [44], waxy rice or
   pea cotyledon [43] amylopectin) [39, 41, 43, 44]
23 Amylose + phosphate (pea cotyledon amylose [43], arsenate can replace phosphate [53]) [41, 43, 53]
24 Maltoheptaose + phosphate (poor substrate) [42]

Product spectrum

1 (1,4-alpha-D-Glucosyl)$_n$ + phosphate [13]
2 ?
3 ?
4 Starch + phosphate [29, 30]
5 Amylopectin + phosphate [30]
6 ?
7 Glycogen + phosphate [21, 31, 37]
8 ?
9 ?
10 ?
11 ?
12 ?
13 ?
14 ?
15 ?
16 Maltodextrin + phosphate [14]
17 Dextrin + phosphate [21]
18 Glycogen + glucose 1-phosphate [18, 21]
19 Dextrin + glucose 1-phosphate [18]
20 ?
21 Starch + glucose 1-phosphate [18, 30]
22 Amylopectin + glucose 1-phosphate [39, 41, 43, 44]
23 Amylose + glucose 1-phosphate [41, 43]
24 ?

Inhibitor(s)

Phosphorylase phosphatase (characterization [61], reactivation by phosphorylase kinase [25]) [25, 61]; beta-Amylose (not rabbit muscle enzyme) [46, 47]; alpha-D-Glucopyranose-1,2-cyclic phosphate (strong, kinetics) [52]; Adenine [34]; ATP (strong, at non-saturating levels of glucose 1-phosphate [34], rabbit muscle enzyme [29], additive inhibition together with tyrosine [34], phosphorylase B [33], kinetics [30], less effective than nucleotide sugars [16], not phosphorylase A or C [33], not [28, 39, 41, 44]) [1, 4–7, 9, 16, 18, 29, 30, 33, 34, 46, 58]; Mg-ATP (reversible by AMP) [19]; ADP (allosteric inhibition, kinetics [30], not [16, 34, 39, 41, 44]) [1, 4–6, 18, 21, 30]; AMP (weak [21, 23], a-isozyme [23], above 2 mM [27], competitive to glucose 1-phosphate [28], yeast or potato enzyme [29], reversible by Mg^{2+} [21], not [30, 34, 39, 41, 44]) [21, 23, 27–29]; cAMP (weak, 5 mM) [27]; ADPglucose (strong (phosphorolysis, isozyme I) [44], weak [52], competitive (phosphorolysis [30]) [13, 20], kinetics [30, 41], reversible by Mg^{2+} [21]) [1, 3, 4, 13, 16, 20, 21, 27, 30, 39, 41, 44, 46, 52]; ADPmannose [16]; ADP-ribose [16]; GDPglucose (kinetics [17], competitive (phosphorolysis) [20]) [16, 17, 20]; UDPglucose (kinetics [20, 30], strong (rabbit or yeast enzyme

[29]) [28, 29], weak (potato enzyme [29]) [5, 29, 52], competitive to glucose 1-phosphate, non-competitive to phosphate [22], at non-saturating levels of glucose 1-phosphate [34], reversible by Mg^{2+} [21], not [24, 39]) [3, 5, 6, 13, 16, 20–22, 28–30, 33, 34, 44, 46, 52, 60]; TDPglucose (competitive) [13]; dTDPglucose [16]; GDP [30]; Guanine [9, 34]; Guanosine [9]; UDP (at high glucose 1-phosphate concentration, glycogen synthesis, not phosphorolysis [22], not [30]) [22, 29]; UTP (not [39]) [30]; Pyridoxal 5'-phosphate (addition of external amounts [44], not [34]) [44]; Pyridoxal (phosphorylase I, not II) [44]; Pyridine 3-aldehyde (weak, phosphorylase I, not II) [44]; Pyridine 4-aldehyde (phosphorylase I, not II) [44]; Deoxypyridoxine (weak, phosphorylase I, not II) [44]; Pyridoxamine (weak, phosphorylase I, not II) [44]; Pyridoxamine 5'-phosphate (weak, phosphorylase I, not II) [44]; Glucose (weak [17, 27], kinetics [62], non-competitive [13], not Atriplex spongiosa [12] or yeast enzyme [29], not [16, 28, 31, 34, 41]) [5, 6, 11–13, 17, 27, 29, 62, 63]; Glucose 6-phosphate (above 5 mM [26], strong [28], weak [52], not potato enzyme [29], not [16, 20, 23, 34]) [6, 11, 17, 21, 26, 28, 29, 52, 58, 62]; Glucose 2-phosphate (weak) [52]; Fructose 1-phosphate (not [16, 44]) [17]; Fructose 6-phosphate (rabbit enzyme: weak [29], not [16, 34, 39, 41, 44]) [6, 17, 29]; Trehalose (weak) [5]; 1,5-Gluconolactone (strong, muscle isozyme a and b) [60]; Glucose 1-phosphate (above 2 mM [13], substrate inhibition [13], kinetics [17, 20], completely reversible by 5'-AMP [17], not [16, 43]) [13, 17, 20]; Fructose 1,6-diphosphate (yeast enzyme, not [16, 17, 34, 39, 41, 44]) [29]; Mannose (weak) [29]; 2,3-Diphosphoglycerate (not [44]) [39]; alpha-Methylglucoside (rabbit, not yeast enzyme) [29]; alpha-Amylose [13]; beta-Amylose [13]; alpha-D-Glucopyranosyl fluoride (phosphorylase b, kinetics, strong: rabbit, weak: potato enzyme) [60]; Cyclodextrin (i.e. structure analogue of alpha-1,4-linked starch molecule, potato enzyme: strong, yeast or rabbit enzyme: weak [32], alpha- or beta-, not gamma-cyclodextrin [28]) [28, 32]; Cyclodextrin-dialdehyde (kinetics) [32]; Cyclohexaamylose [3]; Phosphoenolpyruvate (not Atriplex spongiosa [12], not [41, 44]) [12]; Aromatic compounds (overview, e.g. p-nitrophenol) [59]; Aromatic amino acids (phosphorylase B, not A or C [33]) [9, 33, 34]; L-Tyrosine (additive inhibition together with ATP [34]) [8, 34]; Phenolic compounds (overview [33, 34], 10 mM, no inhibition at 1 mM [34]) [33, 34]; Caffeine (weak [28], rabbit muscle: strong [29]) [28, 29]; Mercuriacetate [21]; NADPH (also oxidized form, less effective than nucleotide sugars, not NAD(H)) [16]; $NaHSO_3$ (kinetics, completely reversible by dilution or dialysis) [53]; Sulfate (weak [24], not [28, 34, 53]) [1, 24]; Phosphate (1 mM) [12]; KNO_3 [13]; CN^- (not [53]) [13]; $K_2S_2O_8$ [13]; KCl (50% loss of activity at 0.08 M and 90% loss of activity at 0.2 M) [23]; NaF [28]; Cl^- (weak) [24]; Bicarbonate (weak) [53]; Ba^{2+} (not [15]) [18]; Ca^{2+} (weak [13, 15, 45], 1 mM, stimulation at 10 mM [18], not [49]) [4, 13, 15, 18, 45]; Mg^{2+} (slight [15], not [13, 41, 45, 40, 49]) [4, 15]; Ag^{2+} (strong) [13, 34, 49]; Co^{2+} (1 mM, not 10 mM [18], not [49]) [15, 18]; Cu^{2+} (strong [13, 15], not [49]) [6, 13, 15, 18, 21,

Enzyme Handbook © Springer-Verlag Berlin Heidelberg 1996
Duplication, reproduction and storage in data banks are only
allowed with the prior permission of the publishers

28, 34]; Fe^{2+} (weak [13], 10 mM, stimulation at 1 mM [18]) [13, 18, 34, 49]; Hg^{2+} (strong [13, 34], phosphorylase A and B, not C [33]) [6, 13, 28, 33, 34, 45, 49]; Mn^{2+} (reaction without glucan primer [50], not [13, 49]) [15, 18, 50]; Sn^{2+} [15]; Zn^{2+} (strong, not [49]) [13, 15, 18, 28, 34, 45]; High salt concentration [23]; N-Acetylimidazole (glucose 1-phosphate prevents) [45]; PCMB (reversible by 2-mercaptoethanol [34], not [18, 33]) [13, 34, 49]; p-Hydroxy-mercuribenzoate (2 mM [21], glucose 1-phosphate prevents [45], not [13, 15]) [21, 45]; Dithiothreitol (not [41]) [18]; Cysteine (strong [18], not [33, 49]) [18]; GSH (not [15, 33, 49]) [18]; Iodoacetate (weak) [13, 18]; IAA (weak, not [33, 49]) [13]; NEM (weak, not [33]) [13]; DTNB (1 mM, weak [15]) [15, 57]; Diethyldithiocarbamate (strong) [18]; 2,2'-Dipyridyl [18]; EDTA (strong, not [15, 21, 34]) [18]; EGTA (i.e. ethylene glycol bis(beta-aminoethyl-ether)-N,N'-tetraacetic acid, strong, not [15]) [18]; Sodium dodecylsulfate [6]; Polyethyleneglycol [19]; Antiserum of purified phosphorylase [34]; Antibodies to skeletal muscle phosphorylase (heart enzyme) [58]; More (no inhibitors: agarose [49], fructose, sucrose, dihydroxyacetone phosphate, 6-phosphogluconate, 2-phosphoglycollate, 2-phosphoglycerate, pyruvate [41], adenosine, cytosine [34], GMP [30, 34, 39], IMP [28], TMP [30], GTP [39], UMP, CMP, CTP [30, 39], o-phenanthroline [21], 2-mercaptoethanol [15, 33, 49], Ni^{2+} [49], Sr^{2+} [15], K^+, Na^+ [45], F^-, ClO_4^- [34], azide [53], acetate, acetylphosphate or butyrylphosphate, 6-phosphogluconate [16], 3-phosphoglycerate [16, 34], maltose, maltotriose or maltotetraose [31], succinate, 2-oxoglutarate, malate [34], p-coumaric acid, caffeic acid or cinnamic acid [33], non-aromatic amino acids [33, 34], gibberellic acid, indolyl-3-acetic acid [49]) [15, 16, 21, 28, 30, 31, 33, 34, 39, 41, 45, 49, 53]

Cofactor(s)/prosthetic group(s)/activating agents

5'-AMP (activation [1, 4–6, 13, 16, 17, 23, 27, 29, 33, 35, 54–56, 58, 62, 63], of b- not a-isozyme [23], slight [16, 17, 27], in both directions [13], required for dephospho-phosphorylase (kinetics) [63], rabbit muscle enzyme (activator site) [29], significant activation by synergism with NaF (glycogen as substrate) [27], 0.01 mM, 1.0 mM or 2 mM, isozyme a, ab or b, respectively [5], phosphorolysis: inhibits at 2 mM and above [27], not (phosphorylase A or C [33]) [19, 20, 24, 28, 33]) [1, 4–6, 13, 16, 17, 23, 27, 29, 33, 35, 54–56, 58, 62, 63]; Phosphate (activation [25, 26, 28, 55, 56, 62, 63], yeast and mammalian enzymes are activated by covalent phosphorylation (phosphorylase kinase) [26], converts phosphorylase b into phosphorylase a [55, 56], in the presence of ATP-Mg [55]. Phosphorylation site: a single Thr-residue in the N-terminal region [26, 28]. P-content correlates with activity [25], 0.1–0.74 mol covalently bound phosphate per mol subunit [25]) [25, 26, 28, 55, 56, 62, 63]; Phosphorylase kinase (activation [25, 26, 28, 55, 56, 62, 63], yeast and mammalian enzymes are activated by covalent phosphorylation of a single Thr-residue in the N-terminal region [26], incorporates 1 mol phosphate per mol subunit [25]) [25, 26, 28, 55, 56, 62, 63]; Pyridoxal 5'-phosphate (requirement [9, 10, 13, 17, 19, 22, 25, 38, 39, 44, 53, 58], 2 mol per mol enzyme [9, 58], 1 mol/mol subunit [17, 38, 39], additional pyridoxal-phosphate inhibits [44], not [33, 34]) [9, 10, 13, 17, 19, 22, 25, 38, 39, 44, 53, 58]; IMP (activation, isozyme a and ab, not b) [5, 62]; cAMP (activation [27, 35], in the absence of AMP, inhibits at 5 mM [27]) [27, 35]; ADP (activation, in the absence of AMP) [27]; ATP (activation [27, 33], not synthesis [27], not phosphorylase A or C [33]) [27, 33]; Albumin (activation, reaction without primer glucan) [50]; Protamine (stimulation, dephospho-phosphorylase, kinetics) [63]; DTT (required to maintain activity, human) [57]; GSH (slight stimulation) [49]; 2-Mercaptoethanol (activation, 5 mM [15], slight [49]) [15, 49]; Cysteine (activation, slight [49]) [15, 49]; p-Hydroxymercuribenzoate (slight activation, 0.1–1.0 mM) [15]; EDTA (activation, phosphorylase C, not A or B [33]) [33, 35]; More (no activation by 2'-AMP, 3'-AMP, ADP, adenosine [13], ATP, cAMP [13, 20] or ATP-regenerating systems [20], no activation by serine, leucine, isoleucine, methionine, ornithine, L-arginine [33]) [13, 20, 33]

Metal compounds/salts
Ca^{2+} (stimulates at 10 mM, inhibits at 1 mM [18], not [33, 49]) [18, 21]; Co^{2+} (activation, not [49]) [21]; Fe^{2+} (stimulates at 1 mM, inhibits at 10 mM) [18]; Mg^{2+} (activation [1, 21, 35, 49], slight [49], not [13, 33]) [1, 21, 35, 49]; Mn^{2+} (activation [21, 35, 49], slight [49], not [13]) [21, 35, 49]; NaF (activation, maximal at 0.2 M [13], slight [27], significant activation by synergism with AMP (glycogen as substrate) [27]) [13, 27]; Ni^{2+} (activation [21], not [49]) [21]; SO_4^{2-} (activation) [13, 55]; CNO^- (slight activation) [13]; Fumarate (activation) [34]; Maleic acid (activation) [34]; Pyruvate (stimulation) [35]; Mercaptoacetic acid (activation, 1 mM, inhibits at 5 mM) [15]; More (no divalent cations required [16], no activation by indoleacetic acid, gibberellic acid [34, 49], Cl^- [13], Cu^{2+} [33, 49], Zn^{2+} [49] or Na^+ [21]) [13, 16, 21, 33, 34, 49]

Turnover number (min^{-1})

Specific activity (U/mg)
More [41, 45]; 0.005 [15]; 0.118 [13]; 0.16 (chloroplastic enzyme, phosphorolysis) [37]; 0.222 [18]; 0.36 [4]; 1.61 (fast isozyme) [49]; 1.76 [24]; 1.9 [16]; 2.67 (slow isozyme) [49]; 3.3 (b-isozyme) [23]; 4.2 [17]; 4.3 (a-isozyme) [23]; 4.7 (non-phosphorylated enzyme) [26]; 5 (phosphorylase II) [40]; 7.14 [20]; 11.7 [42]; 15.9 [38]; 17.7 (maltoheptaose) [14]; 35–49 [39]; 42.3 (isozyme P-1 [36]) [28, 36]; 43.5 (cytoplasmic enzyme, synthesis) [37]; 46 (heart) [56]; 47.5 [25]; 51.6 (skeletal muscle) [56]; 65 [19]; 71 [55]; 85–90 (phosphorylated enzyme) [26]; 300 (isozyme DC2) [1]; 400 (isozyme DC1) [1]; 659 (phosphorylase I) [40]

K_m-value (mM)

More (kinetic mechanism [10], kinetic parameters of 3 starch phosphorylases [33], kinetic data [1, 17, 29, 31, 35, 36, 39–44, 49, 54, 55, 57, 59, 60,
62], comparison of muscle phosphorylases of various animals [55, 57], 0.13
mg/ml (starch (+ phosphate)) [34], 0.19 mg/ml (starch (+ glucose 1-phos-
phate)) [34], 0.24 mg/ml (starch, phosphorylase A) [33], 0.264 mg/ml (amylopectin), 0.277 mg/ml (glycogen), 0.285 mg/ml (amylose) [3], 0.36 mg/ml
(glycogen) [18], 0.453 mg/ml (maltodextrin) [3], 0.65 mg/ml (glycogen) [21],
1.1 mg/ml (glycogen, phosphorylase A) [33], 1.21 mg/ml (starch, phosphorylase C) [33], 1.26 mg/ml (glycogen) [17], 2.7 mg/ml (glycogen, a-isozyme) [23], 4.2 mg/ml (glycogen, b-isozyme) [23], 5 mg/ml (oyster glycogen) [20], 20 mg/ml (oyster glycogen, phosphorolysis), 25 mg/ml (endogen
glycogen, synthesis), 27 mg/ml (corn dextrin, phosphorolysis), 40 mg/ml
(corn dextrin, synthesis) [15], 42 (oyster glycogen, synthesis) [15], 0.46%
(w/v) (glycogen, with AMP), 0.67% (w/v) (glycogen, without AMP) [13]) [1, 3,
10, 13, 15–18, 20, 21, 23, 29, 31, 33–36, 39–44, 49, 54, 55, 57, 59, 60, 62];
0.05 (glucose 1-phosphate, synthesis) [16]; 0.1–0.21 (glycogen (isozyme b
[7], calculated on the basis of molar concentration of non-reducing ends) [7,
58], glucose 1-phosphate (+ amylopectin), synthesis [30], maltodecaheptaose, phosphorolysis [31]) [7, 30, 31, 58]; 0.4–0.5 (glucose 1-phosphate, in
the presence of Mg^{2+} [21], maltodecaheptaose, synthesis [31]) [21, 31];
0.5–0.69 (glycogen, isozymes a and ab [7], rabbit or yeast enzyme [29],
amylopectin [31], calculated on the basis of molar concentration of non-reducing ends) [7, 29, 31]; 0.66–0.93 (glucose 1-phosphate (+ starch) [34],
maltohexaose, maltooctaose, maltopentaose, phosphorolysis [31]) [31, 34];
1.0 (glucose 1-phosphate) [13, 21]; 1.15–1.22 (phosphate (a-isozyme [23])
[7, 23], maltoheptaose, phosphorolysis [31]) [7, 23, 31]; 1.4–1.74 (glycogen
(phosphorylated enzyme) [63], glucose 1-phosphate (+ starch), phosphorylase A [33], phosphate (b-isozyme [23]) [17, 21, 23], glucose 1-phosphate,
Agrobacterium tumefaciens or rabbit liver enzyme [22]) [17, 21–23, 33, 63];
2.0–8.3 (maltooctaose, maltoheptaose, maltohexaose, maltopentaose, maltotetraose, synthesis) [31]; 2.2–2.6 (glucose 1-phosphate [6], phosphate (+
amylopectin [30], phosphorolysis [16, 30]) [7, 16, 30]) [6, 7, 16, 30]; 2.7–2.9
(glucose 1-phosphate (+ starch), phosphorylase C [33], phosphate (phosphorylated enzyme) [63], glucose 1-phosphate, in the presence of AMP
[13]) [13, 33, 63]; 3.0–3.6 (glycogen [6], glucose 1-phosphate ((+ starch),
phosphorylase B [33]) [15, 33], phosphate [20]) [6, 15, 20, 33]; 4.0 (glucose
1-phosphate (+ maltose), phosphorylase C) [33]; 5.0 (glucose 1-phosphate
(+ glycogen), phosphorylase A) [33]; 6.0 (glucose 1-phosphate, heart) [58];
10.87 (phosphate (+ starch)) [34]; 16 (glucose 1-phosphate) [24]; 20 (phosphate, isozyme ab [7], glycogen, dephospho-phosphorylase [63]) [7, 63];
100 (phosphate, dephospho-phosphorylase) [63]

pH-optimum

More (pI: 5.0 (slow isozyme) [49], pI: 5.2 (heart isozyme IIb) [56], pI: 5.3 [14], pI: 5.4 [3], pI: 5.5 (fast isozyme) [49], pI: 6.2 (skeletal muscle, heart isozyme Ib) [56]) [3, 14, 49, 56]; 5–5.8 (phosphorolysis) [36]; 5.1–6.8 (synthesis, 2-(N-morpholino)ethanesulfonic acid buffer) [31]; 5.2–5.6 [49]; 5.5–5.6 (isozyme I) [35]; 5.6 (below, synthesis) [41]; 5.8 [28]; 5.9 [14]; 5.9–6.2 [24]; 5.9–6.3 [45]; 6.0 (synthesis [34]) [8, 15, 21, 34]; 6–7.5 (broad [17], cytoplasmic enzyme, buffer dependent optima in this range [37]) [17, 37]; 6.1 [11]; 6.1–6.3 (isozyme II) [35]; 6.3 [55]; 6.4 [16]; 6.4–6.7 (heart isozyme IIIb) [58]; 6.5–6.9 (heart isozyme Ib) [58]; 6.6 (a-isozyme: 2-(N-morpholino)ethanesulfonic acid preferred to imidazole buffer) [23]; 6.7–6.9 (glycogen) [13]; 6.8 [6]; 6.9–7.2 [5, 7]; 7 [18]; 7–7.4 (phosphorolysis) [42]; 7–7.5 (phosphorolysis) [31]; 7.2 (phosphorolysis) [41]; 7.3–8 (synthesis) [36]; 7.5 (broad, chloroplastic enzyme [37], phosphorolysis) [34, 37]

pH-range

4–8 (about half-maximal activity at pH 4 and 8, slow isozyme) [49]; 4.2–7.6 (about half-maximal activity at pH 4.2 and 7.6, fast isozyme) [49]; 5.1–7.5 (about half-maximal activity at pH 5.1 and 7.5) [14]; 5.2–7.2 (about half-maximal activity at pH 5.2 and 7.2) [21, 45]; 5.2–7.5 (about 70% of maximal activity at pH 5.2 and about half-maximal activity at pH 7.5) [24]; 5.4–8.2 (about half-maximal activity at pH 5.4 and 8.2) [17]; 5.5–9 [40]; 5.6–7.5 (about half-maximal activity at pH 5.6 and 7.5, imidazole buffer, phosphorolysis) [31]; 5.6–8 (about half-maximal activity at pH 5.6 and about 65% of maximal activity at pH 8, citrate/phosphate-buffer) [18]; 5.7–7.2 (about half-maximal activity at pH 5.7 and 7.2) [15]; 5.7–8.0 (about half-maximal activity at pH 5.7 and 8) [40]; 5.8–7.2 (about half-maximal activity at pH 5.8 and 7.2, a-isozyme) [23]; 5.8–7.6 (about half-maximal activity at pH 5.8 and 7.6, b-isozyme) [23]; 6–7.5 (about half-maximal activity at pH 6.0 and 7.5) [13]; 6–8.6 (about 70% of maximal activity at pH 6 and about half-maximal activity at pH 8.6, Tris-buffer) [18]; 7–8.1 (maximal activity at pH 7 and about 60% of maximal activity at pH 8.1, glycylglycine buffer, phosphorolysis) [31]

Temperature optimum (°C)

More (ratio of phosphorolysis to synthesis increases significantly at higher temperatures) [31]; 26 [23]; 30 [6, 28]; 30–35 (slow isozyme) [49]; 35 (phosphorylase B [33], fast isozyme [49]) [33, 49]; 37 [11]; 40 [8, 34]; 45 (phosphorylase A [33]) [33, 45]; 50 (phosphorylase C) [33]

Temperature range (°C)

18–35 (about half-maximal activity at 18°C and 35°C) [23]

3 ENZYME STRUCTURE

Molecular weight

119000 (Phymatotrichum omnivorum, gel filtration) [18]
135000 (Solanum tuberosum phosphorylase I, PAGE) [50]
150000 (Zea mays, sucrose density gradient ultracentrifugation) [39]
151000 (Spinacia oleracea, gel filtration) [38]
159000 (Spinacia oleracea, porosity density gradient electrophoresis) [38]
160000 (Pisum sativum cytoplasmic isozyme, PAGE) [42]
165000 (Solanum tuberosum fast isozyme, gel filtration) [49]
170000 (Entosphenus japonicus, PAGE) [55]
172000 (Physarum polycephalum, gel filtration) [17]
174000 (Klebsiella pneumoniae, sedimentation equilibrium (high speed)
[14], Zea mays phosphorylase I, ultracentrifugation [40]) [14, 40]
175000 (human phosphorylase b, sucrose density gradient centrifugation) [57]
180000 (Neurospora crassa, gel filtration or PAGE) [19]
180000–193000 (Locusta migratoria isozymes, gel filtration) [5]
181000 (Klebsiella pneumoniae, sedimentation equilibrium (low speed)) [14]
185000 (rabbit liver isozyme) [58]
186000 (Physarum polycephalum, gradient-PAGE) [17]
188000 (Dioscorea rotundata, gel filtration) [4]
189000 (Klebsiella pneumoniae, gel filtration [14], human phosphorylase b,
high speed sedimentation equilibrium [57]) [14, 57]
190000 (pig, gel filtration [63], Solanum tuberosum phosphorylase II, sucrose density gradient centrifugation or PAGE [50], rabbit phosphorylase b,
high speed sedimentation equilibrium) [57]
190000–220000 (Ipomoea batatas, PAGE) [48]
194000 (Spinacia oleracea cytoplasmic isozyme, sucrose density gradient
centrifugation) [37]
195000 (Zea mays phosphorylase II, ultracentrifugation) [40]
200000 (Sepia pharaonis [10], Dioscorea cayenensis, gel filtration [45], rabbit heart isozymes Ia, Ib, IIIb [58]) [10, 45, 58]
203000 (Saccharomyces cerevisiae, PAGE) [28]
203800 (Spinacia oleracea chloroplastic isozyme, sucrose density gradient
centrifugation) [37]
204000 (Voandzeia subterranea isozyme I, gel filtration) [35]
209000 (Solanum tuberosum slow isozyme, gel filtration) [49]
210000 (Dictyostelium discoideum, gel filtration) [20]
219000 (Saccharomyces cerevisiae, gel filtration) [28]
220000 (Manihot utilissima, gel filtration) [9]
224000 (human phosphorylase a, high speed sedimentation equilibrium) [57]
240000 (Gracilaria sordida, PAGE) [3]
243000 (Gracilaria sordida, gel filtration) [2]
245000 (Gracilaria sordida [3], Patinopecten yessoensis [6], gel filtration) [3, 6]
250000 (E. coli, PAGE [13], human phosphorylase a, rabbit phosphorylase
b, sucrose density gradient centrifugation [57]) [13, 57]

315000 (Zea mays, gel filtration) [39]

335000 (rabbit phosphorylase a, sucrose density gradient centrifugation) [57]

340000 (Ipomoea batatas, gel filtration) [48]

363000 (rabbit phosphorylase a, high speed sedimentation equilibrium) [57]

400000 (rabbit heart isozyme IIIa) [58]

450000 (Musa paradisiaca, gel filtration) [34]

More (amino acid composition of Klebsiella pneumoniae enzyme [14], of Physarum polycephalum enzyme [17], amino acid sequence of Saccharomyces cerevisiae enzyme [26]) [14, 17, 26]

Subunits

? (x × 40000, Solanum tuberosum fast isozyme, SDS-PAGE [49], x × 97400, Gracilaria sordida, SDS-PAGE [2, 3], x × 104000, Solanum tuberosum slow isozyme, SDS-PAGE [49], x × 112000, Typha latifolia, SDS-PAGE [36], x × 145000, Phymatotrichum omnivorum, SDS-PAGE [18]) [2, 3, 18, 36, 49]

Dimer (2 × 87000, Zea mays phosphorylase I, SDS-PAGE [40], 2 × 88000, Klebsiella pneumoniae, SDS-PAGE [14], 2 × 88700, Pisum sativum cytoplasmic isozyme, SDS-PAGE [42], 2 × 89000, Spinacia oleracea, SDS-PAGE [38], 2 × 90000, Klebsiella pneumoniae, sedimentation equilibrium (low speed), after treatment with 6 M guanidine-HCl [14], Neurospora crassa, SDS-PAGE [19], Dictyostelium discoideum, SDS-PAGE [23], 2 × 92000, Spinacia oleracea cytoplasmic isozyme, SDS-PAGE [37], 2 × 92500, Locusta migratoria, SDS-PAGE [5], rat, SDS-PAGE [56], 2 × 93000, Physarum polycephalum, SDS-PAGE [17], 2 × 94000, Entosphenus japonicus, SDS-PAGE [55], 2 × 95000, Dictyostelium discoideum [20], Voandzeia subterranea isozyme I [35], pig [63], SDS-PAGE, 2 × 96000, Solanum tuberosum phosphorylase II, SDS-urea-PAGE [50], 2 × 98000, Ipomoea batatas, SDS-PAGE [48], 2 × 100000, Saccharomyces cerevisiae, SDS-PAGE [25, 28], presumably s- and I-subunit of very slight MW differences, less active: tetramer of I-subunit, inactive: octamer and larger oligomers [25], 2 × 110000, Manihot utilissima, SDS-PAGE [9], 2 × 120000, Patinopecten yessoensis, SDS-PAGE [6]) [5, 6, 9, 14, 17, 19, 20, 23, 25, 28, 35, 37, 38, 40, 42, 48, 50, 55, 56, 63]

Tetramer (4 × 53000, Zea mays phosphorylase II, SDS-PAGE) [40]

Octamer (8 × 55000, Musa paradisiaca, SDS-PAGE) [34]

More (inactive enzyme tends to form larger oligomers [25], rabbit or human phosphorylase a dissociate into active dimers in the presence of high salt concentration, glucose or glycogen, the human enzyme at 5 mg/ml protein concentration, the rabbit enzyme remains a tetramer even at 1 mg/ml [57]) [25, 57]

Glycoprotein/Lipoprotein

–

4 ISOLATION/PREPARATION

Source organism
Atriplex spongiosa [12]; Chlorella vulgaris [30, 31]; Dioscorea cayenensis (yellow yam) [1, 45]; Dioscorea rotundata (white yam) [4]; Elephantopus scabar (cabbage) [8]; Gracilaria sordida (strain (Harv.) W.Nelson [3]) [2, 3]; Ipomoea batatas (sweet potato, Lam. cv. Tainon 65) [46–48]; Kalanchoe daigremontiana [12]; Manihot utilissima (tapioca) [9]; Musa paradisiaca (banana, cooking variety [33]) [33, 34]; Pisum sativum (pea, cv. Kelvedon Wonder [41], Victory Freezer [43, 44], var. Kleine Rheinländerin [42]) [41–44]; Red seaweeds [2]; Solanum tuberosum (cv. Adelheid or Rosa [49]) [29, 32, 49–53, 60]; Spinacia oleracea (spinach, var. Früremona or Kasperik [38]) [37, 38]; Typha latifolia [36]; Voandzeia subterranea (Bamberra groundnut, var. Thouars.) [35]; Zea mays (sweet corn, var. Golden Bantam, Iowa Belle 104 [39], Inrafrüh [40]) [39, 40]; Brewer's bottom yeast [25]; Cryptococcus laurentii (var. flavescens [21], syn. Rhodotorula peneaus, strain 48–23A [24]) [21, 24]; Dictyostelium discoideum (slime mould, strain NC-4 [20] or Ax3 [23]) [20, 23]; Neurospora crassa (strain CM (wild-type)) [19]; Phymatotrichum omnivorum (strain (Shear) Dugg) [18]; Physarum polycephalum M3c [17]; Saccharomyces carlsbergensis [25]; Saccharomyces cerevisiae (strain PH5–3, harbouring plasmid Yep24::GPH1 [26]) [25, 26, 28, 29, 32]; Agrobacterium tumefaciens [22]; Clostridium pasteurianum W-5 [16]; E. coli K-12 [13]; Klebsiella pneumoniae [14]; Streptococcus salivarius [15, 27]; Carcharhinus flaciformis (silky shark) [57]; Entosphenus japonicus (lamprey) [55]; Homarus americanus (American lobster) [57]; Human [57]; Indocibium guttattam (deap-sea fish) [54]; Locusta migratoria (locust) [5, 7]; Patinopecten yessoensis (scallop) [6]; Pig [58, 63]; Rabbit [22, 29, 32, 53, 57–62]; Rat (male Wistar) [56]; Sepia pharaonis (cuttle fish) [10]; Tilapia mosambica (tilapia) [11]

Source tissue
Cotyledons [35, 43]; Fruit [33]; Leaf (mesophyll and bundle sheath, not epidermal cells [40]) [8, 34, 37, 38, 40–42]; Pollen grains (mature) [36]; Seeds [44]; Shoots [41]; Root (parenchymal cells) [46–48]; Tuber [1, 4, 9, 45, 49–51, 60]; Macroplasmodium [17]; Mycelium [18, 19]; Cell (culmination stage [20, 23]) [13–15, 20–26, 29, 30]; Fat body [5]; Heart [56, 58]; Kidney [63]; Liver (rabbit [22]) [22, 58]; Muscle (adductor muscle [6], flight muscle [7], mantle muscle [10], skeletal muscle [55, 56, 62]) [6, 7, 10, 11, 29, 53, 55–57, 59, 60–62]; More (tissue distribution) [55]

Localization in source
Cytoplasm (loosely bound to native granules of its substrate polyglucan [16, 21], mesophyll cell isozyme I [40], predominant form [42]) [16, 18, 21, 23, 37, 38, 40, 42, 43]; Amyloplast [48]; Chloroplast (bundle sheath cell isozyme II [40]) [37, 40–42]

Purification
   Chlorella vulgaris (partial) [30, 31]; Dioscorea rotundata [4]; Dioscorea cayenensis (2 enzymes, DC1 and 2 [1], partial [45]) [1, 45]; Elephantopus scabar (partial) [8]; Gracilaria sordida [2, 3]; Ipomoea batatas [48]; Manihot
   utilissima [9]; Musa paradisiaca (multiple forms [33]) [33, 34]; Pisum sativum
   (partial [41], 2 of 3 isozymes: cytoplasmic (I) and major chloroplastic (III)
   form [42], 2 isozymes [43, 44]) [41–44]; Solanum tuberosum (2 isozymes:
   slow and fast [49], phosphorylase II [50]) [49, 50]; Spinacia oleracea (affinity chromatography, cytoplasmic and chloroplastic (partial) isozyme [37],
   non-chloroplastic phosphorylase I [38]) [37, 38]; Typha latifolia (isozymes
   P-1 and P-2) [36]; Voandzeia subterranea (isozymes I and II) [35]; Zea mays
   (2 isozymes: I and II [40]) [39, 40]; Cryptococcus laurentii (partial [24]) [21,
   24]; Dictyostelium discoideum (isozymes a and b [23]) [20, 23]; Phymatotrichum omnivorum [18]; Physarum polycephalum [17]; Saccharomyces cerevisiae [25, 26, 28]; Agrobacterium tumefaciens (partial) [22]; Clostridium
   pasteurianum (partial) [16]; E. coli [13]; Klebsiella pneumoniae [14]; Streptococcus salivarius [15]; Entosphenus japonicus [55]; Locusta migratoria
   (isozymes a, ab and b) [5, 7]; Patinopecten yessoensis [6]; Pig [63]; Rabbit
   (phosphorylase ab hybrid) [62]; Rat (heart: 2 isozymes) [56]; Sepia pharaonis [10]

Crystallization
   (human [57], Musa paradisiaca [34], rabbit [57, 60], Saccharomyces cerevisiae [26], Tilapia mosambica [11]) [11, 26, 34, 57, 60]

Cloned
   –

Renatured
   –

---

5 STABILITY

pH
   4–9 (stable) [18]; 5–8 (stable) [28]; 5.5–7.5 (stable) [45]; 5.9–10.5 (30 min
   stable at 30°C) [14]; 11.5 (and above, denaturation) [14]

Temperature (°C)
   More (heating enhances sensitivity to tyrosine inhibition [34], phosphate
   protects against thermal inactivation [14]) [14, 34]; 0 (heart isozyme Ib or
   IIIb: $t_{1/2}$: 2.4–2.9 min, heart isozyme Ia: stable in the presence of 2-mercap-
   toethanol) [58]; 25 (stable below) [28]; 30 (15 min, slight stimulation [18],
   30 min stable at pH 5.9–10.5 [14]) [14, 18]; 35 (phosphorylase B: inactivation above) [33]; 36 (20% loss of activity within 5 min) [24]; 37 (2 h stable)
   [15]; 40 (15 min, 16% loss of activity [18], stable below [45]) [18, 45]; 40–50

Enzyme Handbook © Springer-Verlag Berlin Heidelberg 1996
Duplication, reproduction and storage in data banks are only
allowed with the prior permission of the publishers

(inactivation) [28]; 42 (inactivation above) [23]; 43 (10 min stable [14], and above, inactivation within 5 min [24]) [14, 24]; 45 (phosphorylase A or C: 30 min stable [33], crystalline, several h stable [57]) [33, 57]; 50 (inactivation within 5 min, b-isozyme) [23]; 52 ($t_{1/2}$: 10 min) [14]; 63 (inactivation within 10 min) [14]; 55 (about 40% loss of activity within 5 min [34], 10 min, 80% loss of activity [50]) [34, 50]; 70 (15 min, complete inactivation) [18]

Oxidation

Organic solvent

General stability information

Phosphate protects against thermal inactivation [14]; 5'-AMP stabilizes dilute enzyme solutions in media of high ionic strength, e.g. 50 mM ammonium sulfate or $MgCl_2$ [19]; Glycerol, 10%, stabilizes during purification [39]; Sucrose, 10%, stabilizes during purification [39]; Streptomycin sulfate or ammonium sulfate precipitation inactivates b-, not a-isozyme [23]; Freeze-thawing inactivates (b-isozyme [23]) [23, 30]; Freezing inactivates completely [38]; Dialysis for 24 h against water, stable to, phosphorylase A or C [33]; Glycogen, AMP or caffeine prevent proteolytic action of trypsin [29]; DTT stabilizes [57]

Storage

Bovine serum albumin prevents surface inactivation during storage [17]; Glycerol stabilizes purified enzyme during storage [20, 56]; Storage in frozen or lyophilized state, crystalline human, not rabbit, phosphorylase a or b, stable to [57]; −20°C, phosphorylase B, at least 1 month [33]; −20°C, cytoplasmic isozyme, about 30% loss of activity within 12 months [42]; −20°C, 50% glycerol, at least 4 months [56]; −20°C, partially purified, lyophilized, at least 2 months [63]; −20°C, purified, lyophilized, 30% loss of activity within 3 months [63]; −18°C, phosphorylase A or C, at least 1 month [33]; −10°C, 0.2 M NaF, 1 month [15]; 0°C, several days [16]; 0°C, at least 3 months [30]; 0°C, heart isozyme Ia, stable in the presence of 2-mercaptoethanol [58]; 0–4°C, at least 1 month [33]; 4°C, 10% loss of activity within 2 weeks [38]; 4°C, 3.5 M $(NH_4)_2SO_4$ in solution of neutral pH-value, several months [14]; 4°C, 20% loss of activity within 30 days [18]; 4°C, 20–25 mg/ml in 0.1 M succinate buffer, 0.02% NaN_3, 2–3 months [25]; 6°C, 3 months [17]; 7°C, b-isozmye: inactivation within 48 h, a-isozyme: at least 48 h, in crude extract [23]

6 CROSSREFERENCES TO STRUCTURE DATABANKS

PIR/MIPS code

PIR1:PHECGG (Escherichia coli); PIR2:S02178 (Escherichia coli); PIR2:B23093 (human (fragment)); PIR1:PHRBG (rabbit); PIR2:A24302 (rabbit); PIR2:A23093 (rabbit (fragment)); PIR2:A33328 (rat (fragment)); PIR2:C23093 (rat (fragment)); PIR2:S29860 (yeast (Saccharomyces cerevisiae)); PIR3:S40427 (yeast (Saccharomyces cerevisiae)); PIR2:S20595 (1 slime mold (Dictyostelium discoideum)); PIR3:S37300 (B rat); PIR2:A40138 (brain human); PIR2:A29949 (brain human); PIR2:PL0040 (cardiac pig (fragment)); PIR2:B24200 (cardiac muscle rat (fragment)); PIR2:A25518 (hepatic human); PIR2:A24200 (hepatic rat (fragment)); PIR2:PN0489 (hepatic rat (fragment)); PIR2:PH0226 (hepatic rat (fragment)); PIR3:S22338 (liver rat); PIR2:A27335 (muscle human); PIR2:A60521 (muscle mullet (Liza ramada) (fragment)); PIR2:S00596 (Escherichia coli); PIR1:PHECGM (Escherichia coli (fragment)); PIR2:S03773 (Escherichia coli (fragment)); PIR3:S53489 (potato); PIR3:S12033 (Potato (fragment)); PIR3:S47243 (alpha-1,4 glucan L isoform precursor fava bean); PIR3:S15531 (precursor potato); PIR2:PA0097 (fungus (Fusarium sporotrichioides) (fragment)); PIR3:S34189 (potato); PIR2:A40995 (H potato); PIR1:PHPOAG (precursor potato)

Brookhaven code

1ABB (Rabbit (Oryctolagus cuniculus) muscle); 1GPA (Rabbit (Oryctolagus cuniculus) muscle); 1GPB (Rabbit (Oryctolagus cuniculus) muscle); 1GPY (Rabbit (Oryctolagus cuniculus) muscle); 2GPB (Rabbit (Oryctolagus cuniculus) muscle); 3GPB (Rabbit (Oryctolagus cuniculus) muscle); 4GPB (Rabbit (Oryctolagus cuniculus) muscle); 5GPB (Rabbit (Oryctolagus cuniculus) muscle); 6GPB (Rabbit (Oryctolagus cuniculus) muscle); 7GPB (Rabbit (Oryctolagus cuniculus) muscle); 8GPB (Rabbit (Oryctolagus cuniculus) muscle); 9GPB (Rabbit (Oryctolagus cuniculus) muscle); 1PRI (Rabbit (Oryctolagus cuniculus) muscle); 1PRJ (Rabbit (Oryctolagus cuniculus) muscle); 1PYG (Rabbit (Oryctolagus cuniculus) muscle)

7 LITERATURE REFERENCES

[1] Oluoha, U., Ugochukwu, E.N.: Biol. Plant.,33,249–261 (1991)
[2] Yu, S., Pedersen, M.: Plant Physiol. Biochem.,29,341–347 (1991)
[3] Yu, S., Pedersen, M.: Physiol. Plant.,81,149–155 (1991)
[4] Oluoha, U.: Biol. Plant.,32,64–76 (1990)
[5] Van Marrewijk, W.J.A., Van den Broek, A.T.M., Beenakkers, A.M.T.: Insect Biochem., 18,37–44 (1988)
[6] Hata, K., Yokoyama, I., Suda, M., Hata, M., Matsuda, K.: Comp. Biochem. Physiol. B Comp. Biochem.,87B,747–753 (1987)
[7] Vaandrager, S.H., Van Marrewijk, W.J.A., Beenakkers, A.M.Th.: Insect Biochem., 17,695–700 (1987)
[8] Kumar, A.: Indian J. Plant Physiol.,27,209–213 (1984)

[9] Kumar, A., Sanwal, G.G.: Indian J. Biochem. Biophys.,21,241–247 (1984)
[10] Thomas, T.P., Philip, G.: Indian J. Biochem. Biophys.,20,198–202 (1983)
[11] Mukundan, M.K., Nair, M.R.: Fish. Technol.,14,1–6 (1977)
[12] Sutton, B.G.: Aust. J. Plant Physiol.,2,403–411 (1975)
[13] Chen, G.S., Segel, I.H.: Arch. Biochem. Biophys.,127,175–186 (1968)
[14] Linder, D., Kurz, G., Bender, H., Wallenfels, K.: Eur. J. Biochem.,70,291–303 (1976)
[15] Khandelwal, R.L., Spearman, T.N., Hamilton, I.R.: Arch. Biochem. Biophys., 154,295–305 (1973)
[16] Robson, R.L., Morris, J.G.: Biochem. J.,144,513–517 (1974)
[17] Nader, W., Becker, J.-U.: Eur. J. Biochem.,102,345–355 (1979)
[18] Sambandam, T., Gunasekaran, M.: Arch. Biochem. Biophys.,254,579–585 (1987)
[19] Cuppoletti, J., Segel, I.H.: J. Bacteriol.,139,411–417 (1979)
[20] Thomas, D.A., Wright, B.E.: J. Biol. Chem.,254,1253–1257 (1976)
[21] Schultz, J.C., Ankel, H.: Biochim. Biophys. Acta,215,39–51 (1970)
[22] Madsen, N.: Biochem. Biophys. Res. Commun.,6,310–313 (1961)
[23] Cloutier, M.J., Rutherford, C.L.: J. Biol. Chem.,262,9486–9493 (1987)
[24] Foda, M.S., Phaff, H.J.: Z. Allg. Mikrobiol.,18,95–106 (1978)
[25] Becker, J.-U., Wingender-Drissen, R., Schiltz, E.: Arch. Biochem. Biophys.,225, 667–678 (1983)
[26] Rath, V.L., Hwang, P.K., Fletterick, R.J.: J. Mol. Biol.,225,1027–1034 (1992)
[27] Spearman, T.N., Khandelwal, R.L., Hamilton, I.R.: Arch. Biochem. Biophys., 154,306–313 (1973)
[28] Tanabe, S., Kobayashi, M., Matsuda, K.: Agric. Biol. Chem.,51,2465–2471 (1987)
[29] Tanabe, S., Kobayashi, M., Matsuda, K.: Agric. Biol. Chem.,52,757–764 (1988)
[30] Nakamura, Y., Imamura, M.: Phytochemistry,22,835–840 (1983)
[31] Nakamura, Y., Imamura, M.: Phytochemistry,22,2395–2399 (1983)
[32] Kobayashi, M., Takagi, S., Matsuda, K., Ichishima, E.: Agric. Biol. Chem.,52, 2703–2708 (1988)
[33] Singh, S., Sanwal, G.G.: Phytochemistry,15,1447–1451 (1976)
[34] Kumar, A., Sanwal, G.G.: Biochemistry,21,4152–4159 (1982)
[35] Umezurike, G.M., Ekhorutomwen, S.A.: Biochim. Biophys. Acta,567,331–338 (1979)
[36] Iwata, T., Funagama, T., Hara, A.: Agric. Biol. Chem.,52,407–412 (1988)
[37] Preiss, J., Okita, T.W., Greenberg, E.: Plant Physiol.,66,864–869 (1980)
[38] Steup, M., Schächtele, C., Latzko, E.: Planta,148,168–173 (1980)
[39] Lee, E.Y.C., Braun, J.J.: Arch. Biochem. Biophys.,156,276–286 (1973)
[40] Mateyka, C., Schnarrenberger, C.: Plant Physiol.,86,417–422 (1988)
[41] Kruger, N.J., Ap Rees, T.: Phytochemistry,22,1891–1898 (1983)
[42] Conrads, J., Van Berkel, J., Schächtele, C., Steup, M.: Biochim. Biophys. Acta, 882,452–463 (1986)
[43] Myers, D., Matheson, N.K.: Phytochemistry,30,1079–1087 (1991)
[44] Matheson, N.K., Richardson, R.H.: Phytochemistry,17,195–200 (1973)
[45] Hamdan, I., Diopoh, J.: Plant Sci.,76,1–7 (1991)
[46] Chang, T.-C., Su, J.-C.: Plant Physiol.,80,534–538 (1986)
[47] Pan, S.-M., Chang, T.-C., Juang, R.-H., Su, J.-C.: Plant Physiol.,88,1154–1156 (1988)
[48] Chang, T.-C., Lee, S.-C., Su, J.-C.: Agric. Biol. Chem.,51,187–195 (1987)
[49] Shivaram, K.N.: Z. Naturforsch.,31c,424–432 (1976)
[50] Sivak, M.N., Tandecarz, J.S., Cardini, C.E.: Arch. Biochem. Biophys.,212,525–536 (1981)

[51] Sivak, M.N., Tandecarz, J.S., Cardini, C.E.: Arch. Biochem. Biophys.,212,537–545 (1981)
[52] Kokesh, F., Stephenson, R.K., Kakuda, Y.: Biochim. Biophys. Acta,483,258–262 (1977)
[53] Kamogawa, A., Fukui, T.: Biochim. Biophys. Acta,302,158–166 (1973)
[54] Soman, G., Philip, G.: Biochim. Biophys. Acta,482,35–40 (1977)
[55] Yonezawa, S., Hori, S.H.: Arch. Biochem. Biophys.,181,447–453 (1977)
[56] Berndt, N., Rösen, P.: Arch. Biochem. Biophys.,228,143–154 (1984)
[57] Assaf, S.A., Yunis, A.A.: Ann. N.Y. Acad. Sci.,210,139–152 (1973) (Review)
[58] Schliselfeld, L.H.: Ann. N.Y. Acad. Sci.,210,181–191 (1973) (Review)
[59] Soman, G., Philip, G.: Biochim. Biophys. Acta,358,359–362 (1974)
[60] Ariki, M., Fukui, T.: J. Biochem.,78,1191–1199 (1975)
[61] Khandelwal, R.L.: Biochim. Biophys. Acta,485,379–390 (1977)
[62] Vereb, G., Fodor, A., Bot, G.: Biochim. Biophys. Acta,915,19–27 (1987)
[63] Medicus, R., Mendicino, J.: Eur. J. Biochem.,40,63–75 (1973)

1 NOMENCLATURE

EC number
2.4.1.2

Systematic name
1,4-alpha-D-Glucan:1,6-alpha-D-glucan 6-alpha-D-glucosyltransferase

Recommended name
Dextrin dextranase

Synonyms
Glucosyltransferase, dextrin 6-
Dextrin 6-glucosyltransferase
Dextran dextrinase [1]

CAS Reg. No.
9032-13-7

2 REACTION AND SPECIFICITY

Catalysed reaction
(1,4-alpha-D-Glucosyl)$_n$ + (1,6-alpha-D-glucosyl)$_m$ →
→ (1,4-alpha-D-glucosyl)$_{n-1}$ + (1,6-alpha-D-glucosyl)$_{m+1}$ (mechanism [3])

Reaction type
Hexosyl group transfer

Natural substrates

Substrate spectrum
1 (1,4-alpha-D-Glucosyl)$_n$ + (1,6-alpha-D-glucosyl)$_m$ (i.e. dextrin x + dextran
y, converts polymerized 1,4-linked glucose units into polymerized 1,6-lin-
ked units [1], acts on non-reducing terminal glucosyl residues [3], three
transglucosylation action modes: main reaction, disproportional action on
maltooligosaccharides and disproportional action on isomaltooligosaccha-
rides [3], substrates are maltooligosaccharides, e.g. maltotriose (G3),
maltotetraose (G4) [2–4], maltopentaose, maltohexaose, short-chain amylose, soluble starch [2], maltotetraitol [3, 4], poor substrate: maltose (G2)
[2, 3], no substrates are unhydrolyzed amylose, amylopectin fractions of
corn or potato starch, native oyster or rabbit liver glycogens, Schardinger's
beta-dextrin (cycloheptoamylose), maltose, sucrose, glucose 1-phosphate
or glucose 6-phosphate, trehalose, alpha-methylglucoside, raffinose, melezitose [1], O-6-deoxy-6-[(2-pyridyl)amino]alpha-D-glucopyra-
nosyl-1,4-maltotriose [3], maltitol [4]) [1–4]

Product spectrum
   1 (1,4-alpha-D-Glucosyl)$_{n-1}$ + (1,6-alpha-D-glucosyl)$_{m+1}$ (i.e. dextrin$_{x-1}$ +
   dextran$_{y+1}$ [1], successive elongation) [1]

Inhibitor(s)
   Fe^{3+} [2]; More (no inhibition by Ca^{2+}, Co^{2+}, Fe^{2+}, Mg^{2+}, Mn^{2+}, Ni^{2+}, Zn^{2+},
   p-chloromercuribenzoate and EDTA) [2]

Cofactor(s)/prosthetic group(s)/activating agents

Metal compounds/salts
   Mn^{2+} (slight activation) [2]; More (no activation by Ca^{2+}, Co^{2+}, Fe^{2+}, Fe^{3+},
   Mg^{2+}, Ni^{2+}, Zn^{2+} and EDTA) [2]

Turnover number (min^{-1})

Specific activity (U/mg)
   20.8 [2]

K$_m$-value (mM)

pH-optimum
   4–4.2 (broad) [2]

pH-range
   3.0–5.7 (about half-maximal activity at pH 3.0 and 5.7) [2]

Temperature optimum (°C)
   37–45 [2]

Temperature range (°C)
   17–52 (about half-maximal activity at 17°C and 52°C) [2]

3 ENZYME STRUCTURE

Molecular weight

Subunits
   ? (x × 300000, Acetobacter capsulatum, SDS-PAGE) [2]

Glycoprotein/Lipoprotein
   –

4 ISOLATION/PREPARATION

Source organism
   Acetobacter capsulatum [1–4]

Source tissue
Cell [1–4]

Localization in source
Soluble [1]

Purification
Acetobacter capsulatum [2]

Crystallization
–

Cloned
–

Renatured
–

---

5 STABILITY

pH
3.5–5.2 (30 min stable) [2]

Temperature (°C)
45 (30 min stable below) [2]; 55 (inactivation within 30 min) [2]; 70 (inactivation within 10 min) [1]

Oxidation

Organic solvent
n-Butanol, H_2O-saturated, 50% v/v in acetate buffer, stable to [2]; Ethylene glycol, 40% v/v in acetate buffer, stable to [2]

General stability information

Storage

---

6 CROSSREFERENCES TO STRUCTURE DATABANKS

PIR/MIPS code

Brookhaven code

7 LITERATURE REFERENCES

[1] Hehre, E.J.: J. Biol. Chem.,192,161–174 (1953)
[2] Yamamoto, K., Yoshikawa, K., Kitahata, S., Okada, S.: Biosci. Biotechnol. Biochem., 56,169–173 (1992)
[3] Yamamoto, K., Yoshikawa, K., Okada, S.: Biosci. Biotechnol. Biochem.,57,47–50 (1993)
[4] Yamamoto, K., Yoshikawa, K., Okada, S.: Biosci. Biotechnol. Biochem.,57,136–137 (1993)

1 NOMENCLATURE

EC number
2.4.1.4

Systematic name
Sucrose:1,4-alpha-D-glucan 4-alpha-D-glucosyltransferase

Recommended name
Amylosucrase

Synonyms
Glucosyltransferase, sucrose-1,4-alpha-glucan
Sucrose-glucan glucosyltransferase

CAS Reg. No.
9032-11-5

---

2 REACTION AND SPECIFICITY

Catalysed reaction
Sucrose + (1,4-alpha-D-glucosyl)$_n$ →
→ D-fructose + (1,4-alpha-D-glucosyl)$_{n+1}$

Reaction type
Hexosyl group transfer

Natural substrates
Sucrose + (1,4-alpha-D-glucosyl)$_n$ (involved in biosynthesis of amylopec-tin-glycogen type polysaccharide [2], constitutive enzyme [3, 4]) [2–4]

Substrate spectrum
1 Sucrose + (1,4-alpha-D-glucosyl)$_n$ (transfers glucose to growing alpha-1,4-glucan chains [3], needs primer molecule: e.g. glucan from Neisseria sp. [2, 4], mussel or sweet corn glycogen, or corn amylopectin [2], sucrose alone is no substrate [2], alpha-D-galactopyranosyl-beta-D-fructofuranoside (Galsucrose) [1], alpha-D-glucopyranosyl fluoride [6] can replace sucrose, no substrates are beta-D-fructofuranosyl-alpha-D-xyloside (Xylsucrose), melibiose, raffinose [1], melezitose [1, 2], 3-deoxy-sucrose, alpha-D-allopyranosyl beta-fructofuranoside [5]) [1–6]

Product spectrum
1 D-Fructose + (1,4-alpha-D-glucosyl)$_{n+1}$ (product: glycogen-like polysaccharide [5], highly branched [4]) [1–5]

Inhibitor(s)
   6-Deoxysucrose [5]; 6-Deoxy-6-fluorosucrose [5]; 4,6-Dideoxysucrose [5];
   Sucrose (above 0.1 M, substrate inhibition) [3]; Fructose (competitive inhibitor) [3, 4]; Tris-HCl buffer [3]; More (no inhibition by C-3-modified sucrose
   derivatives) [5]

Cofactor(s)/prosthetic group(s)/activating agents
   More (no ADP, UDP, ADPglucose or UDPglucose involved) [2]

Metal compounds/salts

Turnover number (min^{-1})

Specific activity (U/mg)
   More (cell-free extracts from various Neisseria species) [4]; 3.7 [2]

K_m-value (mM)
   26.5 (sucrose) [5]

pH-optimum
   6.7–7.0 [3]

pH-range

Temperature optimum (°C)
   30 (assay at) [2]

Temperature range (°C)

---

3 ENZYME STRUCTURE

Molecular weight

Subunits

Glycoprotein/Lipoprotein
   –

---

4 ISOLATION/PREPARATION

Source organism
   Neisseria perflava [1–6]; Neisseria canis [4]; Neisseria cinerea [4]; Neisseria
   cumiculi (not var. gigantea) [4]; Neisseria sicca [4]; Neisseria subflava [4]

Source tissue
   Cell [1–6]

Localization in source
   Soluble (in cells grown without sucrose [3]) [2–4]; Insoluble (in cells grown
   with sucrose, probably bound to polymerized product) [3]

Purification
   Neisseria perflava [2]

Crystallization
   –

Cloned
   –

Renatured
   –

---

5 STABILITY

pH

Temperature (°C)
   45 (inactivation within 10 min) [2]

Oxidation

Organic solvent

General stability information
   Rapid freezing or deep freezing inactivates [2]

Storage
   –20°C, several months without loss of activity [3]; 2–4°C, $t_{1/2}$: 15–20 h [2]

---

6 CROSSREFERENCES TO STRUCTURE DATABANKS

PIR/MIPS code

Brookhaven code

---

7 LITERATURE REFERENCES

[1] Feingold, D.S., Avigad, G., Hestrin, S.: J. Biol. Chem.,224,295–307 (1957)
[2] Okada, G., Hehre, E.J.: J. Biol. Chem.,249,126–135 (1974)
[3] MacKenzie, C.R., Johnson, K.G., McDonald, I.J.: Can. J. Microbiol.,23,1303–1307 (1977)
[4] MacKenzie, C.R., McDonald, I.J., Johnson, K.G.: Can. J. Microbiol.,24,357–362 (1978)
[5] Tao, B.Y., Reilly, P.J., Robyt, J.F.: Carbohydr. Res.,181,163–174 (1988)
[6] Okada, G., Hehre, E.J.: Carbohydr. Res.,26,240–243 (1973)

1 NOMENCLATURE

EC number
2.4.1.5

Systematic name
Sucrose:1,6-alpha-D-glucan 6-alpha-D-glucosyltransferase

Recommended name
Dextransucrase

Synonyms
Sucrose 6-glucosyltransferase
Dextran-sucrase [5]
SGE [8]
CEP [9]
Glucosyltransferase, sucrose-1,6-alpha-glucan

CAS Reg. No.
9032-14-8

2 REACTION AND SPECIFICITY

Catalysed reaction
Sucrose + $(1,6\text{-alpha-D-glucosyl})_n \rightarrow$
$\rightarrow$ D-fructose + $(1,6\text{-alpha-D-glucosyl})_{n+1}$

Reaction type
Hexosyl group transfer

Natural substrates
Sucrose + $(1,6\text{-alpha-D-glucosyl})_n$ (participates in glucan synthesis) [8]

Substrate spectrum
1 Sucrose + $(1,6\text{-alpha-D-glucosyl})_n$ [1–23]
2 More (purified dextransucrase possesses an invertase-like activity [22], the enzyme possesses enhanced level of sucrose hydrolyzing activity [16]) [16, 22]

Product spectrum
1 D-Fructose + $(1,6\text{-alpha-D-glucosyl})_{n+1}$ [1–23]
2 ?

Inhibitor(s)

EDTA (enzyme I not affected, enzyme N inhibited, inhibition overcome by Ca^{2+} [17], noncompetitive [12], no effect [20], addition of dextran partially protects [16], addition of Ca^{2+} or Co^{2+} restores activity [16]) [2, 12, 16, 17, 19]; Guanidine-HCl (0.025 M, 65% inhibition of enzyme N, no effect on enzyme I) [17]; Urea (0.25 M, enzyme N completely inactivated, enzyme I retains 60% of its activity [17], 8 M [23]) [17, 20, 23]; Methyl-deoxy-alpha-D-glucopyranoside (weak, competitive) [3]; 6-Deoxy-6-fluoro-alpha-D-glucopyranoside (very weak, noncompetitive) [3]; Periodate-oxidized dextrans [6]; Oxidized saccharides (overview) [6]; Ribocitrin (competitive with regard to dextran T10) [9]; Zn^{2+} (inhibition by binding to 2 types of metal ion sites, one type consists of a single site and has a low apparent affinity to Ca^{2+}, at the remaining site(s), Ca^{2+} has a much higher apparent affinity than Zn^{2+}, Ni^{2+} or Co^{2+} and prevents inhibition by these metal ions) [11]; Ni^{2+} (inhibition by binding to 2 types of metal ion sites, one type consists of a single site and has a low apparent affinity to Ca^{2+}, at the remaining site(s), Ca^{2+} has a much higher apparent affinity than Zn^{2+}, Ni^{2+} or Co^{2+} and prevents inhibition by these metal ions) [11]; Co^{2+} (inhibition by binding to 2 types of metal ion sites, one type consists of a single site and has a low apparent affinity to Ca^{2+}, at the remaining site(s), Ca^{2+} has a much higher apparent affinity than Zn^{2+}, Ni^{2+} or Co^{2+} and prevents inhibition by these metal ions [11], 1 mM, stimulates activity of enzyme N, enzyme I is inhibited [17]) [11, 17]; Tris [11]; Ca^{2+} (not required, below 1 mM: activation, above 1 mM: weak competitive inhibition [10], 1 mM, stimulates activity of enzyme N, enzyme I is inhibited [17]) [10, 17]; Mg^{2+} (1 mM, stimulates activity of enzyme N, enzyme I is inhibited [17], no effect [23]) [11, 17]; Sr^{2+} [11]; Ba^{2+} [11]; Mn^{2+} (enzyme I and II [20], no effect [23]) [11, 16, 17, 20]; 6'-Amino-6'-deoxysucrose (competitive) [12]; Glucono-1,5-lactone (competitive) [12]; Glucuronic acid (reduced, noncompetitive) [12]; Phenylmercuric acetate [16, 17, 20]; Fe^{2+} (1 mM, stimulates activity of enzyme N, enzyme I is inhibited) [17]; SDS [17, 20, 23]; Methyl-6-amino-6-deoxyglucoside (competitive) [12]; Methyl-alpha-D-glucoside (competitive [12], 50 mM, activates release of D-fructose (sucrase activity), inhibits synthesis of dextran (transferase activity) [19]) [12, 19]; N-Methyl-D-glucamine (competitive) [12]; Nojirimycin (noncompetitive) [12]; Maltose [12]; 2-Deoxy-2-fluoroglucopyranosyl fluoride (competitive) [13]; 3-Deoxy-3-fluoroglucopyranosyl fluoride (competitive) [13]; 3-Deoxy-3-thioglucopyranosyl fluoride (competitive) [13]; Cu^{2+} [16]; Fe^{3+} (enzyme I [20]) [16, 20]; Dithiothreitol [23]; Fructose (competitive) [23]; $AgNO_3$ [23]; Guanidine-HCl [20]; HgCl [23]

Cofactor(s)/prosthetic group(s)/activating agents

Dextrans (stimulation [16], activity of enzyme N is more effectively stimulated than enzyme I [17], stimulates reducing sugar production 1.3-fold and
dextran synthesis 9.6-fold [19], B-1416 enzyme is activated 4.35-fold by addition of 0.5% exogenous dextran, B-1375 enzyme is activated 2.76-fold
[21], addition of the B-1299 water-soluble dextan stimulates activity of enzyme I, no effect on activity of enzyme II [20]) [16, 17, 19–21]; Methyl-
alpha-D-glucoside (50 mM, activates release of D-fructose (sucrose activity),
inhibits synthesis of dextran (transferase activity)) [19]

Metal compounds/salts

Ca^{2+} (activates [19], restores activity after EDTA treatment [2], not required,
below 1 mM: activation, above 1 mM: weak competitive inhibition [10], 1
mM, activates [16], stimulates activity of enzyme N [17]) [2, 10, 16, 17, 19];
Mg^{2+} (1 mM, stimulates activity of enzyme N, enzyme I is inhibited) [17];
Fe^{2+} (1 mM, stimulates activity of enzyme N, enzyme I is inhibited) [17]; Co^{2+}
(1 mM, stimulates activity of enzyme N, enzyme I is inhibited [17], 1 mM
slight activation of enzyme I and II [20]) [17, 20]; More (2 enzyme forms: I
and N, enzyme I is gradually converted into enzyme N upon ageing, conversion is stimulated in the presence of NaCl) [17]

Turnover number (min^{-1})

Specific activity (U/mg)

4.47 [8]; 90–170 [4]; More [14, 16, 17, 19, 20]

K_m-value (mM)

11.2 (sucrose) [13]; 12–16 (sucrose) [4]; 26 (alpha-D-glucopyranosyl fluoride) [4]; More (effect of temperature [17]) [17, 19]

pH-optimum

5.0 [2]; 5–5.5 [4]; 5.2 (free and immobilized [14]) [14, 19, 21]; 5.2–6.2 [7];
5.5 (enzyme N [17]) [1, 17, 23]; 5.5–6.9 (enzyme II) [20]; 6.0 (enzyme I)
[17]; 6.3–6.5 (enzyme I) [20]; 6.6–7.0 [8]

pH-range

Temperature optimum (°C)

29 [1]; 30 (B-1375 enzyme with dextran [21], below [2], free and immobilized [14]) [2, 14, 16, 19, 21]; 34–42 [23]; 35 (B-1416 enzyme with and without dextran, B-1375 enzyme without dextran [21], enyme N [17]) [17, 21];
35–40 (enzyme II) [20]; 45 (enzyme I) [20]

Temperature range (°C)

8–36 (8°C: 22% of activity maximum, 36°C: 30% of activity maximum) [1];
37–51 (80% of activity maximum at 37°C and 51°C) [20]

3 ENZYME STRUCTURE

Molecular weight
65000 (Leuconostoc mesenteroides, enzyme I and II, disc gel electrophoresis) [16]
94000 (Streptococcus mutans, gel filtration) [23]
100000 (about, Streptococcus sanguis [7], Leuconostoc mesenteroides, enzyme III, disc gel electrophoresis [16]) [7, 16]
130000–133000 (Leuconostoc mesenteroides, disc gel electrophoresis) [21]
More (2 forms: MW 177000 and MW 158000) [4]

Subunits
? (x x 170000, Streptococcus mutans, SDS-PAGE [22], x x 48000, Leuconostoc mesenteroides, enzyme I, SDS-PAGE [17], x x 65000, Leuconostoc mesenteroides, enzyme II, SDS-PAGE [19]) [17, 19, 22]
Monomer (1 x 69000, Leuconostoc mesenteroides, enzyme I, 1 x 79000, Leuconostoc mesenteroides, enzyme II, SDS-PAGE) [20]
Dimer (2 x 64000–68000, Leuconostoc mesenteroides, SDS-PAGE after alkaline treatment, pH 10.5) [21]

Glycoprotein/Lipoprotein
Glycoprotein (17% carbohydrate [16]) [1, 2, 16]

---

4 ISOLATION/PREPARATION

Source organism
Betacoccus arabinosaceus [1]; Leuconostoc mesenteroides (Sikhae [2], NRRI B-512(F) [2, 4, 10, 11, 13, 14, 16, 19], NRRL B-1416 [12, 21], NRRL B-1299 [17, 20], IAM 1046 [18], B-1375 [21], B-512FM [3]) [2–4, 10–14, 16–21]; Streptococcus sanguis (ATCC 10558) [5, 7]; Streptococcus mutans (E49 [8, 9], 6715 [23], HS6 [22]) [6, 8, 9, 15, 22, 23]

Source tissue
Culture fluid [2, 7, 16–18, 22, 23]

Localization in source
Extracellular [16, 17]

Purification
Leuconostoc mesenteroides (partial [2], NRRL B-1299 [17, 20], 2 forms: I and N, enzyme I is gradually converted into enzyme N upon ageing, conversion is stimulated in the presence of NaCl [17], 2 forms: I and II [20], Sikhae [2], NRRL B-512(F) [2, 4, 14, 16, 19], 2 forms: MW 177000 and MW 158000 [4], B-1375 [21], multiple forms: I, II and III [16], IAM 1046 [18], B-1416 [21], 2 forms: I (major form) and II (smaller amount), dextransucrase II purified [19]) [2, 4, 14, 16–21]; Streptococcus sanguis (2 forms) [5, 7]; Streptococcus mutans (6715 [23], HS6 [22], 6715 [23]) [15, 22, 23]

Crystallization
–

Cloned
–

Renatured
–

---

5 STABILITY

pH
4 (B-512(F) enzyme unstable, Sikhae enzyme more stable) [2]; 4.6–7.2 (4°C, 24 h, 50% loss of activity at pH 4.6 and 7.2, enzyme II) [20]; 4.7–8.5 (4°C, 24 h, 50% loss of activity at pH 4.7 and 8.5, enzyme I) [20]; 5.2 (stability optimum, immobilized enzyme) [14]; 5.2–8.5 (4°C, 0.1% bovine serum albumin, 24 h) [21]; 5.3–5.8 (4°C, 24 h, enzyme II stable) [20]; 5.5 (4°C, 24 h, enzyme N stable) [17]; 5.5–6 (stability optimum, free enzyme) [14]; 6.0 (4°C, 24 h, enzyme I stable) [17]; 6.2–6.9 (4°C, 24 h, enzyme I stable) [20]

Temperature (°C)
35 (10 min, enzyme I stable [20], 10 min, enzyme I and N stable [17], B-1375 enzyme stable [21]) [17, 20, 21]; 40 (10 min, 80% loss of activity without dextran, 30% loss of activity with addition of dextran [16], B-1416 enzyme stable [21]) [16, 21]; 45 (10 min, 80% loss of activity, enzyme I) [20]

Oxidation

Organic solvent

General stability information
Dextran T10 stabilizes [11]; Maximal stabilization of soluble enzyme by 5 mM Ca^{2+} [17]; Glycerol, 33%, significantly stabilizes [19]; Bovine serum albumin, 0.1%, significantly stabilizes [19]

Storage
Storage stability decreased by addition of dextranase [2]; 4°C, pH 6.0–9.0, 4 mg/ml dextran, 24 h stable, without dextran the stable range is narrowed at pH 7.0–9.0 [16]; In a deep-freeze, 33% glycerol, 0.1% bovine serum albumin, less than 20% loss of activity after several months [19]; 4°C, 0.1% bovine serum albumin, pH 5.2–8.5, stable for 24 h [21]

6 CROSSREFERENCES TO STRUCTURE DATABANKS

PIR/MIPS code

PIR2:PU0034 (Streptococcus bovis (fragment)); PIR2:B41898 (Streptococcus gordonii (fragment)); PIR2:B39841 (Streptococcus sobrinus (fragment)); PIR2:A45866 (precursor Streptococcus mutans); PIR2:JT0345 (precursor Streptococcus mutans (strain GS-5))

Brookhaven code

7 LITERATURE REFERENCES

[1] Bailey, R.W., Barker, S.A., Bourne, E.J., Stacey, M.: J. Chem. Soc.,3530–3536 (1957)
[2] Rhee, S.H., Lee, C.H.: J. Microbiol. Biotechnol.,1,176–181 (1991)
[3] Tanriseven, A., Robyt, J.F.: Carbohydr. Res.,225,321–329 (1992)
[4] Miller, A.W., Eklund, S.H., Robyt, J.F.: Carbohydr. Res.,147,119–133 (1986)
[5] Grahame, D.A., Mayer, R.M.: Carbohydr. Res.,142,285–298 (1985)
[6] Ono, K., Nuessle, D.W., Smith, E.E.: Carbohydr. Res.,88,119–134 (1981)
[7] Huang, S., Lee, H.C., Mayer, R.M.: Carbohydr. Res.,74,287–300 (1979)
[8] Takashio, M., Okami, Y.: Agric. Biol. Chem.,47,2161–2171 (1983)
[9] Takashio, M., Okami, Y.: Agric. Biol. Chem.,47,2153–2159 (1983)
[10] Miller, A.W., Robyt, J.F.: Biochim. Biophys. Acta,880,32–39 (1986)
[11] Miller, A.W., Robyt, J.F.: Arch. Biochem. Biophys.,248,579–586 (1986)
[12] Kobayashi, M., Yokoyama, I., Matsuda, K.: Agric. Biol. Chem.,50,2585–2590 (1986)
[13] Michiels, A.G., Wang, A.Y., Clark, D.S., Blanch, H.W.: Appl. Biochem. Biotechnol., 31,237–246 (1991)
[14] Kaboli, H., Reilly, P.J.: Biotechnol. Bioeng.,22,1055–1069 (1980)
[15] Russell, R.R.B.: FEMS Microbiol. Lett.,6,197–199 (1979)
[16] Kobayashi, M., Matsuda, K.: Biochim. Biophys. Acta,614,46–62 (1980)
[17] Kobayashi, M., Matsuda, K.: J. Biochem.,79,1301–1308 (1976)
[18] Suzuki, D., Kobayashi, T.: Agric. Biol. Chem.,39,557–558 (1975)
[19] Kobayashi, M., Mihara, K., Matsuda, K.: Agric. Biol. Chem.,50,551–556 (1986)
[20] Kobayashi, M., Matsuda, K.: Biochim. Biophys. Acta,397,69–79 (1975)
[21] Yokoyama, I., Kobayashi, M., Matsuda, K.: Agric. Biol. Chem.,49,1385–1391 (1985)
[22] Fukui, K., Fukui, Y., Moriyama, T.: J. Bacteriol.,118,796–804 (1974)
[23] Chludzinski, A.M., Germaine, G.R., Schachtele, C.F.: J. Bacteriol.,118,1–7 (1974)

1 NOMENCLATURE

EC number
2.4.1.7

Systematic name
Sucrose:orthophosphate alpha-D-glucosyltransferase

Recommended name
Sucrose phosphorylase

Synonyms
Phosphorylase, sucrose
Disaccharide glucosyltransferase [3]
Sucrose glucosyltransferase

CAS Reg. No.
9074-06-0

2 REACTION AND SPECIFICITY

Catalysed reaction
Sucrose + phosphate →
→ alpha-D-glucose 1-phosphate + D-fructose (mechanism and structure [2, 3, 7])

Reaction type
Hexosyl group transfer

Natural substrates
Sucrose + phosphate (inducible bacterial enzyme) [1]

Substrate spectrum
1 Sucrose + phosphate (r [1–3, 6–8], favoured reaction [1], H_2O may replace phosphate, in the absence of inorganic acceptors the glucosyl group can be transferred to a monosaccharide (e.g. fructose or sorbose) [1–3], trans-1,2-cyclohexanediol, cis-1,2-cyclohexanediol, ethylene glycol, methanol, ethanol [7]) [1–3, 6–8]
2 Sucrose + arsenate [1, 2]

Enzyme Handbook © Springer-Verlag Berlin Heidelberg 1996
Duplication, reproduction and storage in data banks are only
allowed with the prior permission of the publishers

    3 alpha-D-Glucose 1-phosphate + D-fructose (specific for D-glucose con-
       figuration [1], broad glucosyl acceptor specificity [5], several alcohols
       [2], pentitols, L-fucose, alpha-methyl-D-glucose and glyceraldehyde [5]
       can act as acceptors, poor acceptors are D-glucose, D-galactose, D-xy-
       lose, D-mannose, D-ribose, D-glucuronic acid, glycerol, D-sorbitol, in-
       ositol and sucrose [5], no acceptors are L-glucose, L-xylose, D/L-rham-
       nose, D-fucose, 2-deoxy-D-glucose, glucosamine, N-acetylglucosamine,
       D-galacturonic acid, ascorbic acid, erythritol, D-mannitol, maltose, iso-
       maltose, trehalose, cellobiose, maltotriose, melibiose, melezitose, pa-
       nose, raffinose [5]) [1–3, 5]
    4 alpha-D-Glucose 1-phosphate + H_2O (ir) [1, 3]
    5 alpha-D-Glucose 1-phosphate + arsenate (ir) [3]
    6 alpha-D-Glucose 1-phosphate + L-sorbose [1–3, 7]
    7 alpha-D-Glucose 1-phosphate + L-arabinose (favoured reaction [1], not
       [5]) [1, 3]
    8 alpha-D-Glucose 1-phosphate + D-xylulose [1, 3]
    9 alpha-D-Glucose 1-phosphate + L-arabinulose [3]
   10 alpha-D-Glucose 1-phosphate + ethylene glycol [2]
   11 alpha-D-Glucose 1-phosphate + methanol [2]
   12 alpha-D-Glucose 1-phosphate + trans-1,2-cyclohexanediol [2]
   13 alpha-D-Glucose 1-phosphate + xylitol [5]
   14 alpha-D-Glucose 1-fluoride + H_2O [2]
   15 alpha-D-Glucose 1-fluoride + phosphate [2]

Product spectrum

    1 alpha-D-Glucose 1-phosphate + D-fructose (mainly pyranose form) [1, 2,
       6]
    2 Glucose + arsenate (via unstable glucose 1-arsenate + glucose) [1, 2]
    3 Sucrose + phosphate [1–3]
    4 ?
    5 ?
    6 alpha-D-Glucosyl-alpha-L-sorbofuranoside + phosphate [3]
    7 ?
    8 alpha-D-Glucopyranosyl-D-xylulofuranoside + phosphate [1]
    9 ?
   10 Hydroxyethylglucoside + phosphate [2]
   11 alpha-Methylglucoside + phosphate [2]
   12 Hydroxycyclohexylglucoside + phosphate [2]
   13 4-O-alpha-D-Glucopyranosyl-xylitol + phosphate [5]
   14 ?
   15 ?

Inhibitor(s)
  Glucose (strong) [1]; Ethylene glycol (phosphorolysis) [2]; trans-1,2-Cyclo-
  hexanediol (phosphorolysis) [2]; Chloroethanol [7]; 2-Mercaptoethanol [7]

Cofactor(s)/prosthetic group(s)/activating agents
  More (no organic or anorganic cofactor required) [1]

Metal compounds/salts

Turnover number (min^{-1})

Specific activity (U/mg)
  More [3]; 173.8 [4]

K_m-value (mM)
  More (kinetic study [2, 3, 7], kinetics of immobilized enzyme [8]) [2, 3, 7, 8];
  13 (D-fructose) [2]; 130 (sorbose) [2]; 270 (trans-1,2-cyclohexanediol) [2]

pH-optimum
  More (pI: 4.6) [4]; 6.3 (hydrolysis) [3]; 6.5 (phosphorolysis, immobilized enzyme) [8]; 6.8 (broad [1], disaccharide synthesis plus glucose 1-phosphate
  hydrolysis, immobilized enzyme [8]) [1]; 7.0 (phosphorolysis) [3]

pH-range
  5.0–7.8 (about 60% of maximal activity at pH 5.0 and 7.8, phosphorolysis,
  immobilized enzyme) [8]; 5.4–8.2 (about half-maximal activity at pH 5.4 and
  8.2, disaccharide synthesis, immobilized enzyme) [8]; 5.5–8.5 (about
  half-maximal activity at pH 5.5 and 8.5, hydrolysis) [3]; 5.5–9.0 (about 60%
  of maximal activity at pH 5.5 and 9.0, phosphorolysis) [3]

Temperature optimum (°C)

Temperature range (°C)

3 ENZYME STRUCTURE

Molecular weight
  55000 (Leuconostoc mesenteroides, HPLC-gel filtration) [4]
  56400 (Leuconostoc mesenteroides, sedimentation equilibrium centrifugation) [4]
  80000–100000 (Pseudomonas saccharophila, gel filtration [3], analytical
  ultracentrifugation [2]) [2, 3]

Subunits
  Monomer (1 × 55000, Leuconostoc mesenteroides, SDS-PAGE) [4]
  Dimer (2 × 50000, Pseudomonas saccharophila, SDS-PAGE) [2]

Glycoprotein/Lipoprotein
  –

4 ISOLATION/PREPARATION

Source organism
   Leuconostoc mesenteroides [1, 4–6]; Pseudomonas saccharophila [1–3, 7,
   8]; Pseudomonas putrefaciens [1]; Streptococcus mutans [9]

Source tissue
   Cell [1–9]

Localization in source
   Cytoplasm [3]

Purification
   Leuconostoc mesenteroides [4]; Pseudomonas saccharophila [3]

Crystallization
   –

Cloned
   (Leuconostoc mesenteroides gene, cloned and expressed in E. coli 1100,
   using a 'sleeper' bacteriophage vector (slp-spl-1)) [6]

Renatured
   –

5 STABILITY

pH

Temperature (°C)
   30 ($t_{1/2}$: 34–38 days, immobilized enzyme) [8]; 35 ($t_{1/2}$: 10.3 days, immobilized enzyme) [8]; 40 ($t_{1/2}$: 5 days, immobilized enzyme) [8]

Oxidation

Organic solvent
   Ethylene glycol, up to 30% v/v, in ethylene-glycol water mixtures, stable to
   [2, 7]; Methanol, 25% v/v, stable to [7]

General stability information

Storage

6 CROSSREFERENCES TO STRUCTURE DATABANKS

PIR/MIPS code
   PIR2:PQ0164 (Leuconostoc mesenteroides (fragment)); PIR2:A27626
   (Streptococcus mutans)

Brookhaven code

4

7 LITERATURE REFERENCES

[1] Doudoroff, M. in "The Enzymes",2nd Ed. (Boyer, P.D., Lardy, H., Myrbäck, K., Eds.) , 5,229–236, Academic Press, New York (1961) (Review)

[2] Mieyal, J.J., Abeles, R.H. in "Enzymes",3rd Ed. (Boyer, P.D., Ed.) ,7,515–532, Academic Press, New York (1972) (Review)

[3] Silverstein, R., Voet, J., Reed, D., Abeles, R.H.: J. Biol. Chem.,242,1338–1346 (1967)

[4]Koga, T., Nakamura, K., Shirokane, Y., Mizusawa, K., Kitao, S., Kikuchi, M.: Agric. Biol. Chem.,55,1805–1810 (1991)

[5] Kitao, S., Sekine, H.: Biosci. Biotechnol. Biochem.,56,2011–2014 (1992)

[6] Kitao, S., Nakano, E.: J. Ferment. Bioeng.,73,179–184 (1992)

[7] Mieyal, J.J., Simon, M., Abeles, R.H.: J. Biol. Chem.,247,532–542 (1972)

[8] Taylor, F., Chen, L., Gong, C.S., Tsao, G.T.: Biotechnol. Bioeng.,24,317–328 (1982)

[9] Russell, R.R.B., Mukasa, H., Shimamura, A., Ferretti, J.J.: Infect. Immun.,56,2763ff. (1988)

1 NOMENCLATURE

EC number
  2.4.1.8

Systematic name
  Maltose:orthophosphate 1-beta-D-glucosyltransferase

Recommended name
  Maltose phosphorylase

Synonyms
  Phosphorylase, maltose

CAS Reg. No.
  9030-19-7

---

2 REACTION AND SPECIFICITY

Catalysed reaction
  Maltose + phosphate →
  → D-glucose + beta-D-glucose 1-phosphate (mechanism [3, 5])

Reaction type
  Hexosyl group transfer

Natural substrates
  Maltose + phosphate (catalyzes narrowly defined set of glycosyl transfer reactions with little hydrolysis) [5]

Substrate spectrum
  1 Maltose + phosphate (r [1–6], no substrates are alpha-methylglucoside,
    dextran from Leuconostoc, cellobiose, gentiobiose, type-l-meningo-
    coccus polysaccharide, soluble starch [3], beta-maltose, alpha-maltosyl-
    fluoride [5], maltitol, maltotriitol, sucrose, lactose, maltobiontic acid,
    maltotriose, maltotetraose [6], trehalose, isomaltose [3, 6]) [1–6]
  2 Maltose + arsenate (no substrates are alpha/beta-maltosyl fluoride, meth-
    yl-alpha-maltoside, maltal [5]) [1, 3, 5]
  3 beta-D-Glucose 1-phosphate + D-glucose (specific, not alpha-D-glucose
    1-phosphate [1, 3], favoured reaction [1], no substrates are D-fructose,
    D-galactose, L-glucose, D-mannose, D-ribose, D/L-arabinose, alpha-meth-
    ylglucoside, maltose, trehalose, cellobiose, D-gluconate [3]) [1, 3, 5]
  4 beta-D-Glucose 1-phosphate + D-xylose [1, 3, 4]
  5 beta-D-Glucose 1-phosphate + D-xylulose [1]

6 beta-D-Glucosylfluoride + alpha-D-glucose (no acceptors are methyl-alpha-D-glucoside, D-glucal, beta-D-glucose, salicin and alpha/beta-D-glucosylfluoride) [5]

Product spectrum
1 beta-D-Glucose 1-phosphate + D-glucose [1]
2 Glucose + arsenate [1, 3, 5]
3 Maltose + phosphate [1, 3, 5]
4 4-O-alpha-D-Glucosyl-D-xylose + phosphate [1, 4]
5 ?
6 alpha-Maltose + HF [5]

Inhibitor(s)
$CuSO_4$ [6]; $HgCl_2$ [6]; PCMB [6]; More (no inhibition by $MgCl_2$, $MnSO_4$, $CaCl_2$, $Pb(CH_3COO)_2$, $Ba(OH)_2$, $ZnSO_4$, glucose 6-phosphate, glucose 1-phosphate, fructose 6-phosphate, AMP, EDTA, IAA) [6]

Cofactor(s)/prosthetic group(s)/activating agents
More (no cofactor required [1], no pyridoxal 5-phosphate required [6]) [1, 6]

Metal compounds/salts
More (no metal ions required) [1]

Turnover number (min⁻¹)

Specific activity (U/mg)
14.9 [5]; 18.0 [6]

K_m-value (mM)
More (kinetic studies) [5]; 1.7 (arsenate) [6]; 1.9 (maltose) [6]; 2.6 (phosphate) [6]

pH-optimum
5.4 [6]; 6.5 [1]

pH-range
4.5–7.2 (about half-maximal activity at pH 4.5 and 7.2) [6]

Temperature optimum (°C)
30 (assay at) [5, 6]; 37 (assay at) [3]

Temperature range (°C)

---

3 ENZYME STRUCTURE

Molecular weight
150000 (Lactobacillus brevis, gel filtration) [6]

Subunits
    Dimer (2 × 80000, Lactobacillus brevis, SDS-PAGE) [6]

Glycoprotein/Lipoprotein
    –

---

4 ISOLATION/PREPARATION

Source organism
    Neisseria meningitides (type-I-meningococcus, non-virulent strain no.69 [3])
    [1–4]; Lactobacillus brevis [5, 6]

Source tissue
    Cell [1–6]

Localization in source

Purification
    Lactobacillus brevis (partial) [5, 6]

Crystallization
    –

Cloned
    –

Renatured
    –

---

5 STABILITY

pH
    5.4–7.2 (stable between) [6]

Temperature (°C)
    35 (30 min stable) [6]; 42 ($t_{1/2}$: 30 min) [6]; 46 ($t_{1/2}$: 10 min) [6]; 50 (and
    above, inactivation) [6]

Oxidation

Organic solvent

General stability information

Storage
    –20°C, several months [6]; 2°C, several months [6]

6 CROSSREFERENCES TO STRUCTURE DATABANKS

PIR/MIPS code

Brookhaven code

---

7 LITERATURE REFERENCES

[1] Doudoroff, M. in "The Enzymes",2nd Ed. (Boyer, P.D., Lardy, H., Myrbäck, K., Eds.), 5,229–236, Academic Press, New York (1961) (Review)
[2] Mieyal, J.J., Abeles, R.H. in "Enzymes",3rd Ed. (Boyer, P.D., Ed.) ,7,515–532, Academic Press, New York (1972) (Review)
[3] Fitting, C., Doudoroff, M.: J. Biol. Chem.,199,153–163 (1952)
[4] Putman, E.W., Fitting Litt, C., Hassid, W.Z.: J. Am. Chem. Soc.,77,4351–4353 (1955)
[5] Tsumuraya, Y., Brewer, C.F., Hehre, E.J.: Arch. Biochem. Biophys.,281,58–65 (1990)
[6] Kamogawa, A., Yokobayashi, K., Fukui, T.: Agric. Biol. Chem.,37,2813–2819 (1973)

1 NOMENCLATURE

EC number
2.4.1.9

Systematic name
Sucrose:2,1-beta-D-fructan 1-beta-D-fructosyltransferase

Recommended name
Inulosucrase

Synonyms
Sucrose 1-fructosyltransferase

CAS Reg. No.
9030-16-4

2 REACTION AND SPECIFICITY

Catalysed reaction
Sucrose + (2,1-beta-D-fructosyl)$_n$ $\rightarrow$
$\rightarrow$ glucose + (2,1-beta-D-fructosyl)$_{n+1}$

Reaction type
Hexosyl group transfer
Transfructosidation [1]

Natural substrates

Substrate spectrum
1 Inulin + sucrose (inulin i.e. linear polyfructan of ca. 30 fructan units, beta-1,2-linked) [1]

Product spectrum
1 (Fructosyl)$_{n-1}$ + glucose-(fructose)$_2$ (further fructose units may be transferred to the trisaccharide) [1]

Inhibitor(s)

Cofactor(s)/prosthetic group(s)/activating agents

Metal compounds/salts

Turnover number (min^{-1})

Specific activity (U/mg)

K_m-value (mM)

pH-optimum
   6.0–6.5 [1]

pH-range

Temperature optimum (°C)

Temperature range (°C)

---

3 ENZYME STRUCTURE

Molecular weight

Subunits

Glycoprotein/Lipoprotein
   –

---

4 ISOLATION/PREPARATION

Source organism
   Helianthus tuberosus (artichoke) [1]

Source tissue
   Tuber [1]

Localization in source

Purification

Crystallization
   –

Cloned
   –

Renatured
   –

5 STABILITY

pH

Temperature (°C)

Oxidation

Organic solvent

General stability information

Storage

6 CROSSREFERENCES TO STRUCTURE DATABANKS

PIR/MIPS code

Brookhaven code

7 LITERATURE REFERENCES

[1] Edelmann, J., Bacon, J.S.D.: Biochem. J.,49,529–540 (1951)

1 NOMENCLATURE

EC number
2.4.1.10

Systematic name
Sucrose:2,6-beta-D-fructan 6-beta-D-fructosyltransferase

Recommended name
Levansucrase

Synonyms
Fructosyltransferase, sucrose 6-
Sucrose 6-fructosyltransferase
beta-2,6-Fructosyltransferase
beta-2,6-Fructan:D-glucose 1-fructosyltransferase [1]

CAS Reg. No.
9030-17-5

---

2 REACTION AND SPECIFICITY

Catalysed reaction
Sucrose + (2,6-beta-D-fructosyl)$_n$ $\rightarrow$
$\rightarrow$ glucose + (2,6-beta-D-fructosyl)$_{n+1}$ (ping-pong mechanism [4])

Reaction type
Hexosyl group transfer

Natural substrates
Sucrose + (2,6-beta-D-fructosyl)$_n$ [1]

Substrate spectrum
1 Sucrose + (2,6-beta-D-fructosyl)$_n$ (r [5], at lower temperatures such as 5°C
and 15°C the transfructosylation is preferentially catalyzed rather than the
hydrolysis of sucrose, but inversely at higher temperatures such as 30°C
and 40°C the hydrolysis is preferentially catalyzed [5], other D-fructosyl
acceptors: D-xylose, D-arabinose, L-arabinose, lactose, maltose, maltotriose, cellobiose, melibiose [6], purified enzyme exhibits beta-D-fructofura-
nosidase activity [2], enzyme catalyzes the liberation of reducing sugars
from substrates having 2-beta-D-fructofuranose residues on a terminal
such as sucrose, raffinose, levan from Aerobacter levanicum and inulin
with a relative activity of 100:104:1:0.01 [5], polymerase and hydrolase
activity can be separately modulated by site-directed mutagenesis [18])
[1–18]

Product spectrum

1 Glucose + (2,6-beta-D-fructosyl)n+1 (high molecular weight branched levan [15], the levan synthesized on raffinose contains one mol of galactosylglucose per mol as one of the 2 terminal glycosyl moieties [9]) [1–18]

Inhibitor(s)

Mannose [7]; 6-Phosphogluconate [7]; Glucose [7]; Ethanol [7]; Ag^+ [5]; Hg^{2+} (not: $HgCl_2$, 5 mM [7]) [5]; Cu^{2+} [5]; Mersalyl acid [5]; High ionic strength (suppresses transfer of fructosyl residues, especially in synthesis of high molecular weight levan, suppresses hydrolysis of levan, effect occurs with almost the same degree in hydrolysis of both low and high molecular weight levans [9], under conditions of high ionic strength, only levan with an average degree of polymerization of 120 is synthesized with a reasonable yield) [13]; More (not: EDTA [5, 6], 2-mercaptoethanol [6], hydroxylamine hydrochloride [6]) [5, 6]

Cofactor(s)/prosthetic group(s)/activating agents

Levan (accelerates the rate of polymerization of levan, effective only under conditions of low ionic strength) [13]; Fructose 1,6-biphosphate (5 mM, activates partially purified enzyme) [7]

Metal compounds/salts

Aspartic acid (5 mM, activates partially purified enzyme) [7]; Glucosamine-HCl (5 mM, activates partially purified enzyme) [7]

Turnover number (min^{-1})

Specific activity (U/mg)

1692000 [1]; 731 [6]; 203 (Bacillus subtilis, extracellular) [14]; 415 (Bacillus amyloliquefaciens, extracellular) [14]; More [7]

K_m-value (mM)

12 (sucrose) [15]; 28 (sucrose) [1]; 50 (sucrose) [6]; 122 (sucrose) [5]

pH-optimum

5.0 [5]; 6.0 (glucose release from sucrose [12]) [1, 6, 12, 17]; 6.5 (sucrose hydrolysis to glucose and fructose) [7]

pH-range

More (15°C, 10% sucrose, pH 4–6: transfructosylation preferentially occurs and oligo- and polysaccharides are produced, pH 7–8: hydrolysis of sucrose is enhanced, oligosaccharides are produced rather than polysaccharides as transfructosylation products) [5]

Temperature optimum (°C)

0 (levan synthesis) [12]; 25 (assay at) [14]; 30 (assay at) [5]; 40 (synthesis of levan) [6]; 50 [5]; 55 [17]; 55–60 (sucrose + D-xylose) [6]

Temperature range (°C)

3 ENZYME STRUCTURE

Molecular weight
20000 (Bacillus subtilis, gel filtration) [12]
50000–52000 (Bacillus subtilis, Bacillus amyloliquefaciens, extracellular and membrane-associated enzyme forms, SDS-PAGE, gel filtration) [14]
94000 (Zymomonas mobilis, gel filtration) [5]
120000 (Rahnella aquatilis, gel filtration [6], Zymomonas mobilis, gel filtration [7]) [6, 7]
220000 (Actinomyces viscosus, extracellular, gel filtration) [15]
250000 (Actinomyces viscosus, cell-wall-associated enzyme form, gel filtration) [15]
More (tertiary structure) [11]

Subunits
Monomer (1 x 48000, Erwinia herbicola, SDS-PAGE, native form occurs as an aggregate [2], 1 x 52000, Bacillus subtilis and Bacillus amyloliquefaciens, extracellular and membrane-associated form, SDS-PAGE [14]) [2, 14]
Dimer (2 x 56000, Zymomonas mobilis, SDS-PAGE [5], 2 x 64000, Rahnella aquatilis, SDS-PAGE [6]) [5, 6]

Glycoprotein/Lipoprotein
–

---

4 ISOLATION/PREPARATION

Source organism
Bacillus amyloliquefaciens [14]; Rahnella aquatilis (JCM-1683) [6]; Aerobacter levanicum [1]; Erwinia herbicola (NRRL B-1678) [2]; Bacillus subtilis (BS5C4 constitutive strain [8], QB2010 [10], var. saccharolyticus [12, 13], QB127 [14]) [3, 4, 8–14, 18]; Zymomonas mobilis [5, 7]; Actinomyces viscosus [15]; Streptococcus mutans (FA-1 and 6715) [16]; Pseudomonas syringae pv. phaseolicola [17]

Source tissue
Membrane vesicles [14]; Culture fluid [2, 12, 14]; Cell [1, 6]

Localization in source
Extracellular [2, 5, 15, 17]; Intracellular [17]; Membrane-associated [14]; Cell-wall-associated [15]

Purification
Aerobacter levanicum (precipitation with levan) [1]; Erwinia herbicola (NRRL B-1678) [2]; Zymomonas mobilis (partial [7]) [5, 7]; Rahnella aquatilis (JCM-1683) [6]; Bacillus subtilis [11, 12]; Actinomyces viscosus [15]; Streptococcus mutans (partial) [16]

Enzyme Handbook © Springer-Verlag Berlin Heidelberg 1996
Duplication, reproduction and storage in data banks are only
allowed with the prior permission of the publishers

Crystallization
[5, 8, 11]

Cloned
(Bacillus subtilis QB 2010 gene cloned and expressed in E. coli, precursor form) [10]

Renatured
–

5 STABILITY

pH
3.5–6.0 (30°C, 24 h, stable) [5]; 3.5–7.5 (4°C, stable) [12]; 5.5–9.0 (30°C, 2 h, stable) [6]

Temperature (°C)
–18 – +52 (stable) [1]; 37 (pH 6.0, 30 min) [5]; 40 (20 min, stable below) [6]; 45 (15 min, pH 7.0, loss of activity) [12]; 60 (20 min, 50% loss of activity) [6]; 70 (20 min, complete loss of activity) [6]

Oxidation

Organic solvent

General stability information
Enzyme possesses a tertiary structure wholly dependent on the presence of Fe^{3+} or Ca^{2+} [3]

Storage
4°C, stable for 2 weeks [1]

6 CROSSREFERENCES TO STRUCTURE DATABANKS

PIR/MIPS code
PIR2:JQ0802 (precursor Bacillus amyloliquefaciens); PIR2:A25040 (precursor Bacillus subtilis); PIR2:B28551 (precursor Streptococcus mutans (strain GS-5)); PIR2:JC2519 (precursor Zymomonas mobilis)

Brookhaven code

7 LITERATURE REFERENCES

[1] Reese, E.T., Avigad, G.: Biochim. Biophys. Acta,113,79–83 (1966)
[2] Cote, G.L., Imam, S.H.: Carbohydr. Res.,190,299–307 (1989)
[3] Chambert, R., Petit-Glatron, M.-F.: FEBS Lett.,275,61–64 (1990)
[4] Chambert, R., Treboul, G., Dedonder, R.: Eur. J. Biochem.,41,285–300 (1974)

[5] Yanase, H., Iwata, M., Nakahigashi, R., Kita, K., Kato, N., Tonomura, K.: Biosci. Bio-
     technol. Biochem.,56,1335–1337 (1992)
[6] Ohtsuka, K., Hino, S., Fukushima, T., Ozawa, O., Kanematsu, T., Uchida, T.: Biosci.
     Biotechnol. Biochem.,56,1371–1377 (1992)
[7] Lyness, E.W., Doelle, H.W.: Biotechnol. Lett.,5,345–350 (1983)
[8] Berthou, J., Laurent, A., Lebrun, E., van Rapenbusch, R.: J. Mol. Biol.,82,111–113
     (1974)
[9] Yamamoto, S., Iizuka, M., Tanaka, T., Yamamoto, T.: Agric. Biol. Chem.,49,343–349
     (1985)
[10] Fouet, A., Arnaud, M., Klier, A., Rapoport, G.: Biochem. Biophys. Res. Commun.,
     119,795–800 (1984)
[11] LeBrun, E., van Rapenbusch, R.: J. Biol. Chem.,255,12034–12036 (1980)
[12] Tanaka, T., Oi, S., Iizuka, M., Yamamoto, T.: Agric. Biol. Chem.,42,323–326 (1978)
[13] Tanaka, T., Oi, S., Yamamoto, T.: J. Biochem.,85,287–293 (1979)
[14] Mäntsälä, P., Puntala, M.: FEMS Microbiol. Lett.,13,395–399 (1982)
[15] Pabst, M.J., Cisar, J.O., Trummel, C.L.: Biochim. Biophys. Acta,566,274–282 (1979)
[16] Figures, W.R., Edwards, J.R.: Biochim. Biophys. Acta,577,142–146 (1979)
[17] Sauerstein, J., Reuter, G.: J. Basic Microbiol.,28,667–672 (1988)
[18] Chambert, R., Petit-Glatron, M.-F.: Biochem. J.,279,35–41 (1991)

1 NOMENCLATURE

EC number
2.4.1.11

Systematic name
UDPglucose:glycogen 4-alpha-D-glucosyltransferase

Recommended name
Glycogen (starch) synthase

Synonyms
UDPglucose-glycogen glucosyltransferase
Glucosyltransferase, uridine diphosphoglucose-glycogen
Glycogen (starch) synthetase
UDP-glucose-glycogen glucosyltransferase
UDP-glycogen synthase
UDPG-glycogen synthetase
UDPG-glycogen transglucosylase
Uridine diphosphoglucose-glycogen glucosyltransferase
More (a similar enzyme, EC 2.4.1.21 utilizes ADPglucose)

CAS Reg. No.
9014-56-6

---

2 REACTION AND SPECIFICITY

Catalysed reaction
UDPglucose + (1,4-alpha-D-glucosyl)$_n$ $\rightarrow$
$\rightarrow$ UDP + (1,4-alpha-D-glucosyl)$_{n+1}$

Reaction type
Hexosyl group transfer

Natural substrates
UDPglucose + glycogen (glycogen production) [4]

Substrate spectrum
1 UDPglucose + glycogen (glycogen can be substituted by glucose [1], starch [2]) [1–34]

Product spectrum
1 Glycogen + UDP [18]

Inhibitor(s)
UDPpyridoxal [1]; LiBr [1]; $CaCl_2$ [4]; $MnCl_2$ [4]; $CuCl_2$ [4]; $HgCl_2$ [4]; UDP
(competitive to UDPglucose, noncompetitive to glycogen [17]) [11, 13, 14,
17, 18, 26, 29]; ADP [13]; GDP [13]; ATP (not inhibitory [33]) [13, 18, 19, 21,
26, 32]; Adenine nucleotides [14]; Phosphate (inhibition of synthase D [16])
[16, 22, 26, 33]; SO_4^{2-} [18, 22, 33]; Co^{2+} [20]; Cu^{2+} [20]; Zn^{2+} [20]; Hg^{2+}
[20]; SDS [20]; Nucleoside phosphates [22]; $(NH_4)_2SO_4$ [25]; Potassium
phosphate [25]; Phenylmercuric acetate [20]; p-Chloromercuribenzoate
[28]; N-Ethylmaleimide [28]; Iodobenzoate [28]; Iodoacetate (not inhibitory
[20]) [28]; 1,5-Gluconolactone [29]; More (not inhibitory: EDTA, iodoacetic
acid, thioglycolate) [20]

Cofactor(s)/prosthetic group(s)/activating agents
Glucose 6-phosphate (synthase D, i.e. phosphorylated form is glucose
6-phosphate dependent, synthase I, i.e. dephosphorylated form is glucose
6-phosphate independent, interconversion of D and I forms and vice versa)
[3, 5, 7–11, 13, 19, 21, 24, 25]; Galactose 6-phosphate (stimulates interconversion of D to I form) [19]; Glucosamine 6-phosphate (stimulates interconversion of D to I form) [19]; 2-Deoxyglucose 6-phosphate (stimulates interconversion of D to I form) [19]; Ribose 5-phosphate (stimulates interconversion of D to I form) [19]

Metal compounds/salts
Mg^{2+} (stimulates interconversion of D to I form [19], no influence on activity
[4, 20], activation in presence of glucose 6-phosphate [25], activation [26])
[19, 25, 26]; Ca^{2+} (stimulates interconversion of D to I from [19], no influence
on activity [20], activation in presence of glucose 6-phosphate [25]) [19,
25]; Mn^{2+} (stimulates interconversion of D to I form [19], no influence on activity [20], activation in presence of glucose 6-phosphate [25]) [19, 25]; KCl
(activation in presence of glucose 6-phosphate) [25]; NaCl (activation in
presence of glucose 6-phosphate) [25]; NH_4Cl (activation in presence of
glucose 6-phosphate) [25]; Li^+ (activation) [22]; Na^+ (activation) [22]; K^+
(activation) [22]; Cl^- (activation) [22]; F^- (activation) [22]; Phosphates (e.g.
sugar phosphates, inorganic phosphate, nucleoside phosphates, activation
of synthase I) [15]; More (no divalent cation requirement, no activation by
Fe^{3+}) [20]

Turnover number (min^{-1})

Specific activity (U/mg)
40 [3]; 1514 [4]; 63 [8]; More (assay method [34]) [6, 7, 12, 20, 21, 23–25,
27, 30, 34]

K_m-value (mM)

0.0005 (UDPglucose, presence of glucose 6-phosphate) [1]; 0.0015 (UDP-glucose, absence of glucose 6-phosphate) [1]; 0.033 (UDPglucose, synthase I, 25°C, presence of glucose 6-phosphate) [27]; 0.082 (UDPglucose, synthase I, 37°C, presence of glucose 6-phosphate) [27]; 0.113 (UDPglucose, synthase I, 25°C, absence of glucose 6-phosphate) [27]; 0.26–0.34 (UDPglucose) [26]; 0.31–0.32 (UDPglucose, value independent of glucose 6-phosphate concentration [7], UDPglucose, synthase D [16]) [7, 16]; 0.415–0.416 (UDPglucose, synthase D, value independent of presence of glucose 6-phosphate) [27]; 0.508 (UDPglucose, synthase I, 37°C, absence of glucose 6-phosphate) [27]; 0.57–1.25 (UDPglucose, value dependent on the presence of glucose 6-phosphate) [25]; 0.9 (UDPglucose) [10]; 2.8 (UDPglucose, synthase I) [16]; More (overview [2], dependence on glucose 6-phosphate concentration [3], glycogen: 3.5 mg/ml [4], values for phosphorylated enzyme forms [10], regulation by phosphorylation [11, 12, 15]) [2–4, 7, 10–15]

pH-optimum

5.5–9.2 (synthase D, presence of glucose 6-phosphate) [25]; 6.7 (synthase I, enzyme form 1, absence of Na_2SO_4) [30]; 6.8–9.2 (synthase I, presence of glucose 6-phosphate) [21]; 7.0–8.5 (synthase D, absence of glucose 6-phosphate) [25]; 7–9 (synthase I) [3]; 7.1 (synthase I, enzyme form 2, absence of Na_2SO_4) [30]; 7.2–8.0 (synthase I, absence of glucose 6-phosphate) [21]; 7.4 (glucose 6-phosphate independent form) [13]; 7.5 (absence of glucose 6-phosphate [7]) [7, 32]; 7.7 (synthase I, enzyme form 1, presence of Na_2SO_4) [30]; 7.8 (absence of glucose 6-phosphate) [7]; 8.0 (brain enzyme [2], synthase I, enzyme form 2, presence of Na_2SO_4 [30]) [2, 30]; 8.2–9.0 [20]; 8.3 (glucose 6-phosphate dependent form) [13]; 8.4 (muscle enzyme) [2]; 8.8 (synthase D [3], presence of glucose 6-phosphate [7]) [3, 7]

pH-range

6.0–9.0 [7]

Temperature optimum (°C)

30 [20]; 30–40 [3]; 37 [4, 21]

Temperature range (°C)

More (human enzyme has no activity above 50°C, rat enzyme is active up to at least 60°C) [3]

3 ENZYME STRUCTURE

Molecular weight

420000 (rabbit, gel filtration) [23]
410000 (Phymatotrichum omnivorum, polyacrylamide electrophoresis under native conditions [4], human, gel filtration [21]) [4, 21]
390000 (human, sucrose gradient centrifugation [21], rabbit, sucrose gradient centrifugation [23]) [21, 23]
377000 (rabbit, high speed sedimentation equilibrium centrifugation) [23]
300000–310000 (Saccharomyces cerevisiae, gel filtration [5, 6], ultracentrifugation [28]) [5, 6, 28]
270000 (Neurospora crassa, sucrose density gradient centrifugation) [20]
183000 (rabbit, synthase D, sucrose density gradient centrifugation) [24]
170000 (rabbit, synthase I, sucrose density gradient centrifugation) [24]

Subunits

Tetramer (4 × 99000, Phymatotrichum omnivorum, SDS-PAGE [4],
4 × 76000–83000, Saccharomyces cerevisiae, SDS-PAGE [5, 6, 28],
4 × 88000, rabbit, SDS-PAGE [23]) [4–6, 23, 28]
Trimer (3 × 88000–90000, Neurospora crassa, SDS-PAGE) [20]
Dimer (2 × 85000, rabbit, SDS-PAGE) [24]
? (x × 85000, Ascaris suum, SDS-PAGE [7], rabbit, synthase I, SDS-PAGE [9], x × 90000, rabbit, synthase D, SDS-PAGE [9], x × 80000–85000, human, SDS-PAGE [21], x × 77000–80000, rat, SDS-PAGE [29]) [7, 9, 21, 29]

Glycoprotein/Lipoprotein

More (no covalently attached amino sugars, 2% carbohydrate) [23]

---

4 ISOLATION/PREPARATION

Source organism

Rabbit [1, 2, 9, 11, 14, 23, 24, 31]; Rat [2, 3, 29, 34]; Yeast [2]; Human [3, 17, 18, 21, 25, 32]; Phymatotrichum omnivorum [4]; Saccharomyces cerevisiae [5, 6, 28]; Ascaris suum [7]; Bovine [8, 10, 12, 26, 30]; Biomphalaria glabrata (snail) [13]; Hymenolepis diminuta [15]; Mytilus edulis [16]; Rana ridibunda (frog) [19]; Neurospora crassa [20]; Dictyostelium discoideum [22]; Pig [27]; Rainbow trout [33]

Source tissue

Skeletal muscle [1, 2, 11, 23, 31]; Liver [2, 3, 9, 24, 28, 33, 34]; Brain [2, 27]; Mycelium [4]; Muscle [7]; Heart [8, 10, 26, 30]; Cephalopedal region [13]; Renal medulla [14]; Polymorphonuclear leukocytes [17, 18, 21]; Leg muscle [19]; Placenta [25]; Erythrocytes [32]

Localization in source

Soluble (muscle [2]) [2, 7]; Microsomes (liver) [2]; Particle-bound (liver) [2]

Purification
    Human (synthases D and I [3], synthase I [21], synthase D [25]) [3, 17, 21,
    25]; Phymatotrichum omnivorum [4]; Saccharomyces cerevisiae (from gly-
    cogen deficient strain, separation of synthases D and I [5], synthase D [28])
    [5, 6, 28]; Ascaris suum (synthases D and I) [7]; Bovine (synthases D and I
    [8], synthase I [12], synthase I, two kinetic forms [30]) [8, 12, 30]; Rabbit
    (synthases D and I [9], synthase I [14, 23], synthase D [24], synthase I, two
    kinetic forms [31]) [9, 14, 23, 24, 31]; Mytilus edulis (synthases D and I)
    [16]; Rana ridibunda (synthases D and I) [19]; Neurospora crassa [20]; Pig
    (partial, synthases D and I) [27]; Rat [29]; Rainbow trout [33]

Crystallization
    –

Cloned
    –

Renatured
    –

---

5 STABILITY

pH
    6.8–7.6 [21]; 7.4–8.2 [20]; More (overview: stability in various buffer sys-
    tems) [3]

Temperature (°C)
    25 (stable in imidazole or beta-glycerophosphate buffer, instable in Tris or
    glycylglycine buffer) [24]; 30 (instable above) [7]; 37 (inactivation in a few h)
    [21]; 50 (inactivation above, human) [3]; 60 (inactivation above, rat) [3]

Oxidation

Organic solvent

General stability information
    Phosphate stabilizes [21]; Sulfate stabilizes [21, 31]; Glycogen stabilizes
    [21, 33]; UDP stabilizes [21]; Glucose 6-phosphate stabilizes [21]; DTT sta-
    bilizes [21]; Albumin is essential during lyophilization [21]; UDPglucose sta-
    bilizes [33]

Storage
    –80°C, 50 mM glycerophosphate buffer, pH 7.0, 2 mM EDTA, 40 mM mer-
    captoethanol, 10% sucrose [7]; –70°C, 50 mM Tris/HCl buffer, pH 7.8, 25%
    v/v glycerol, 5 mM EDTA, 2 mM EGTA, 1 mM DTT, 6 months [6]; –70°C,
    beta-glycerophosphate buffer, 1 year [24]; –20°C, glycogen-free or glyco-
    gen-containing enzyme, lyophilized, more than 3 years, solubilized enzyme
    several weeks [21]; –20°C, 1% w/v bovine serum albumin, 2 weeks [22];

−20°C, 45 mM Tris/HCl buffer, pH 7.5, 1 mM DTT, 10% glycerol, several months [25]; −20°C, 45 mM Tris/HCl buffer, pH 7.5, 1 mM mercaptoethanol, 10% glycerol, several months [28]; −20°C, 3 weeks, 55–100% loss of activity [33]; Liquid N_2, 50 mM Tris/HCl buffer, pH 7.0, 1 mM DTT, 50% glycerol [8]; 4°C, 24 h stable [20]

6 CROSSREFERENCES TO STRUCTURE DATABANKS

PIR/MIPS code

PIR3:S11481 (rice); PIR3:S56270 (isoform 1 yeast (Saccharomyces cerevisiae)); PIR2:A32156 (human); PIR2:A13221 (rabbit (fragment)); PIR2:A14050 (rabbit (fragment)); PIR2:S02269 (rat (fragment)); PIR2:B25348 (P-2 peptide rabbit (fragment)); PIR2:B28991 (hepatic rabbit (fragment)); PIR2:A35362 (hepatic rat); PIR2:A33369 (skeletal muscle rabbit); PIR2:A26321 (skeletal muscle rabbit (fragment)); PIR2:A31566 (skeletal muscle rabbit (fragment)); PIR2:JQ0703 (rice); PIR2:S22874 (yeast (Saccharomyces cerevisiae) (fragments)); PIR2:A38326 (1 yeast (Saccharomyces cerevisiae)); PIR2:S51396 (2 yeast (Saccharomyces cerevisiae)); PIR1:YUBHY (precursor barley); PIR2:S07314 (precursor maize); PIR2:S22519 (precursor rice); PIR1:YUWTY (precursor wheat)

Brookhaven code

7 LITERATURE REFERENCES

[1] Pitcher, J., Smythe, C., Cohen, P.: Eur. J. Biochem.,176,391–395 (1988)
[2] Leloir, L.F., Cardini, C.E. in "The Enzymes",2nd ed. (Boyer, P.D., Lardy, H., Myrbäck, K., eds.) 6,317–326, Academic Press, N.Y. (1962) (Review)
[3] Westphal, S.A., Nuttall, F.Q.: Arch. Biochem. Biophys.,292,479–486 (1992)
[4] Sangan, P., Gunasekaran, M.: Mycologia,83,669–673 (1991)
[5] Peng, Z.-Y., Trumbly, R.J., Reimann, E.M.: J. Biol. Chem.,265,13871–13877 (1990)
[6] Carabaza, A., Arino, J., Fox, J.W., Villar-Palasi, C., Guinovart, J.J.: Biochem. J.,268, 401–407 (1990)
[7] Hannigan, L.L., Donahue, M.J., Masaracchia, R.A.: J. Biol. Chem.,260, 16099–16105 (1985)
[8] Dickey-Dunkirk, S., Killilea, S.D.: Anal. Biochem.,146,199–205 (1985)
[9] Camici, M., DePaoli-Roach, A.A., Roach, P.J.: J. Biol. Chem.,259,3429–3434 (1984)
[10] Mitchell, J.W., Thomas, J.A.: J. Biol. Chem.,256,6160–6169 (1981)
[11] Brown, D.F., Hegazy, M., Reimann, E.M.: Biochem. Biophys. Res. Commun.,134, 1129–1135 (1986)
[12] Rasmussen, L.H., Pedersen, K.M., Juhl, H.: Biochimie,67,615–623 (1985)
[13] Schwartz, C.F.W., Carter, C.E.: J. Parasitol.,68,228–235 (1982)
[14] Schlender, K.K., Doster, C.M.: Comp. Biochem. Physiol. B Comp. Biochem.,71B, 423–430 (1982)
[15] Mied, P.A., Bueding, E.: J. Parasitol.,65,14–24 (1979)

[16] Cook, P.A., Gabbot, P.A.: Comp. Biochem. Physiol. B Comp. Biochem.,60B, 419–421 (1978)

[17] Plesner, L., Plesner, I., Esmann, V.: Mol. Cell. Biochem.,12,45–61 (1976)

[18] Solling, H.: Eur. J. Biochem.,94,231–242 (1979)

[19] Itarte, E., Castineiras, M.-J., Guinovart, J.J., Prerz, M.R.: Biochim. Biophys. Acta, 524,305–315 (1978)

[20] Takahara, H., Matsuda, K.: Biochim. Biophys. Acta,522,363–374 (1978)

[21] Solling, H., Esmann, V.: Eur. J. Biochem.,81,119–128 (1977)

[22] Saunders, D.A., Wright, B.E.: J. Gen. Microbiol.,100,89–97 (1977)

[23] Nimmo, H.G., Proud, C.G., Cohen, P.: Eur. J. Biochem.,68,21–30 (1976)

[24] Killilea, S.D., Whelan, W.J.: Biochemistry,15,1349–1356 (1976)

[25] Huang, K.P., Robinson, J.C.: Arch. Biochem. Biophys.,175,583–589 (1976)

[26] Nakai, C., Thomas, J.A.: J. Biol. Chem.,250,4081–4086 (1975)

[27] Passonneau, J.V., Schwartz, J.P., Rottenberg, D.A.: J. Biol. Chem.,250,2287–2292 (1975)

[28] Huang, K.-P., Cabib, E.: J. Biol. Chem.,249,3851–3857 (1974)

[29] McVerry, P.H., Kim, K.-H.: Biochemistry,13,3505–3511 (1974)

[30] Thomas, J.A., Larner, J.: Biochim. Biophys. Acta,293,62–72 (1973)

[31] Schlender, K.K., Larner, J.: Biochim. Biophys. Acta,293,73–83 (1973)

[32] Moses, S.W., Bashan, N., Gutman, A.: Eur. J. Biochem.,30,205–210 (1972)

[33] Lin, D.C., Segal, H.L., Massaro, E.J.: Biochemistry,11,4466–4471 (1972)

[34] Thomas, J.A., Schlender, K.K., Larner, J.A.: Anal. Biochem.,25,486–499 (1968)

1 NOMENCLATURE

EC number
2.4.1.12

Systematic name
UDPglucose:1,4-beta-D-glucan 4-beta-D-glucosyltransferase

Recommended name
Cellulose synthase (UDP-forming)

Synonyms
GS-I [5]
UDP-glucose:1,4-beta-D-glucan 4-beta-D-glucosyltransferase [6]
beta-1,4-Glucosyltransferase [5]
Glucosyltransferase, uridine diphosphoglucose-1,4-beta-glucan
beta-1,4-Glucan synthase
beta-1,4-Glucan synthetase
beta-Glucan synthase
1,4-beta-D-Glucan synthase
1,4-beta-Glucan synthase
Glucan synthase
UDP-glucose-1,4-beta-glucan glucosyltransferase
Uridine diphosphoglucose-cellulose glucosyltransferase
UDPglucose-beta-glucan glucosyltransferase
UDPglucose-cellulose glucosyltransferase

CAS Reg. No.
9027-19-4

2 REACTION AND SPECIFICITY

Catalysed reaction
UDPglucose + (1,4-beta-D-glucosyl)$_n$ →
→ UDP + (1,4-beta-D-glucosyl)$_{n+1}$

Reaction type
Hexosyl group transfer

Natural substrates
UDPglucose + (1,4-beta-D-glucosyl)$_n$ (involved in the synthesis of cellulose)
[1]

Substrate spectrum

1 UDPglucose + (1,4-beta-D-glucosyl)$_n$ (reversibility not demonstrated [1])
   [1–8]

Product spectrum

1 UDP + (1,4-beta-D-glucosyl)$_{n+1}$ [1–8]

Inhibitor(s)

More (degree of inhibition: nucleotide-triphosphate, nucleotide-diphosphate, nucleotide-monophosphate) [4]; UDP (non-competitive [4], low concentration inhibits strongly, high concentration inhibits weakly [5]) [4, 5]; ADP [4, 5]; GDP (enhances activity [5]) [4]; UTP [4, 5]; ATP [4, 5]; GTP (enhances activity [5]) [4]; UMP [4, 5]; AMP [4]; GMP (enhances activity [5]) [4]; Amphomycin (in presence of 0.1% digitonin) [4]; Tunicamycin (in presence of 0.1% digitonin) [4]; Bacitracin (in presence of 0.1% digitonin) [4]; TTP [5]; TMP [5]; TDP (inhibits at low concentrations, stimulates at high concentrations) [5]; Congo red (non-competitive) [3]

Cofactor(s)/prosthetic group(s)/activating agents

Cellodextrins (stimulate cellulose synthesis, cellulose synthesis is dependent on presence of a soluble primer) [1]; Cyclic diguanylic acid (stimulation) [2]; Polyethylene glycol (activation by polyethylene glycol and GTP is cooperative and involves association of the enzyme with a protein factor essential for high rates of enzyme activity) [6, 7]; GTP (activation by polyethylene glycol and GTP is cooperative and involves association of the enzyme with a protein factor essential for high rates of enzyme activity [6, 7], enhances activity [5]) [5–7]; TDP (inhibits at low concentrations, stimulates at high concentrations) [5]; GDP (enhances activity) [5]; GMP (enhances activity) [5]; CTP (enhances activity) [5]; CDP (enhances activity) [5]; D-Glucose (1 M, 7-fold stimulation) [8]; Methyl-beta-D-glucoside (1 M, 11-fold stimulation) [8]; Cellobiose (0.2 M, 3-fold stimulation) [8]; More (regulatory properties [6, 7], of membrane-bound enzyme [7]) [6, 7]

Metal compounds/salts

Mg^{2+} (divalent cation required, Ca^{2+} and Mg^{2+}, 0.5–1.0 mM are effective for full activation [5], enhances activity [8]) [5, 8]; Ca^{2+} (divalent cation required, Ca^{2+} and Mg^{2+}, 0.5–1.0 mM are effective for full activation) [5]

Turnover number (min^{-1})

Specific activity (U/mg)

K_m-value (mM)

0.0044 (UDPglucose) [5]; 1–2.5 (UDPglucose, depending on individual membrane preparation) [4]

pH-optimum

6.0 (assay at) [2]; 7.2–8.5 [8]; 7.5 (assay at) [4]; 8.0 [5]; 8.3 [1]

pH-range
   6.5–8.3 (6.5: 13% of activity maximum, 7.5: 39% of activity maximum, 8.3:
   activity maximum, enzyme is not tested under more alkaline conditions because of instability of UDPglucose) [1]

Temperature optimum (°C)
   25 (assay at) [4]; 27 (assay at) [2]; 29 (assay at) [1]; 30 (assay at) [5]

Temperature range (°C)

---

3 ENZYME STRUCTURE

Molecular weight

Subunits

Glycoprotein/Lipoprotein

   –

---

4 ISOLATION/PREPARATION

Source organism
   Oryza sativa (L. cv. Nipponkai) [5]; Acetobacter xylinum (ATCC 10821) [1];
   Saprolegnia monoica [2, 3]; Nicotiana tabacum [4]; Acetobacter xylinum [6,
   7]; Lupinus albus [8]

Source tissue
   Seeds [8]; Cell [1]; Mycelium [2, 3]; Callus [4]; Cultured cells [5]

Localization in source
   Plasma membrane vesicles [4]; Cytoplasmic organelles [5]

Purification

Crystallization
   –

Cloned
   –

Renatured
   –

---

5 STABILITY

pH

Temperature (°C)
   30 (rather labile) [6]

Oxidation

Organic solvent

General stability information
Repeated freezing and thawing, substantial loss of activity [6]

Storage
0°C, stable for several days [5]; 4°C, 1% digitonin, solubilized enzyme is stable for at least several h [6]; –70°C, in liquid N_2, stable [6]

---

6 CROSSREFERENCES TO STRUCTURE DATABANKS

PIR/MIPS code
PIR2:S16266 (93K protein precursor Acetobacter pasteurianus); PIR2:S13732 (catalytic chain Acetobacter pasteurianus)

Brookhaven code

---

7 LITERATURE REFERENCES

[1] Glaser, L.: J. Biol. Chem.,232,627–636 (1958)
[2] Girard, V., Fevre, M., Mayer, R., Benziman, M.: FEMS Microbiol. Lett.,82,293–296 (1991)
[3] Nodet, P., Girard, V., Fevre, M.: FEMS Microbiol. Lett.,69,225–228 (1990)
[4] Haass, D., Hackspacher, G., Franz, G.: Plant Sci.,41,1–9 (1985)
[5] Kuribayashi, I., Kimura, S., Morita, T., Igaue, I.: Biosci. Biotechnol. Biochem.,56, 388–393 (1992)
[6] Aloni, Y., Cohen, R., Benziman, M., Delmer, D.: J. Biol. Chem.,258,4419–4423 (1983)
[7] Aloni, Y., Delmer, D.P., Benziman, M.: Proc. Natl. Acad. Sci. USA,79,6448–6452 (1982)
[8] Larsen, G.L., Brummond, D.O.: Phytochemistry,13,361–365 (1974)

1 NOMENCLATURE

EC number
2.4.1.13

Systematic name
UDPglucose:D-fructose 2-alpha-D-glucosyltransferase

Recommended name
Sucrose synthase

Synonyms
Glucosyltransferase, uridine diphosphoglucose-fructose
Sucrose synthetase
Sucrose-UDP glucosyltransferase
Sucrose-uridine diphosphate glucosyltransferase
UDP-glucose-fructose glucosyltransferase
Uridine diphosphoglucose-fructose glucosyltransferase

CAS Reg. No.
9030-05-1

---

2 REACTION AND SPECIFICITY

Catalysed reaction
UDPglucose + D-fructose →
→ UDP + sucrose (mechanism [3, 8, 15, 25, 28])

Reaction type
Hexosyl group transfer

Natural substrates
UDP + sucrose (involved in sucrose metabolism) [1, 15, 17]
UDPglucose + D-fructose (major enzyme of sucrose synthesis) [10]

Substrate spectrum
1 UDPglucose + D-fructose (r [1–28], sucrose cleavage preferred [8], main
   substrate is UDPglucose [3, 11]. ADPglucose [1, 14, 15, 19, 21, 22], TDP-
   glucose [1, 15], GDPglucose [11, 15, 19, 21], CDPglucose [11, 15, 19]
   and IDPglucose [11] can replace UDPglucose. L-Sorbose, 5-keto-D-fruc-
   tose, D-tagatose, fructose 6-phosphate, levanbiose can replace fructose
   to a small extent [1]. No glucosyl acceptors are D-xylulose, L-rhamnulose,
   D-glucoheptulose, D-mannoheptulose, turanose, inulobiose, melibiulose,
   lactulose, cellobiulose, 3,4-di-O-methylfructose, dihydroxyacetone, pyruvate [1]) [1–28]

2 UDP + sucrose (ADP (less effective [5, 19, 20]) [1, 5, 11, 14–17, 19, 20, 24], TDP (57% (SS2–isozyme), 91% (SS1-isozyme) as effective as UDP [20]) [1, 11, 20], GDP (poor [1], not [5, 24]) [1], CDP (poor [1, 5, 24]) [1, 5, 19, 24] or UTP (poor) [24] can replace UDP) [1–28]

Product spectrum
1 UDP + sucrose [1–28]
2 UDPglucose + D-fructose [1–28]

Inhibitor(s)
Cu^{2+} (0.01 mM [1], 0.4 mM [7]) [1, 7]; Hg^{2+} (0.01 mM [1], 0.4 mM [7]) [1, 7]; Zn^{2+} (strong) [26]; Ag^+ (0.4 mM) [7]; Ni^{2+} [3]; Mn^{2+} (sucrose cleavage [5], 3 mM [18]) [3, 5, 18]; Fe^{3+} (0.4 mM) [7]; Iodoacetic acid (partially reversible by GSH or DTT [7]) [1, 7]; N-Ethylmaleimide (partially reversible by GSH or DTT [7]) [1, 7]; p-Hydroxymercuribenzoate (partially reversible by GSH or DTT [7]) [1, 7]; UDPglucose (allosteric inhibition [5], not competitive to UDP [8]) [5, 8, 19, 28]; AMP (not [19, 26]) [1, 3]; ADP (high concentrations [14], sucrose cleavage, not synthesis [26]) [14, 26]; ATP (strong [5], sucrose synthesis [19], cleavage, not synthesis [26]) [1, 5, 19, 26]; UMP [1]; UDP (strong [1, 4, 8, 17, 19, 22, 26, 28], high concentrations [26], sucrose synthesis, partially reversible by $MgCl_2$ [4], competitive to UDPglucose [17], not competitive to UDPglucose [8], kinetics [28]) [1, 4, 8, 17, 19, 22, 26, 28]; UTP (sucrose cleavage [26], synthesis (weak [26]) [5, 19, 26]) [1, 5, 19, 26]; TDP [1]; GTP (sucrose cleavage) [5]; Salicine (kinetics) [28]; Imidazole [19]; Tris (sucrose cleavage [19], ADP + sucrose [24]) [19, 24]; Glucose (uncompetitive to UDPglucose [26]) [19, 26]; Fructose (competitive to sucrose [8], uncompetitive to UDPglucose [28], high concentration, sucrose synthesis [14], kinetics [25]) [8, 14, 19, 24, 25, 28]; NH_4^+ [3]; Ba^{2+} [3]; Ca^{2+} (sucrose cleavage [5]) [3, 5, 26]; Mg^{2+} (sucrose cleavage [5]) [5, 26]; F^- [3]; CN^- [3]; More (no inhibition by K^+, Na^+ [26], DTT, GSH, 2-mercaptoethanol, EDTA [19], fructose or glucose 6-phosphate, fructose 1,6- [4, 19] or 2,6-diphosphate [4], glucose 1- or 6-phosphate [4, 19] or cAMP [4], galactose, mannose, maltose, raffinose, 3-phosphoglycerate, phosphoenolpyruvate, ethanol, succinate, 2-oxoglutarate, glutamine, NAD^+, diphosphate [19]) [4, 19, 26]

Cofactor(s)/prosthetic group(s)/activating agents
Allantoin (activation, sucrose synthesis) [19]; UDP-D-xylose (activation of epicotyl and cotyledon isozymes) [7]

Metal compounds/salts
Mn^{2+} (activation [5, 10, 17–19, 22], 0.1 mM [18], sucrose synthesis [5], together with Mg^{2+} [10], inhibits at 3 mM [18]) [5, 10, 17–19, 22]; Mg^{2+} (activation [4, 5, 10, 18, 19, 25, 26], sucrose synthesis, 0.1–10.0 mM [18], slight activation of sucrose cleavage [19], inhibition of sucrose cleavage [26]) [4, 5, 10, 18, 19, 25, 26]; Ca^{2+} (activation of sucrose synthesis [5, 19, 26], slight

activation of sucrose cleavage (10 mM) [19], inhibition [3, 5, 26]) [5, 19, 26];
More (no cation [3] or ionic activator [7] requirement, no activation by K^+,
Na^+, NH_4^+, Cl^-, Br^-, F^-, NO_3^-, phosphate, sulfate, borate, acetate, citrate
[19]) [3, 7, 19]

Turnover number (min^{-1})

Specific activity (U/mg)
More [15, 21]; 0.89 [11]; 0.91 [1]; 2.8 (Solanum tuberosum) [16]; 3.27 [17];
3.54 (Oryza sativa) [16]; 3.85 [18]; 4.35 [12]; 6 (isozyme SS2) [20]; 8.34
[24]; 10 (isozyme SS1) [20]; 15.1 [19]; 128 (sucrose cleavage) [26]; 180
(sucrose synthesis) [26]

K_m-value (mM)
More (kinetic study) [8, 13, 17]; 0.005 (UDP) [19]; 0.012–0.033 (UDPglucose) [2, 19]; 0.061–0.094 (UDP (pH 6 [1]) [1, 2], TDP, pH 6 [1]) [1, 2]; 0.11
(UDPglucose, Populus tremuloides) [21]; 0.13 (ADP) [19]; 0.14–0.15 (UDP,
Tris buffer [24]) [24, 26]; 0.17–0.20 (GDP [15], UDP, ADP, HEPES buffer
[24]) [15, 24]; 0.3–0.38 (UDPglucose (pH 7.2 [1], isozyme SS2 [22]), TDP
[15]) [1, 15]; 0.44–0.9 (CDP [15], UDP (pH 7.2 [1]) [1, 9, 14]) [1, 9, 14, 15];
0.98–1.66 (fructose (Solanum tuberosum [16]) [8, 16], CDP [19], UDP [3],
ADPglucose [19], UDPglucose, ADP [26]) [3, 8, 16, 19, 26]; 1.92–2.7 (UDP-
glucose) [8, 16]; 2.08–3.1 (fructose, pH 7.2 [1]) [1, 16, 26]; 3.3–3.8 (ADP,
ADPglucose) [14]; 3.7–5.0 (fructose, (Populus tremuloides [21])) [2, 5, 19,
21]; 5 (sucrose) [3]; 5.3–6.0 (UDPglucose [14], fructose [9], UDPfructose,
isozyme SS2 [22]) [9, 14, 22]; 11.5 (UDPfructose, isozyme SS1) [22];
16.9–40 (sucrose (+ ADP [15])) [8, 15, 19, 24, 26]; 40 (fructose (+ ADPglucose)) [14]; 52–290 (sucrose (pH 6 [1], + UDP [14])) [1, 2, 5, 9, 14, 20]

pH-optimum
More (pI: 5.4 (isozyme SS2), pI: 5.5 [24], pI: 5.8 (isozyme SS1 [20]), pI: 6.16
[11]) [11, 20, 24]; 6.0 (sucrose cleavage) [14, 18, 19, 22]; 6.0–6.5 (sucrose
cleavage) [1]; 6.4 (sucrose cleavage) [24]; 6.5 (sucrose cleavage) [3, 5,
26]; 7.0 (sucrose synthesis [1], cleavage [2]) [1, 2]; 7.0–9.5 (broad, sucrose
synthesis) [14]; 7.5 (sucrose synthesis, isozyme SS2) [22]; 8.0 (sucrose synthesis) [21, 26]; 8.5 (sucrose synthesis) [2]; 9.0 (sucrose synthesis (isozyme
SS1 [22])) [5, 22]; 9.5 (sucrose synthesis) [19]

pH-range
5.0–8.0 (about half-maximal activity at pH 5.0 and 8.0, sucrose cleavage)
[22]; 5.0–8.2 (about 90% of maximal activity at pH 5.0 and about half-maximal activity at pH 8.2, sucrose cleavage) [19]; 5.0–10.0 (about half-maximal
activity at pH 5.0 and 10.0, sucrose synthesis) [22]; 5.5–9.0 (about 80% of
maximal activity at pH 5.5 and about half-maximal activity at pH 9.0, sucrose
cleavage) [14]; 5.6–10.0 (about half-maximal activity at pH 5.6 and about
90% of maximal activity at pH 10.0, sucrose synthesis) [14]; 6.2–6.6 (96% of

maximal activity at pH 6.2 and 6.6) [24]; 7.5–10.0 (about half-maximal activity at pH 7.5 and about 90% of maximal activity at pH 10.0, sucrose synthesis) [19]

Temperature optimum (°C)
37 [3]; 55 [5]

Temperature range (°C)

---

3 ENZYME STRUCTURE

Molecular weight
353000 (Vicia faba, gel filtration) [24]
360000 (Prunus persica, gel filtration [2], Zea mays, analytical ultracentrifugation in the presence of Mg^{2+} [18]) [2, 18]
362000 (Oryza sativa, gel filtration) [11]
370000 (Triticum aestivum germ, gel filtration) [4]
375000 (Phaseolus aureus, low speed sedimentation equilibrium centrifugation) [15]
380000 (Triticum aestivum leaf, gel filtration [4], Glycine max, PAGE [19]) [4, 19]
405000 (Phaseolus aureus, gel filtration) [15]
410000 (Pisum sativum, gel filtration [7], Oryza sativa, PAGE [14]) [7, 14]
412000 (Glycine max, gel filtration) [19]
420000 (Pharbitis nil [5], banana [6], gel filtration) [5, 6]
440000 (Oryza sativa, gel filtration) [14]
540000 (Ipomea batatas [17], Cucumis sativus [22], gel filtration) [17, 22]
More (Zea mays, amino acid composition) [18, 20]

Subunits
Oligomer (x × 87000, Zea mays, SDS-PAGE [20], x × 35000 + x × 70000, Pharbitis nil, SDS-PAGE [5]) [5, 20]
Tetramer (4 × 87000, Prunus persica, SDS-PAGE [2], 4 × 88000, Zea mays, SDS-PAGE (depending on the ionic species and ionic strength of the solution the Zea mays enzyme can assume catalytically active, tetrameric, octameric and other higher aggregated forms of which the tetramer is the predominant form) [18], 4 × 90000, Glycine max, SDS-PAGE [19], 4 × 92000, Oryza sativa, SDS-PAGE [11], 4 × 92600, Vicia faba, SDS-PAGE [24], 4 × 94000, Phaseolus aureus, SDS-PAGE [15], 4 × 100000, Oryza sativa, SDS-PAGE [14], 4 × 110000, banana, SDS-PAGE [6]) [2, 6, 11, 14, 15, 18, 19, 24]

Glycoprotein/Lipoprotein
–

4 ISOLATION/PREPARATION

Source organism

Sugar beet [1]; Sugar cane [10]; Ipomea batatas (sweet potato, cv. Okinava no.100) [17]; Solanum tuberosum (potato, var. Norium no.1) [16]; Cucumis sativus (cucumber) [22]; Vicia faba (field or faba bean, cv. Maris Bead) [24]; Pisum sativum (pea) [7]; Phaseolus aureus [15]; Glycine max (soy bean, cv. Prize [23], Merr cv. Williams [19]) [19, 23]; Oryza sativa (rice, var. Nihonbare [16]) [11, 14, 16]; Zea mays (maize, sh1/sh1- (Black Mexican Sweet) or sh1bz1-m4-genotype [20], hybrid B27xB14 [25]) [12, 18, 20, 25, 26]; Leleba oldhami (bamboo) [8]; Triticum aestivum (wheat, cv. San Augustin INTA [4]) [3, 4]; Pharbitis nil (morning-glory) [5]; Dianthus caryophyllus (carnation) [27]; Gladiolus sp. (gladiolus) [27]; Anigozanthos manglesii (kangaroo paw) [27]; Clianthus formosus (sturt pea) [27]; Helianthus tuberosus (Jerusalem artichoke) [13, 28]; Banana [6]; Populus tremuloides (quaking aspen) [21]; Prunus persica (peach) [2]; Prunus dulcis (almond) [27]; Chlorella vulgaris (green alga) [9]; Scenedesmus obliquus (green alga) [9]

Source tissue

Root [1, 17]; Leaf [4, 10]; Tubers [13, 16, 28]; Fruit [2, 6, 22]; Seed (immature [3], ripening [14]) [3, 11, 14, 16, 18]; Germ [4]; Endosperm [12, 20, 25, 26]; Seedling (cotyledons, epicotyls and roots of etiolated seedlings [7]) [7, 15, 20]; Cotyledons [24]; Shoot [8]; Petals [27]; Flower stalk (Dianthus caryophyllus, Gladiolus sp., Anigozanthos manglesii) [27]; Callus [5]; Cell suspension culture [20]; Nodules (nodulaid inoculum: Rhizobium japonicum CB 1809 [19], Bradyrhizobium japonicum strains [23]) [19, 23]

Localization in source

Cytosol (soluble) [1, 11, 19]; Membrane-bound [21]

Purification

Sugar beet [1]; Prunus persica [2]; Triticum aestivum (partial [4]) [3, 4]; Pharbitis nil [5]; Banana [6]; Pisum sativum (at least 3 isozymes) [7]; Chlorella vulgaris (partial) [9]; Scenedesmus obliquus (partial) [9]; Oryza sativa (no isozymes detected by isoelectric focusing [11]) [11, 14, 16]; Zea mays (isolation of corresponding mRNA [12], 2 isozymes: SS1 and SS2 [20]) [12, 18, 20, 26]; Helianthus tuberosus (partial, 2 isozymes) [13]; Phaseolus aureus (amino acid composition) [15]; Solanum tuberosum [16]; Glycine max [19, 23]; Cucumis sativus (2 isozymes: SS1 and SS2, partial) [22]; Vicia faba (partial) [24]

Crystallization

--

Cloned

--

Renatured

--

5 STABILITY

pH

5.0 (below, complete inactivation) [18]; 5.5–8.0 (stable) [18]; 6.5 ($t_{1/2}$: 5 h, 37°C [1], $t_{1/2}$: 1 min, 55°C, complete inactivation within 10 min [26]) [1, 26]; 8.0 (most stable [1], above, gradual loss of activity [18], 55°C: 20% loss of activity within 1 min, 80% loss of activity within 10 min [26]) [1, 18, 26]

Temperature (°C)

37 ($t_{1/2}$: 5 h, pH 6.5) [1]; 55 (pH 6.5: $t_{1/2}$: 1 min, complete inactivation after 10 min, pH 8: 20% loss of activity after 1 min, 80% loss of activity within 10 min) [26]; 60 (and above, 5 min, rapid loss of activity) [18]

Oxidation

Organic solvent

General stability information

2-Mercaptoethanol (0.1 mM [1] or 5 mM [19]) stabilizes [1, 19]; EDTA, 0.1 mM, stabilizes [1, 15]; DTT, 0.1 mM, stabilizes [15]; Protamin sulfate stabilizes [1]; High salt concentrations inactivate [1]; Freeze-thawing inactivates [1, 15]; Repeated freeze-thawing, stable to if foaming is avoided [18]; Repeated freeze-thawing leads to slight decrease of activity, even in the presence of PMSF [20]

Storage

–80°C, 20% glycerol, 0.1–1.0 mM DTT, stable [20]; –20°C, 20% loss of activity within 2 weeks, $t_{1/2}$: 1 month [11]; –20°C, 50% glycerol, stable [12]; –20°C, about 75% of activity retained after 4 months [18]; 4°C, 0.1 mM EDTA and DTT, more than 50% of activity retained after 1 month [15]; 4°C, 20 mM potassium phosphate buffer, pH 7.0, 5 mM 2-mercaptoethanol, 4 weeks stable [19]; 4°C, isozyme SS1, 4 months, isozyme SS2, $t_{1/2}$: 3 weeks [22]; 4°C, storage leads to gradual precipitation of denatured protein [18]; 4°C, prolonged storage leads to slight decrease of activity, even in the presence of PMSF [20]

---

6 CROSSREFERENCES TO STRUCTURE DATABANKS

PIR/MIPS code

PIR1:YUMU (Arabidopsis thaliana); PIR2:S24966 (barley (fragment)); PIR2:S37560 (carrot); PIR2:S31479 (fava bean); PIR1:YUZMS (maize); PIR1:YUPOS (potato); PIR2:S25526 (rice); PIR2:S23543 (rice); PIR2:A29484 (soybean (fragment)); PIR3:S22131 (sugarcane); PIR2:S22535 (1 rice (fragment)); PIR2:JT0280 (1 wheat (fragment)); PIR2:S22536 (2 rice (fragment)); PIR2:JT0281 (2 wheat (fragment)); PIR2:S22537 (3 rice (fragment)); PIR2:S29242 (Ss1 barley); PIR2:S32451 (Ss2 barley)

Brookhaven code

7 LITERATURE REFERENCES

[1] Avigad, G., Milner, Y.: Methods Enzymol.,8,341–345 (1966)
[2] Moriguchi, T. Yamaki, S.: Plant Cell Physiol.,29,1361–1366 (1988)
[3] Anand, S., Singh, R.: J. Plant Sci. Res.,2,1–10 (1986)
[4] Larsen, A.E., Salerno, G.L:, Pontis, G.: Physiol. Plant.,67,37–42 (1986)
[5] Hisajima, S., Ito, T.: Biol. Plant.,23,356–364 (1981)
[6] Yang, C.-L., Su, J.-C.: J. Chin. Biochem. Soc.,9,100–101 (1980)
[7] Sung, H.-Y., Su, J.-C.: J. Chin. Biochem. Soc.,6,22–37 (1977)
[8] Yang, C.-L., Su, J.-C.: Proc. Natl. Sci. Counc. Part2 (Taiwan) ,10,271–284 (1977)
[9] Duran, W.R., Pontis, H.G.: Mol. Cell. Biochem.,16,149–152 (1977)
[10] Patil, B.A., Joshi, G.V.: Proc. Indian Natl. Sci. Acad. Part B,38,50–54 (1972)
[11] Elling, L., Kula, M.-R.: J. Biotechnol.,29,277–286 (1993)
[12] Wöstemeyer, J., Behrens, U., Merckelbach, A., Müller, M., Starlinger, P.: Eur. J. Biochem.,114,39–44 (1981)
[13] Wolosiuk, R.W., Pontis, H.G.: FEBS Lett.,16,237–240 (1971)
[14] Nomura, T., Akazawa, T.: Arch. Biochem. Biophys.,156,644–652 (1973)
[15] Delmer, D.P.: J. Biol. Chem.,247,3822–3828 (1972)
[16] Murata, T.: Agric. Biol. Chem.,36,1815–1818 (1972)
[17] Murata, T.: Agric. Biol. Chem.,35,1441–1448 (1971)
[18] Su, J.-C., Preiss, J.: Plant Physiol.,61,389–393 (1978)
[19] Morell, M., Copeland, L.: Plant Physiol.,78,149–154 (1985)
[20] Echt, C.S., Chourey, P.S.: Plant Physiol.,79,530–536 (1985)
[21] Graham, L.L., Johnson, M.A.: Phytochemistry,17,1231–1233 (1978)
[22] Gross, K.C., Pharr, D.M.: Phytochemistry,21,1241–1244 (1982)
[23] Thummler, F., Verma, D.P.S.: J. Biol. Chem.,262,14730–14736 (1987)
[24] Ross, H.A., Davies, H.V.: Plant Physiol.,100,1008–1013 (1992)
[25] Doehlert, D.C.: Plant Sci.,52,153–157 (1987)
[26] Tsai, C.-Y.: Phytochemistry,13,885–891 (1974)
[27] Hawker, J.S., Walker, R.R., Ruffner, H.P.: Phytochemistry,15,1441–1443 (1976)
[28] Wolosiuk, R.W., Pontis, H.G.: Arch. Biochem. Biophys.,165,140–145 (1974)

1 NOMENCLATURE

EC number
2.4.1.14

Systematic name
UDPglucose:D-fructose-6-phosphate 2-alpha-D-glucosyltransferase

Recommended name
Sucrose-phosphate synthase

Synonyms
UDPglucose-fructose-phosphate glucosyltransferase
Sucrosephosphate-UDP glucosyltransferase
SPS [2]
Glucosyltransferase, uridine diphosphoglucose-fructose phosphate
Sucrose 6-phosphate synthase
Sucrose phosphate synthetase
Sucrose phosphate-uridine diphosphate glucosyltransferase

CAS Reg. No.
9030-06-2

2 REACTION AND SPECIFICITY

Catalysed reaction
UDPglucose + D-fructose 6-phosphate →
→ UDP + sucrose 6-phosphate (enzyme operates either by a ordered bi-bi
or a Theorell-Chance mechanism [26], ordered mechanism [27])

Reaction type
Hexosyl group transfer

Natural substrates
UDPglucose + D-fructose 6-phosphate (key enzyme regulating the sucrose
synthesis [13, 14, 16, 26], involved in regulation of carbon partitioning in the
leaves [14]) [13, 14, 16, 26]

Substrate spectrum
1 UDPglucose + D-fructose 6-phosphate (r [1], equilibrium lies far towards
UDP and sucrose 6-phosphate [1], ADPglucose is inactive [24]) [1–27]

Product spectrum
1 UDP + sucrose 6-phosphate [1–27]

Inhibitor(s)

gamma-ATP (phosphorylates and inactivates) [12]; Tris-maleate buffer [17]; Citrate (stimulation [4]) [17]; EDTA [17]; p-Chloromercuribenzoate (inhibition is reversed by DTT or 2-mercaptoethanol) [17]; ADP (slight) [17]; CTP [17]; ATP [17]; GTP [17]; Fructose 1,6-bisphosphate (stimulates [25]) [26]; NaCl (enzyme is not halophilic or halotolerant, at 0.2 M: remaining activity of 10%) [16]; EDTA [4]; Sucrose (not [25]) [20]; UDP (competitive [1, 6, 17], reversal of inhibition by divalent cations [19]) [1, 4, 6, 17, 19, 22, 24–27]; UTP [4, 17]; Inorganic phosphate (in presence of 5 mM glucose 6-phosphate: partial competitive inhibitor with respect to both substrates, in absence of glucose 6-phosphate: more complex inhibition pattern [23]) [8, 17, 18, 21–23, 25–27]; 5-Azidouridine 5'-diphosphate-glucose [10]; Ca^{2+} (inhibition of sweet potato root and barley leaf enzyme) [24]; Glucono-1,5-lactone [17]; Pyruvate (slight) [17]; Maleate [17]; Isocitrate [17]; Tartrate (slight) [17]; Succinate (slight) [17]; Chloride (slight) [17]; Iodide (slight) [17]; Nitrate (slight) [17]; Sulfate [17]; Malonate [17]; Fumarate (slight) [17]; Sucrose 6-phosphate (not [25]) [18]; Tris-HCl buffer (slight) [18]; Diphosphate (slight) [22]; Potassium fluoride [22]; Mg^{2+} (slight stimulation of sweet potato root and potato tuber enzyme, inhibits barley leaf enzyme) [24]; More (Mg^{2+} can restore activity to control values when inhibited by nucleoside triphosphates, citrate or phosphate [17], enzyme is inactivated by protein phosphorylation in vitro, which appears to be the mechanism of light modulation in vitro) [2]

Cofactor(s)/prosthetic group(s)/activating agents

Casein (1%, causes 100% increase in activity over the no-casein control) [3]; 1,5-Anhydroglucitol 6-phosphate (stimulates at subsaturation concentration of fructose 6-phosphate) [18]; Glucose 6-phosphate (activates, inorganic phosphate antagonizes glucose 6-phosphate activation by competiting with the activator for binding to the modifier site [23]) [21–23, 25]; Glucose 1-phosphate (weak activation) [22]; Fructose 1-phosphate (weak activation) [22]; Fructose 1,6-bisphosphate (stimulates [25], inhibits [26]) [25]

Metal compounds/salts

Mg^{2+} ($MgCl_2$ stimulates [4], stimulates [6, 17, 25], slight stimulation of sweet potato root and potato tuber enzyme, inhibits barley leaf enzyme [24], can restore activity to control values when inhibited by nucleoside triphosphates, citrate or phosphate [17]) [4, 6, 17, 24, 25]; Mn^{2+} ($MnCl_2$ stimulates [4], stimulates [6, 17], stimulation of sweet potato and potato tuber enzyme [24]) [4, 6, 17, 24]; Citrate (stimulates [4], inhibition [17]) [4]; Fe^{2+} (activates, K_m 0.3 mM) [6]; Fe^{3+} (activates, K_m 2.0 mM) [6]; Co^{2+} (activates) [6]; $Co(NH_3)_6^{3+}$ (activates) [6]; K^+ (activates at high concentration) [6]; Inorganic phosphate (stimulates particularly in the presence of higher concentrations of fructose 6-phosphate) [16]; More (enzyme is inactivated by protein phosphorylation in vitro, which appears to be the mechanism of light modulation in vitro, activity of the inactivated enzyme is strongly stimulated by high ionic strength, salt stimulation is reversible and antagonized by the presence of ethylene glycol in the assay mixture, salt stimulation of deactivated enzyme is observed for a variety of C-4 plants, not for any of the C-3 species tested [2], highly activated enzyme contains phosphorylated residues that increase activation state, spontaneous inactivation occurs by removal of these phosphate groups [12]) [2, 12]

Turnover number (min^{-1})

Specific activity (U/mg)
More [1, 9, 16, 17, 26]; 4.22 [25]; 25 [13]; 57 [10]

K_m-value (mM)
1.3 (fructose 6-phosphate) [4]; 2.1 (UDPglucose) [4]; 3.0 (fructose 6-phosphate) [1, 5]; 7.4 (UDPglucose) [1]; 10 (UDPglucose) [5]; More [6, 8, 10, 16, 18, 21, 24–26]

pH-optimum
6–8 (sweet potato, barley) [24]; 6.5 [4, 17]; 6.5–7.5 [18]; 7.0 (in presence of 10 mM inorganic phosphate [22]) [22, 25]; 7.5 (in absence of effectors [22]) [6, 22]; More (piperazine-N,N'-bis(2-ethane-sulfonic acid), HEPES and MES buffers are optimal for activity) [17]

pH-range
6.5–8.0 (50% of activity maximum at pH 6.5 and 8.0, absence of effectors) [22]

Temperature optimum (°C)
45 [4]; 46 [16]

Temperature range (°C)
20–46 (20°C: about 15% of activity maximum, 30°C: about 50% of activity maximum, 46°C: activity maximum) [16]

3 ENZYME STRUCTURE

Molecular weight
253000 (Spinacia oleracea, zonal sedimentation data) [10]
380000 (morning glory, gel filtration [4], Triticum aestivum, gel filtration, sedimentation velocity method [17]) [4, 17]
450000 (rice) [6]
456000 (Pisum sativum, gel filtration) [25]
460000 (Spinacia oleracea, gel filtration) [22]

Subunits

? (x x 138000, Zea mays, SDS-PAGE [13], x x 120000, Spinacia oleracea, SDS-PAGE [9]) [9, 13]

Dimer (2 x 120000, Spinacia oleracea, SDS-PAGE) [10]

Glycoprotein/Lipoprotein

–

4 ISOLATION/PREPARATION

Source organism

Triticum aestivum (wheat) [1, 17, 19, 20, 27]; Zea mays (maize) [2, 3, 7, 13, 14]; Morning glory (genera Ipomoea and Convovulus) [4]; Chlorella vulgaris [5]; Glycine max (soybean) [7]; Scenedesmus obliquus [5]; Rice [6]; Chlorella vulgaris [5]; Spinacia oleracea (spinach) [7–12, 15, 18, 21–23, 26]; Pisum sativum [25]; Dunaliella tertiolecta [16]; Ipomoea batatas (sweet potato) [24]; Solanum tuberosum (potato) [24]; Hordeum vulgare (barley) [24]; Brassica napus (rape) [24]; Trifolium repens (ladino clover) [24]

Source tissue

Root [24]; Tuber [24]; Germ [1, 17, 19, 20, 27]; Leaf [2, 3, 7–13, 18, 21–24, 26]; Callus cells [4]; Germinating seeds (scutellum [6]) [6, 25]; Cells [16]

Localization in source

Purification

Rice (partial) [6]; Triticum aestivum [1, 17]; Morning glory [4]; Chlorella vulgaris (partial) [5]; Scenedesmus obliquus (partial) [5]; Zea mays (partial) [13]; Spinacia oleracea (2 forms which differ in kinetic properties [8], by immunoprecipitation [9], partial [18, 21, 22]) [8–10, 18, 21, 22, 26]; Dunaliella tertiolecta (partial) [16]; Ipomoea batatas [24]; Solanum tuberosum [24]; Pisum sativum [25]

Crystallization

Cloned

(expression of maize enzyme in tomato [14]) [14, 15]

Renatured

–

5 STABILITY

pH

Temperature (°C)

Oxidation

Organic solvent

General stability information
  Casein, 1%, stabilizes the extremely labile enzyme even at 15°C preincubation conditions [3]; Glycerol, 20%, stabilizes [17]; Fructose 6-phosphate, 5 mM, stabilizes [17]

Storage
  0–4°C, stable in presence of 5 mM fructose 6-phosphate or 20% glycerol [17]; –10°C, 2 months, 90% of original activity retained [26]; –80°C, for at least 4 weeks [10]; –20°C, no loss of activity after 1 month, 50–60% loss of activity after 1 year [17]; 0°C, 4 weeks, 50% loss of activity [18]; Frozen in liquid N_2, stable for at least 11 months [18]

6 CROSSREFERENCES TO STRUCTURE DATABANKS

PIR/MIPS code
  PIR2:JQ1329 (maize); PIR2:S34172 (potato); PIR2:JQ2277 (spinach); PIR3:S39784 (spinach)

Brookhaven code

7 LITERATURE REFERENCES

[1] Mendicino, J.: J. Biol. Chem.,235,3347–3352 (1960)
[2] Huber, S.C., Huber, J.L.: Plant Cell Physiol.,32,327–333 (1991)
[3] Polisetty, R., Sicher, R.C.: Indian J. Plant Physiol.,30,390–392 (1987)
[4] Hisajima, S., Hasegawa, T., Suzuki, T.: Denpun Kagaku,27,167–172 (1980)
[5] Duran, W.T., Pontis, H.G.: Mol. Cell. Biochem.,16,149–152 (1977)
[6] Nomura, T., Akazawa, T.: Plant Cell Physiol.,15,477–483 (1974)
[7] Kerr, P.S., Kalt-Torres, W., Huber, S.C.: Planta,170,515–519 (1987)
[8] Siegl, G., Stitt, M.: Plant Sci.,66,205–210 (1990)
[9] Walker, J.L., Huber, S.C.: Plant Physiol.,89,518–524 (1989)
[10] Salvucci, M.E., Drake, R.R., Haley, B.E.: Arch. Biochem. Biophys.,281,212–218 (1990)
[11] Huber, J.L.A., Huber, S.C., Nielsen, T.H.: Arch. Biochem. Biophys.,270,681–690 (1989)
[12] Huber, J.L., Hite, D.R.C., Outlaw, W.H., Huber, S.C.: Plant Physiol.,95,291–297 (1991)

[13] Bruneau, J.-M., Worrell, A.C., Cambou, B., Lando, D., Voelker, T.A.: Plant Physiol., 96,473–478 (1991)
[14] Worrell, A.C., Bruneau, J.-M., Summerfelt, K., Boersig, M., Voelker, T.A.: Plant Cell, 3,1121–1130 (1991)
[15] Klein, R.R., Crafts-Brander, S.J., Salvucci, M.E.: Planta,190,498–510 (1993)
[16] Müller, W., Wegmann, K.: Planta,141,159–163 (1978)
[17] Salerno, G.L., Pontis, H.G.: Planta,142,41–48 (1978)
[18] Amir, J., Preiss, J.: Plant Physiol.,69,1027–1030 (1982)
[19] Salerno, G.L., Pontis, H.G.: FEBS Lett.,64,415–418 (1976)
[20] Salerno, G.L., Pontis, H.G.: FEBS Lett.,86,263–267 (1978)
[21] Doehlert, D.C., Huber, S.C.: FEBS Lett.,153,293–297 (1983)
[22] Doehlert, D.C., Huber, S.C.: Plant Physiol.,73,989–994 (1983)
[23] Doehlert, D.C., Huber, S.C.: Plant Physiol.,76,250–253 (1984)
[24] Murata, T.: Agric. Biol. Chem.,36,1877–1884 (1972)
[25] Lunn, J.E., Ap Rees, T.: Phytochemistry,29,1057–1063 (1990)
[26] Harbron, S., Foyer, C., Walker, D.: Arch. Biochem. Biophys.,212,237–246 (1981)
[27] Salerno, G.L., Pontis, H.G.: Arch. Biochem. Biophys.,180,298–302 (1977)

1 NOMENCLATURE

EC number
2.4.1.15

Systematic name
UDPglucose:D-glucose-6-phosphate 1-alpha-D-glucosyltransferase

Recommended name
alpha,alpha-Trehalose-phosphate synthase (UDP-forming)

Synonyms
UDPglucose-glucose-phosphate glucosyltransferase
alpha,alpha-Trehalose phosphate synthase (UDP-forming)
Phosphotrehalose-uridine diphosphate transglucosylase
Trehalose 6-phosphate synthase
Trehalose 6-phosphate synthetase
Trehalose phosphate synthase
Trehalose phosphate synthetase
Trehalose phosphate-uridine diphosphate glucosyltransferase
Trehalose-P synthetase [6]
Transglucosylase [7]
Glucosyltransferase, uridine diphosphoglucose phosphate
Trehalosephosphate-UDP glucosyl transferase
More (cf. EC 2.4.1.36)

CAS Reg. No.
9030-07-3

---

2 REACTION AND SPECIFICITY

Catalysed reaction
UDPglucose + D-glucose 6-phosphate $\rightarrow$
$\rightarrow$ UDP + alpha,alpha-trehalose 6-phosphate (ordered bi-bi mechanism [1])

Reaction type
Hexosyl group transfer

Natural substrates

Substrate spectrum
1 UDPglucose + glucose 6-phosphate (r [5], specific for glucose 6-phosphate [4, 5], UDPglucose can be substituted by GDPglucose [3–5], ADPglucose [4], UDPglucose cannot be substituted by ADPglucose or GDPglucose [1]) [1, 3–6, 8, 9, 13, 14]

Enzyme Handbook © Springer-Verlag Berlin Heidelberg 1996
Duplication, reproduction and storage in data banks are only
allowed with the prior permission of the publishers

Product spectrum
1 Trehalose 6-phosphate + UDP [6, 9, 14]

Inhibitor(s)
K^+ (above 300 mM) [1]; SDS [3]; Deoxycholate [3]; UMP [5]; UDP [5]; UTP [5]; AMP (slight) [5]; ATP (slight) [5]; Zn^{2+} [6]; Trehalose (noncompetitive to UDPglucose) [6]; Cellobiose [6]; Polyribonucleotide inhibitor from Mycobacterium tuberculosis [7]; Poly-DL-ornithine [10]; Poly-D-lysine [10]; Poly-DL-lysine [10]; Poly-L-ornithine [10]; Phosphate [12]; UDPglucose (competitive to GDPglucose) [14]; GDPglucose (noncompetitive to UDPglucose) [14]; Mycoribin [14]

Cofactor(s)/prosthetic group(s)/activating agents
Heparin (activation) [3, 4, 10]; Tween 80 (activation) [3]; Triton X-100 (activation) [3]; Chondroitin sulfate (activation) [4, 10, 14]; TPS-activator protein [11]; Dermatan sulfate (activation) [10]; Heparan sulfate (activation) [10]; gamma-Carragenan (activation) [10]; Polynucleotides (activation) [10]; Polyanions (activation) [14]

Metal compounds/salts
Mg^{2+} (dependent on 3–6 mM $MgCl_2$ [1], activation [4–6, 9, 14]) [1, 4–6, 9, 14]; $NaHCO_3$ (100–300 mM, activation) [1]; $KHCO_3$ (100–300 mM, activation) [1]; KCl (activation) [4]; Mn^{2+} (activation) [6]; Ca^{2+} (activation) [6]; Ba^{2+} (activation) [6]; Fe^{2+} (activation) [14]; Co^{2+} (activation) [14]; Zn^{2+} (activation) [14]; Cd^{2+} (activation) [14]; Ni^{2+} (activation) [14]; Monovalent cations (activation) [13]

Turnover number (min^{-1})

Specific activity (U/mg)
317.7 (GDPglucose) [3]; 127 (UDPglucose) [1]; 157.3 (UDPglucose) [3]; 15 [11]; More [2, 14]

K_m-value (mM)
0.3 (UDPglucose) [6]; 1.0 (glucose 6-phosphate (+ GDPglucose)) [3]; 1.2 (glucose 6-phosphate (+ UDPglucose)) [3]; 1.5 (UDPglucose) [1]; 1.6 (UDPglucose) [3]; 2.0 (glucose 6-phosphate [1], GDPglucose [3]) [1, 3]; 5 (glucose 6-phosphate) [6]; More (complex bimodal kinetics [4]) [4, 12, 14]

pH-optimum
6.5 (potassium phosphate buffer [1]) [1, 4]; 6.6 [9]; 7.0 (Tris/HCl buffer [1]) [1, 14]

pH-range

Temperature optimum (°C)
50 [1]

Temperature range (°C)

3 ENZYME STRUCTURE

Molecular weight
 630000 (Saccharomyces cerevisiae, gel filtration) [2]
 300000 (Saccharomyces cerevisiae, complex of EC 2.4.1.15/EC 3.1.3.12,
 gel filtration) [11]
 63000 (Ectothiorhodospira halochloris, gel filtration) [1]
 45000 (Mycobacterium smegmatis, gel filtration) [3]

Subunits
 Oligomer (x × 115000 + x × 57000, Saccharomyces cerevisiae, complex of
 EC 2.4.1.15/EC 3.1.3.12, SDS-PAGE) [11]
 Multimer (x × 56200, Saccharomyces cerevisiae, calculation from gene sequence, x × 55000, isozyme TPS1, x × 100000, isozyme TPS2, x × 105000,
 isozyme TPS3, Saccharomyces cerevisiae, SDS-PAGE) [2]
 ? (x × 45000, or x × 90000, Mycobacterium smegmatis, aggregation to dimer
 or oligomer, SDS-PAGE) [3]

Glycoprotein/Lipoprotein
 –

4 ISOLATION/PREPARATION

Source organism
 Ectothiorhodospira halochloris [1]; Saccharomyces cerevisiae [2, 11, 12];
 Mycobacterium smegmatis [3, 10, 14]; Dictyostelium discoideum [4];
 Neurospora crassa [5]; Hyalophora cecropia [6]; Mycobacterium tuberculo-
 sis [7]; Schistocera gregaria (locust) [8]; E. coli [13]; Salmonella typhimuri-
 um [13]; Saccharomyces carlsbergensis [9]

Source tissue
 Fat body (of larvae [6], of adult locusts [8]) [6, 8]

Localization in source
 Soluble [1]

Purification
 Ectothiorhodospira halochloris (partial) [1]; Saccharomyces cerevisiae
 (trehalose-6-phosphate synthase/phosphatase complex, EC 2.4.1.15/EC
 3.1.3.12 [11]) [2, 11]; Mycobacterium smegmatis [3]; Dictyostelium
 discoideum (partial) [4]

Crystallization
 –

Cloned
 (isozyme TPS1) [2]

Renatured

–

5 STABILITY

pH

Temperature (°C)
  44 (up to) [4]

Oxidation

Organic solvent

General stability information
  Stable to at least 4 successive freeze/thaw cycles [4]

Storage
  –20°C, 1 mM Bistris buffer, pH 6.3 [2]; –18°C, 2–3 d, no loss of activity [1];
  –12°C, at least 11 weeks [4], 4–6°C, 2 d, 25% loss of activity, 3 d, 75% loss
  of activity [4]

6 CROSSREFERENCES TO STRUCTURE DATABANKS

PIR/MIPS code

Brookhaven code

7 LITERATURE REFERENCES

[1] Lippert, K., Galinski, E.A., Trüper, H.G.: Antonie Leeuwenhoek,63,85–91 (1993)
[2] Bell, W., Klaassen, P., Ohnacker, M., Boller, T., Herweijer, M., Schoppink, P., van der
    Zee, P., Wiemken, A.: Eur. J. Biochem.,209,951–959 (1992)
[3] Pan, Y.T., Mitchell, M., Elbein, A.D.: Arch. Biochem. Biophys.,186,392–400 (1978)
[4] Killick, K.A.: Arch. Biochem. Biophys.,196,121–133 (1979)
[5] Betz, R., Holldorf, A.W.: Biochem. Soc. Trans.,3,988–989 (1975)
[6] Murphy, T.A., Wyatt, G.R.: J. Biol. Chem.,240,1500–1508 (1965)
[7] Lornitzo, F.A., Goldman, D.S.: J. Biol. Chem.,239,2730–2734 (1964)
[8] Candy, D.J., Kilb, B.A.: Biochem. J.,78,531–536 (1961)
[9] Cabib, E., Leloir, L.F.: J. Biol. Chem.,231,259–275 (1958)
[10] Elbein, A.D., Mitchell, M.: Carbohydr. Res.,37,223–238 (1974)
[11] Londesborough, J., Vuorio, O.: J. Gen. Microbiol.,137,323–330 (1991)
[12] Vandercammen, A., Francois, J., Hers, H.-G.: Eur. J. Biochem.,182,613–620 (1989)
[13] Giaever, H.M., Styrvold, O., Kaasen, I., Strom, A.R.: J. Bacteriol.,170,2841–2849
     (1988)
[14] Lapp, D., Patterson, B.W., Elbein, A.D.: J. Biol. Chem.,246,4567–4579 (1971)

1 NOMENCLATURE

EC number
2.4.1.16

Systematic name
UDP-N-acetyl-D-glucosamine:chitin 4-beta-N-acetylglucosaminyltransferase

Recommended name
Chitin synthase

Synonyms
Chitin-UDP N-acetylglucosaminyltransferase
Acetylglucosaminyltransferase, chitin-uridine diphosphate
Chitin synthetase
trans-N-Acetylglucosaminosylase

CAS Reg. No.
9030-18-6

---

2 REACTION AND SPECIFICITY

Catalysed reaction
UDP-N-acetyl-D-glucosamine + $[1,4-(\text{N-acetyl-beta-D-glucosaminyl})]_n$ →
→ UDP + $[1,4-(\text{N-acetyl-beta-D-glucosaminyl})]_{n+1}$ (mechanism [15], the
enzyme itself is capable both of initiating chitin chains without a primer and
of determining their chain length [34])

Reaction type
Hexosyl group transfer

Natural substrates
UDP-N-acetyl-D-glucosamine + $[1,4-(\text{N-acetyl-beta-D-glucosaminyl})]_n$ (key
enzyme in chitin biosynthesis [13], isozyme Chs 2 is the essential enzyme
for primary septum formation in Saccharomyces cerevisiae [14], isozyme
Chs 1 is a repair enzyme [14], chitin synthase II is responsible for chitin synthesis [33], the enzyme itself is capable both of initiating chitin chains without a primer and of determining their chain length [34]) [13, 14, 33, 34]

Substrate spectrum
1 UDP-N-acetyl-D-glucosamine + $[1,4-(\text{N-acetyl-beta-D-glucosaminyl})]_n$
[1–36]

Product spectrum
1 UDP + $[1,4-(\text{N-acetyl-beta-D-glucosaminyl})]_{n+1}$ [1–36]

Inhibitor(s)

Polyoxin D (isozyme Chs 2 is more resistant to these antibiotic than isozyme Chs 1 [14], only at high concentrations [8], not [30], Hyalophora cecropia enzyme is almost insensitive, Trichoplusia enzyme is inhibited [7]) [1–3, 7–10, 12–14, 31–33, 35, 36]; NaCl (0.5 M, 86% inhibition of Chs 1 and 29% inhibition of Chs 2) [1]; Nikkomycin (almost insensitive to [7]) [2, 10, 16]; 1-Geranyl-2-methylbenzimidazole (weak) [2]; Cu^{2+} [3]; Zn^{2+} [3, 24, 31, 32]; $ZnCl_2$ [33]; Fe^{2+} [24, 31]; Ba^{2+} [24]; Ca^{2+} [24]; 3',5'-AMP (2 mM, slight) [35]; UDPglucose [25]; UDPmannose [35]; UDPxylose [35]; UDPglucuronic acid [35]; UDPgalacturonic acid [35]; Captan [2, 7]; Benzoylphenyl urea derivatives [2]; Primulin [4]; Calcofluor white (noncompetitive) [5]; 3,5-Dichloro-4-methoxybenzyl alcohol [11]; 2,3,5,6-Tetrachloro-4-methoxyphenol (Drosophilin A) [11]; Pentachlorophenol [11]; 2,2'-Methylenebis[3,4,6-trichlorophenol] (hexachlorophene) [11]; 3,5-Dichlorobenzyl alcohol (weak) [11]; 3,5-Dichloro-4-methoxybenzaldehyde (weak) [11]; Strobilurin analogues [11]; Amphotericin B (noncompetitive) [23]; Nystatin [23]; Filipin (weak) [23]; Soluble chitodextrins (enzymes from some organisms activated [28]) [24]; ATP (2 mM, stimulation [35]) [26]; ADP [26]; Unsaturated fatty acids (linoleicacid, oleic acid) [29]; Nikkomycin X (isozyme Chs 2 is more resistant to these antibiotic than Chs 1 [14]) [6, 14]; Nikkomycin Z (isozyme Chs 2 is more resistant to these antibiotic than Chs 1 [14]) [6, 14]; Amphotericin B methyl ester [6]; UDP [7, 8, 10, 13, 16, 20, 26, 30, 36]; 5'-UDP [9]; UTP [7, 8, 26]; Polyoxin B (Trichoplusia ni [7]) [7, 16]; N-Acetyl-D-glucosamine-N,N'-diacetylchitobiose [7]; Glycerol [8]; EDTA [20, 22, 33]; Diflubenzuron (not [7, 8]) [30]; More (soluble protein inhibitor isolated from cytoplasm of Mucor rouxii [21], pH-dependent, heat-stable inhibitor in the soluble cytoplasm from the mycelium of Mortierella vinacea [24]) [21, 24]

Cofactor(s)/prosthetic group(s)/activating agents

Proteases (apparently exists as a zymogen, requires proteolytic activation [2, 10, 12, 13, 17, 29, 36], activation by trypsin [3, 22], no activation by trypsin [30], proteinase B, EC 3.4.22.9 activates chitin synthase zymogen [19], isozyme Chs 2 is stimulated severalfold by treatment with different proteases [1], papain is a good activator for isozyme Chs 1, poor activator for Chs 2 [1], proteinase K and pronase stimulate Chs 1 and Chs 2 [1], Phascolomyces articulosus enzyme is activated by cell proteases, Choanephora cucurbitarum enzyme is activated by acid protease, slightly activated by trypsin and inhibited by neutral protease [31], acid protease is an excellent activator, weak activation by trypsin and neutral protease [32], chitin synthase II is not activated by proteolysis, chitin synthase I is strongly activated by trypsin treatment [33]) [1–3, 10, 12, 13, 17, 19, 22, 29–33, 36]; Activating factor (from Saccharomyces cerevisiae, protein inhibitor of yeast protease activates chitin synthetase zymogen) [18]; 2-Deoxyglucose (stimulates) [26]; Soluble chitodextrins (activate enzymes from: Allomyces macrogynus, Venturia inaequalis, Neurospora crassa, Aspergillus flavus, Coprinus cinereus,

Mucor rouxii, inhibit enzymes from: Mortierella vinacea, Cunninghamella elegans, Phycomyces blakesleeanus, Saccharomyces cerevisiae, Saccharomyces carlsbergensis [28], inhibition [24]) [28]; N-Acetyl-D-glucosamine (stimulates [1–3, 9, 10, 12, 13, 24, 26, 31–36], K_m: 4.7 mM [10], required [28], slightly increases activity [3], no effect [30]) [1–3, 9, 10, 12, 13, 24, 26, 28, 31–36]; UDP-N-acetylglucosamine (substrate is an allosteric activator) [26]; Digitonin (strong stimulation [6], chitin synthase II is maximally stimulated, nearly 2-fold at digitonin to protein ratio of 0.042, chitin synthase I at a ratio of 0.3–0.75 [33]) [6, 33]; ATP (2 mM, stimulation [35], slight stimulation [8], inhibition [26]) [8, 35]; Glucose (slight stimulation [8], stimulation [26]) [8, 26]; Phosphatidylinositol (stimulates) [20]; Phosphatidylserine (stimulates [20], phospholipid required, phosphatidylserine and lysophosphatidylserine are the best activators [29]) [20, 29]; Phosphatidylethanolamine (stimulates) [20]; Lysophosphatidylserine (phospholipid required, phosphatidylserine and lysophosphatidylserine are the best activators) [29]

Metal compounds/salts
Mg^{2+} (required for maximal activity [7, 12, 13, 28, 30], Mn^{2+} or Mg^{2+} required, Mg^{2+} best activator [12, 13, 33], stimulates isozyme Chs 2 [1], stimulates [2, 9, 10, 20, 22, 24, 25, 31, 32, 35]) [1–3, 7, 9, 10, 12, 13, 20, 22, 24, 25, 28, 30–33, 35]; Mn^{2+} (stimulates [1, 10, 12, 20, 31, 32], slight stimulation [13, 30]) [1, 10, 12, 13, 20, 30–32]; Co^{2+} (stimulates [1, 12, 20, 31, 32], inhibits isozyme Chs 1 [1], no effect [3, 13], best stimulator of Chs 2 [14], inhibits chitin synthase I [33]) [1, 12, 14, 20, 31, 32]; NH_4^+ (slight stimulation) [31]; K^+ (slight stimulation) [31]; Ca^{2+} (slight stimulation [31, 32], no effect [3]) [31, 32]

Turnover number (min^{-1})

Specific activity (U/mg)
42.7 [10]; 4.7 [27]; More [13, 15, 21, 34]

K_m-value (mM)
0.0317 (UDP-GlcNAc) [8]; 0.7 (UDP-GlcNAc) [10, 13]; 0.8–0.9 (UDP-GlcNAc) [1]; 2 (UDP-GlcNAc) [12]; More [6, 20, 22, 24, 28, 29, 31–35]

pH-optimum
5.8–6.2 [24]; 6.0 (Mortierella pusilla) [32]; 6.2 (Mortierella candelabrum [32]) [31, 32]; 6.4 [2]; 6.5 [13, 35]; 6.5–7.5 [10]; 7.0 (chitin synthase I [33]) [30, 33]; 7.5 (secondary peak at pH 6.2 [28]) [22, 28]; 7.5–8.5 [3]; 8.0 (particulate preparation [24], chitin synthase II [33]) [24, 33]; 8.0–8.2 [20]; 8.5 [12]

pH-range
5.5–8.0 (5.5: about 40% of activity maximum, 8.0: about 5% of activity maximum) [13]; 7.1–8.3 (about 50% of activity maximum at pH 7.1 and 8.3) [22]

Temperature optimum (°C)
25 (assay at [12], chitin synthase I [33]) [12, 33]; 28 (Mortierella pusilla [32]) [32, 35]; 29 [28]; 30 (Mortierella candelabrum [32]) [2, 13, 24, 32]; 30.5 [22]; 31–33 [24]; 32 [12]; 40 (chitin synthase II) [33]; 42–46 [3]

Temperature range (°C)
9–43 (about 50% of activity maximum at 9°C and 43°C, particulate preparation) [24]; 17–42 (about 50% of activity maximum at 17°C and 42°C) [35]; 17.5–41 (about 50% of activity maximum at 17.5°C and 41°C) [22]; 21–29.5 (21°C: about 50% of activity maximum, 29.5°C: activity maximum) [28]; More (chitin synthase I retains high activity at 45–50°C, chitin synthase II is inactive) [33]

3 ENZYME STRUCTURE

Molecular weight
520000 (Absidia glauca, gel filtration) [13]
570000 (Saccharomyces cerevisiae, sedimentation coefficient determination) [34]
More (Coprinus cinereus, reversible aggregation into large multimolecular units from 150000 to several million) [27]

Subunits
? (x × 67000, Coprinus cinereus, SDS-PAGE, reversible aggregation into large multimolecular units from 150000 to several million [27], x × 30000, Absidia glauca, SDS-PAGE, aggregation to large complexes [13], x × 63000 (major band) + x × 63000 (weaker band), Saccharomyces cerevisiae, SDS-PAGE [34]) [13, 27, 34]

Glycoprotein/Lipoprotein
–

4 ISOLATION/PREPARATION

Source organism
Allomyces macrogynus [28]; Venturia inaequalis [28]; Neurospora crassa (wall-less variant, slime [20]) [4, 5, 16, 20, 28]; Aspergillus flavus [28]; Mortierella vinacea [24, 28]; Cunninghamella elegans [28]; Phycomyces blakesleeanus [28, 35]; Saccharomyces cerevisiae (chitin synthase II [33], isozymes Chs 1 and Chs 2 [1, 14], are products of different genes [1]) [1, 10, 14, 15, 18, 19, 29, 33, 34]; Saccharomyces carlsbergensis [28]; Sclerotium rolfsii [2]; Apodachlya sp. [3]; Agaricus bisporus [6]; Trichoplusia ni [7]; Hyalophora cecropia [7]; Absidia glauca [13]; Candida albicans [17]; Stomyxs calcitrans [7]; Tribolium castaneum [8]; Tribolium confusium [9]; Tribolium brevicornis [9]; Tenebrio molitor [9]; Galleria mellonella [9]; Coprinus

cinereus [11, 25–28]; Neocallimastix frontalis [12]; Mucor rouxii (IM-80 [23])
[21, 23, 28, 36]; Aspergillus nidulans [22]; Artemia salina (brine shrimp)
[30]; Choanephora cucurbitarum [31]; Phascolomyces articulosus [31];
Mortierella candelabrum [32]; Mortierella pusilla [32]

Source tissue
Hyphae [5, 6, 31]; Cells [1, 10, 17, 20]; Mycelium [2, 11, 12, 17, 28, 32, 35];
Larvae [7, 9, 30]; Wing tissue [7]; Pupae [7, 8]; Gut [9]; Peritrophic mem-
brane [9]; Stipe [25–27]

Localization in source
Plasma membrane (isozyme Chs 2, associated with [1], integral membrane
protein [10]) [1, 10, 29, 35]; Microsomes [9, 13, 25]; Membrane (bound [13])
[13, 22]; More (activity localized in mixed membrane fraction [3, 31, 32],
74000 x g pellet [3], enzyme present in microsomal fraction, cell-wall fraction and mitochondrial fraction [24]) [3, 24, 31, 32]

Purification
Agaricus bisporus [6]; Saccharomyces cerevisiae [10, 29, 34]; Absidia
glauca [13]; Mucor rouxii [21, 36]; Coprinus cinereus [27]; Artemia salina
[30]

Crystallization
–

Cloned
–

Renatured
–

---

5 STABILITY

pH

Temperature (°C)
4 (24 h, more than 90% loss of activity [20], 45 h, Phascolomyces articulosus, gradual decrease of activity with time, Choanephora cucurbitarum, rapid decrease within the first 5 h [31]) [20, 31]; 25 (2 h, solubilized
preparation, retains 80% of activity) [25]; 30 (24 h, more than 90% loss of
activity) [20]; 60 (5 min, complete loss of activity) [13]

Oxidation

Organic solvent

General stability information
Glycerol stabilizes [13]; Lyophilization, 70% of the activity is recovered [34]

Storage

−80°C, stable for several months [34]; 4°C, 1 month, 10% loss of activity [29]; −20°C, solubilized preparation stable for 5 weeks [24]; −80°C, stable for several weeks [10]; −80°C, 20% glycerol, stable [13]; 4°C, 20% glycerol, 50% loss of activity after several days [13]; −70°C, high-speed pellets quick-frozen in liquid N_2, stable for up to 2 weeks [20]; Enzyme activity from Mortierella pusilla increases during low temperature storage, enzyme from Mortierella candelabrum not [32]

6 CROSSREFERENCES TO STRUCTURE DATABANKS

PIR/MIPS code

PIR3:S20538 (imperfect fungus (Candida albicans)); PIR2:A41638 (Neurospora crassa); PIR2:JT0767 (Phycomyces blakesleeanus (fragment)); PIR2:S11808 (yeast (Candida albicans)); PIR2:A23944 (yeast (Saccharomyces cerevisiae)); PIR2:JC2308 (1 Rhizopus oligosporus); PIR2:JC2309 (2 Rhizopus oligosporus); PIR2:S45167 (2 yeast (Saccharomyces cerevisiae)); PIR2:S45879 (3 yeast (Saccharomyces cerevisiae)); PIR2:JC2314 (chsA Emericella nidulans); PIR2:JC2315 (chsB Emericella nidulans); PIR2:JC2408 (class I Emericella nidulans); PIR2:A44427 (membrane-bound pin mould (Absidia glauca) (fragment))

Brookhaven code

7 LITERATURE REFERENCES

[1] Sburlati, A., Cabib, E.: J. Biol. Chem.,261,15147–15152 (1986)
[2] Cohen, E., Elster, I., Chet, I.: Pestic. Sci.,17,175–182 (1986)
[3] Huizar, H.E., Aronson, J.M.: Exp. Mycol.,9,302–309 (1985)
[4] Selitrennikoff, C.P.: Exp. Mycol.,9,179–182 (1985)
[5] Selitrennikoff, C.P.: Exp. Mycol.,8,269–272 (1984)
[6] Haenseler, E., Nyhlen, L.E., Rast, D.M.: Exp. Mycol.,7,17–30 (1983)
[7] Cohen, E., Casida, J.E.: Pestic. Biochem. Physiol.,17,301–306 (1982)
[8] Mayer, R.T., Chen, A.C., DeLoach, J.R.: Insect Biochem.,10,549–556 (1980)
[9] Cohen, E., Casida, J.E.: Pestic. Biochem. Physiol.,13,121–128 (1980)
[10] Cabib, E., Kang, M.S., Au-Young, J.: Methods Enzymol.,138,643–649 (1987) (Review)
[11] Pfefferle, W., Anke, H., Bross, M., Steglich, W.: Agric. Biol. Chem.,54,1381–1384 (1990)
[12] Gay, L., Hebraud, M., Girard, V., Fevre, M.: J. Gen. Microbiol.,135,279–283 (1989)
[13] Machida, S., Saito, M.: J. Biol. Chem.,268,1702–1707 (1993)
[14] Cabib, E.: Antimicrob. Agents Chemother.,35,170–173 (1991)
[15] Fähnrich, M., Ahlers, J.: Eur. J. Biochem.,121,113–118 (1981)
[16] Gow, L.A., Selitrennikoff, C.P.: Curr. Microbiol.,11,211–216 (1984)
[17] Hardy, J.C., Gooday, G.W.: Curr. Microbiol.,9,51–54 (1983)
[18] Ulane, R.E., Cabib, E.: J. Biol. Chem.,249,3418–3422 (1974)

[19] Ulane, R.E., Cabib, E.: J. Biol. Chem.,251,3367–3374 (1976)
[20] Selitrennikoff, C.P.: Biochim. Biophys. Acta,571,224–232 (1979)
[21] Lopez-Romero, E., Ruiz-Herrera, J., Bartnicki-Garcia, S.: Biochim. Biophys. Acta, 525,338–345 (1978)
[22] Ryder, N.S., Peberdy, J.F.: J. Gen. Microbiol.,99,69–76 (1977)
[23] Rast, D.M., Bartnicki-Garcia, S.: Proc. Natl. Acad. Sci. USA,78,1233–1236 (1981)
[24] Peberdy, J.F., Moore, P.M.: J. Gen. Microbiol.,90,228–236 (1975)
[25] Gooday, G.W., de Rousset-Hall, A.: J. Gen. Microbiol.,89,137–145 (1975)
[26] de Rousett-Hall, A., Gooday, G.W.: J. Gen. Microbiol.,89,146–154 (1975)
[27] Montgomery, G.W.G., Adams, D.J., Gooday, G.W.: J. Gen. Microbiol.,130,291–297 (1984)
[28] Moore, P.M., Peberdy, J.F.: Can. J. Microbiol.,22,915–921 (1976)
[29] Duran, A., Cabib, E.: J. Biol. Chem.,253,4419–4425 (1978)
[30] Horst, M.N.: J. Biol. Chem.,256,1412–1419 (1981)
[31] Manocha, M.S., Begum, A.: Can. J. Microbiol.,31,6–12 (1985)
[32] Adjimani, J.P., Manocha, M.S.: Can. J. Microbiol.,31,1035–1040 (1985)
[33] Orlean, P.: J. Biol. Chem.,262,5732–5739 (1987)
[34] Kang, M.S., Elango, N., Mattia, E., Au-Young, J., Robbins, P.W., Cabib, E.: J. Biol. Chem.,259,14966–14972 (1984)
[35] Nung Jan, Y.: J. Biol. Chem.,249,1973–1979 (1974)
[36] Ruiz-Herrera, J., Lopez-Romero, E., Bartnicki-Garcia, S.: J. Biol. Chem.,252, 3338–3343 (1977)

1 NOMENCLATURE

EC number
2.4.1.17

Systematic name
UDPglucuronate beta-D-glucuronosyltransferase (acceptor-unspecific)

Recommended name
Glucuronosyltransferase

Synonyms
1-Naphthol glucuronyltransferase [4]
1-Naphthol-UDP-glucuronosyltransferase
17beta-Hydroxysteroid UDP-glucuronosyltransferase
3alpha-Hydroxysteroid UDP-glucuronosyltransferase [3, 24]
4-Hydroxybiphenyl UDP-glucuronosyltransferase
4-Methylumbelliferone UDP-glucuronosyltransferase
4-Nitrophenol UDPglucuronosyltransferase [36]
4-Nitrophenol UDP-glucuronyltransferase
4-Nitrophenol UDPGT [31]
17-OH Steroid UDPGT [31]
3-OH Androgenic UDPGT [31]
Bilirubin uridine diphosphoglucuronyltransferase
Bilirubin UDPglucuronosyltransferase [26]
Bilirubin UDP-glucuronosyltransferase
Bilirubin monoglucuronide glucuronyltransferase
Bilirubin UDPGT [17]
Bilirubin glucuronyltransferase
Ciramadol UDP-glucuronyltransferase
Estriol UDPglucuronosyltransferase
Estrone UDPglucuronosyltransferase [37]
Glucuronosyltransferase, uridine diphospho-
Glucuronosyltransferase, uridine diphosphoglucuronate-bilirubin glucuronoside
Glucuronosyltransferase, uridine diphosphoglucuronate-bilirubin
Glucuronosyltransferase, uridine diphosphoglucuronate-estriol
Glucuronosyltransferase, uridine diphosphoglucuronate-estradiol
Glucuronosyltransferase, uridine diphosphoglucuronate-4-hydroxybiphenyl
Glucuronosyltransferase, uridine diphosphoglucuronate-1,2-diacylglycerol
Glucuronosyltransferase, uridine diphosphoglucuronate-estriol 16alpha-
Glucuronyltransferase, uridine diphospho-
GT [1, 10]

Morphine glucuronyltransferase [4]
p-Hydroxybiphenyl UDP glucuronyltransferase
p-Nitrophenol UDP-glucuronosyltransferase [24]
p-Nitrophenol UDP-glucuronyltransferase
p-Nitrophenylglucuronosyltransferase
p-Phenylphenol glucuronyltransferase
Phenyl-UDP-glucuronosyltransferase
PNP-UDPGT [31]
UDP glucuronate-estradiol-glucuronosyltransferase
UDP glucuronosyltransferase
UDP glucuronate-estriol glucuronosyltransferase
UDP glucuronic acid transferase
UDP glucuronyltransferase
UDP-glucuronate-4-hydroxybiphenyl glucuronosyltransferase
UDP-glucuronate-bilirubin glucuronyltransferase
UDP-glucuronosyltransferase [1, 7]
UDP-glucuronyltransferase
UDPGA transferase
UDPGA-glucuronyltransferase
UDPglucuronosyltransferase [15]
UDPGT [9]
Uridine diphosphoglucuronyltransferase
Uridine diphosphoglucuronate-bilirubin glucuronosyltransferase
Uridine diphosphate glucuronyltransferase
Uridine 5'-diphosphoglucuronyltransferase
Uridine diphosphoglucuronosyltransferase
EC 2.4.1.42 (formerly)
EC 2.4.1.59 (formerly)
EC 2.4.1.61 (formerly)
EC 2.4.1.76 (formerly)
EC 2.4.1.77 (formerly)
EC 2.4.1.84 (formerly)
EC 2.4.1.107 (formerly)
EC 2.4.1.108 (formerly)
More (this entry denotes a family of enzymes accepting a wide range of substrates, including phenols, alcohols, amines and fatty acids. Some of the activities catalyzed were previously listed separately as EC 2.4.1.42, 59, 61, 76, 77, 84, 107 and 108. A temporary nomenclature for the various forms, whose delineation is in a state of flux, is suggested in [1])

CAS Reg. No.
9030-08-4; 37277-52-4; 37277-66-0; 37205-52-0; 61969-98-0; 62213-47-2; 62213-43-8

2 REACTION AND SPECIFICITY

Catalysed reaction
UDPglucuronate + acceptor →
→ UDP + acceptor beta-D-glucuronoside

Reaction type
Hexosyl group transfer

Natural substrates
Bilirubin + UDPglucuronate (physiological elimination of bilirubin) [6]
More (elimination of lipid-soluble endogenous compounds and xenobiotics)
[1, 72–75]

Substrate spectrum
1 4-Nitrophenol + UDPglucuronate (no activity [68]) [1–3, 8, 10–12, 14, 18, 20, 21, 24, 25, 28, 31, 32, 34–37, 41, 43–46, 51, 53, 54, 56, 57, 62, 65–67, 71, 77, 80, 81]
2 1-Naphthol + UDPglucuronate [1, 2, 4, 7, 8, 10, 12, 18, 19, 21, 25, 28, 30, 31, 36, 43, 63, 66]
3 Morphine + UDPglucuronate (no activity [37, 62, 66]) [2, 4, 19, 28, 36, 43, 68]
4 N-Hydroxy-2-naphthylamine + UDPglucuronate [2, 4, 80]
5 3-Hydroxybenzo[a]pyrene + UDPglucuronate [2, 4, 28, 80]
6 4-Hydroxybiphenyl + UDPglucuronate (very low activity [18], no activity [68]) [2, 18, 21, 62]
7 Testosterone + UDPglucuronate (isozyme 17-OH steroid UDPGT [31], no activity [21, 37, 62, 66–68], UDPglucuronate can be substituted by UDP-galacturonate yielding 83–88% of the activity [33]) [2, 4, 12, 19, 24, 25, 27, 28, 31, 33, 36, 40, 47, 63]
8 Bilirubin + UDPglucuronate (no activity [8, 21, 46, 68], very low activity [18], Gunn rat liver enzyme inactive [9, 42, 77], cat liver enzyme active [42]) [2, 5, 6, 9, 12, 16–18, 22, 25–27, 29, 41, 42, 55, 58, 66, 70, 76, 77, 79]
9 Bilirubin monoglucuronide + UDPglucuronate [6, 27, 29]
10 Bilirubin 7,7,7-triphenylheptanoic acid + UDPglucuronate [9]
11 4-Methylumbelliferone + UDPglucuronate [10, 15, 18, 25, 28, 31, 67, 70]
12 3'-Azido-3'-deoxythymidine + UDPglucuronate [13]
13 Chloramphenicol + UDPglucuronate (very low activity [18], no activity [62]) [18]
14 2-Methoxy-4-(2-propenyl)phenol + UDPglucuronate (i.e. eugenol) [21]
15 Serotonin + UDPglucuronate (i.e. 5-hydroxytryptamine, cat liver enzyme inactive, Gunn rat liver enzyme active [42]) [21, 42]
16 Estradiol-17beta + UDPglucuronate [25, 59–61]
17 Phenolphthalein + UDPglucuronate [27, 28, 51]
18 beta-Estradiol + UDPglucuronate [31]

19 alpha-Naphthylamine + UDPglucuronate [15, 67]
20 Estriol + UDPglucuronate [15, 59, 61, 67]
21 Etiocholanone + UDPglucuronate [24]
22 17beta-Hydroxy-5alpha-androstan-3-one + UDPglucuronate [24]
23 Lithocholic acid + UDPglucuronate [24, 69, 78]
24 Chenodeoxycholic acid + UDPglucuronate [24, 33, 69, 78]
25 Estrone + UDPglucuronate (r [48], estrone-specific activity inactive towards 4-nitrophenol [34]) [34, 40, 47, 48, 59]
26 2-Aminophenol + UDPglucuronate [51, 54, 56, 57]
27 8-Hydroxyquinoline + UDPglucuronate [51]
28 4-Nitrothiophenol + UDPglucuronate [52]
29 Thiophenol + UDPglucuronate [52]
30 2-Aminobenzoate + UDPglucuronate [54]
31 Estradiol-17alpha + UDPglucuronate [61]
32 Benzo[a]pyrene 3,6-quinone + UDPglucuronate [39]
33 Benzo[a]pyrene-3,6-quinol + UDPglucuronate [64]
34 Deoxycholic acid + UDPglucuronate [69]
35 Ursodeoxycholic acid + UDPglucuronate [69, 78]
36 Glycolithocholic acid + UDPglucuronate [69]
37 Cholic acid + UDPglucuronate [78]
38 More (specificity not influenced by lysophosphatidylcholine [80], specificity of cloned human enzyme [84]) [80, 84]

Product spectrum
   1 ?
   2 1-Naphthol glucuronide + UDP [1]
   3 ?
   4 ?
   5 ?
   6 ?
   7 ?
   8 Bilirubin monoglucuronoside + bilirubin diglucuronoside + UDP [5, 6]
   9 Bilirubin diglucuronide + UDP [6, 29]
  10 ?
  11 ?
  12 3'-Azido-3'-deoxythymidine 5'-O-glucuronate + UDP [13]
  13 ?
  14 ?
  15 Tryptamine 5-O-beta-D-glucuronide + UDP [42]
  16 ?
  17 ?
  18 ?
  19 ?
  20 ?
  21 ?

22 ?
23 ?
24 ?
25 ?
26 ?
27 4-Nitrothiophenol thio-beta-D-glucuronide + UDP [52]
28 ?
29 Thiophenol thio-beta-D-glucuronide + UDP [52]
30 ?
31 ?
32 Benzo[a]pyrene 3,6-quinol glucuronide + UDP (no information to which
    positions 3-, 6-, or both the glucuronate is attached) [39]
33 Benzo[a]pyrene 3,6-quinol mono- and diglucuronide + UDP [64]
34 ?
35 ?
36 ?
37 ?
38 ?

Inhibitor(s)
omega,omega,omega-Triphenylalkyl-UDP derivatives (e.g. 7,7,7-triphenyl-
heptyl-UDP) [7]; 4-Chlorophenyl-UDP [12]; 4-Bromophenyl-UDP [12]; 4-Io-
dophenyl-UDP [12]; 4-Isopropylphenyl-UDP [12]; 4-tert-Butylphenyl-UDP
[12]; 2-Chlorophenyl-UDP [12]; 2-Bromophenyl-UDP [12]; [12]; 2,5-Dichloro-
phenyl-UDP [12]; 2,6-Dimethoxyphenyl-UDP [12]; 3-Methyl-2-nitrobenzyl-
UDP [12]; 2-(4-Bromophenyl)ethyl-UDP [12]; 2-(4-Nitrophenyl)ethyl-UDP [12];
2-(1-Naphthyl)ethyl-UDP [12]; 2-(2-Naphthyl)ethyl-UDP [12]; 2,2,2-(Triphenyl)-
ethyl-UDP [12]; 2-(1-Naphthyl)ethyl-UMP [12]; 2-(2-Naphthyl)ethyl-UMP [12];
2-(4-Bromophenyl)ethyl-UMP [12]; p-[di-n-Propylsulfamoyl]benzoic acid (i.e.
probenecid) [13]; Triphenylacetic acid (competitive to bilirubin) [17, 26];
3,3,3-Triphenylpropionic acid (competitive to bilirubin) [17]; 4,4,4-Triphenyl-
butanoic acid (competitive to bilirubin) [17]; 5,5,5-Triphenylpentanoic acid
(competitive to bilirubin) [17]; 6,6,6-Triphenylhexanoic acid (competitive to
bilirubin) [17]; 7,7,7-Triphenylheptanoic acid (competitive to bilirubin) [17];
8,8,8-Triphenyloctanoic acid (competitive to bilirubin) [17]; 9,9,9-Triphenyl-
nonanoic acid (noncompetitive to bilirubin) [17]; UDP (mixed type inhibition
towards UDPglucuronic acid and 1-naphthol [30], product inhibition be-
comes negligible in presence of UDP-N-acetylglucosamine, is amplified in
phospholipase A treated enzyme [49]) [30, 49, 57, 59]; UDP-N-acetylglucosamine (competitive to UDPglucuronic acid, mixed-type inhibition towards
1-naphthol [30], inhibits phospholipase A treated enzyme [49]) [30, 49];
Phospholipase A (in whole microsomes: inhibition of glucuronidation of testosterone, stimulation of glucuronidation of estradiol and estrone) [47]; EDTA
(decreases rate of glucuronidation if enzyme is stimulated by UDP-N-ace-

---

tylglucosamine and divalent metal ions [54], not [48]) [54]; UMP [57]; UDP-glucose (in microsomal preparation, below 16°C) [49]; UTP [57]; ATP [59]; Estrone (competitive to estradiol-17beta) [59, 61]; Zn^{2+} [58, 78]; p-Chloromercuribenzoate [48]; Triton X-100 (above 0.01% v/v) [78]; Digitonin (inhibition above 0.6 mg/mg protein, activation below) [78]; Ni^{2+} [78]; Fe^{2+} [78]; 4-Nitrophenol (uncompetitive to chenodeoxycholic acid) [78]; Testosterone (noncompetitive to chenodeoxycholic acid) [78]; Bilirubin (competitive to chenodeoxycholic acid [78], not [36]) [78]; Estradiol (competitive to chenodeoxycholic acid) [78]; Hg^{2+} (inhibition of 4-methylumbelliferone UDP-glucuronyltransferase, activation of bilirubin UDP-glucuronyltransferase) [70]; Codeine [68]; More (immunochemical experiments [28], product inhibition [48], not inhibitory: N-ethylmaleimide [36, 48], Lubrol 12A9 [36], dithioerythritol [48]) [28, 36, 48]

Cofactor(s)/prosthetic group(s)/activating agents

Lysophosphatidylcholine (slight stimulation of highly purified preparations [4], activation [10]) [4, 10]; Phosphatidylcholine (required by purified preparations for optimal activity) [8, 18, 21, 24]; Dilauroylphosphatidylcholine (increases activity of liver enzyme towards 4-methylumbelliferone, but decreases this activity of the kidney enzyme) [10]; UDP-N-acetylglucosamine (presumed physiological activator [16], activation [49, 52, 56], activation in presence of divalent metal ions [54]) [16, 49, 52, 54, 56]; Cholate (activation) [18]; Phospholipids (no effect on glucuronidation of 1-naphthol, 4-nitrophenol, testosterone, 4-methylumbelliferone, morphine, twofold activation of glucuronidation of 3-hydroxybenzo[a]pyrene, 3–4-fold activation of glucuronidation of phenolphthalein [28], effects of various phospholipids [24], estrone UDP-glucuronosyltransferase is inactive in absence of phospholipids, 300-fold stimulation by phospholipids. 4-Nitrophenol UDP-glucuronosyltransferase is active in absence of phospholipids, but stimulated 3–4-fold in their presence [71]) [24, 28, 71]; Triton X-100 (0.01%, 1.6-fold activation [78], activation [30]) [30, 78]; Mersalyl (activation) [47]; Phospholipase A (stimulation of glucuronidation of estradiol and estrone, inhibition of glucuronidation of testosterone in whole microsomes [47], in presence of Ca^{2+} activation of glucuronidation of 4-nitrophenol [50]) [47, 50]; Phospholipase C (increase of activity in whole microsomes) [56]; Diethylnitrosamine (10 mM, 3.1–3.7-fold activation) [46]; Cysteine (10 mM, slight activation) [59]; Digitonin (0.6 mg/mg liver protein, 1.8-fold activation, inhibition above [78], activation [58]) [58, 78]; Alkyl ketones (activation) [36]; Choline (required as phospholipid (polar head group)) [71]; Lecithins (activation, unsaturated lecithins are more efficient than saturated) [71]

Metal compounds/salts

Mg^{2+} (10 mM, activation [48], activation in presence of UDP-N-acetylglucosamine [54], 5–20 mM essential [55], no effect [52], activation, can be
partially replaced by Mn^{2+}, Ca^{2+}, Co^{2+}, Cd^{2+}, Fe^{2+}, Pb^{2+} [58], as $MgCl_2$: 5
mM, activation [4], 30 mM, activation [8], 50 mM, activation [30]) [4, 8, 30,
48, 54, 55, 58]; Mn^{2+} (1 mM, activation [47], increase of glucuronidation of
4-nitrophenol and 2-aminobenzoate [54]) [47, 54]; Ca^{2+} (activation in presence of UDP-N-acetylglucosamine [54], 20 mM, activation [59], can partially
replace Mg^{2+} [58]) [54, 58, 59]; Co^{2+} (activation in presence of UDP-N-acetylglucosamine [54], can partially replace Mg^{2+} [58]) [54, 58]; Zn^{2+} (activation in presence of UDP-N-acetylglucosamine [54], can partially replace
Mg^{2+} [58]) [54, 58]; Cu^{2+} (activation in presence of UDP-N-acetylglucosamine [54], can partially replace Mg^{2+} [58]) [54, 58]; EDTA (activation at 2
mM, no effect at 10 mM [59], no effect [36]) [59]; Hg^{2+} (activation of bilirubin
UDP-glucuronyltransferase, inhibition of 4-methylumbelliferone
UDP-glucuronyltransferase) [70]; More (alkaline pretreatment of microsomes
increases activity) [14]

Turnover number (min^{-1})

Specific activity (U/mg)

6.06 (1-naphthol, Wistar rat treated with 3-methylcholanthrene) [4]; 5.35
(1-naphthol, Wistar rat treated with phenobarbital) [4]; 0.67 (morphine,
Wistar rat treated with phenobarbital) [4]; 4.0 (1-naphthol, Wistar rat treated
with 3-methylcholanthrene) [21]; 5.0 (4-nitrophenol, Wistar rat treated with
3-methylcholanthrene) [21]; 1.8 (eugenol, Wistar rat treated with 3-methyl-
cholanthrene) [21]; 1.7 (5-hydroxytryptamine, Wistar rat treated with 3-me-
thylcholanthrene) [21]; 0.092 (4-hydroxybiphenyl, Wistar rat treated with
3-methylcholanthrene) [21]; 0.142 (bilirubin, non-treated Wistar rat, activity
absent from Gunn rat) [9]; 0.168 (chenodeoxycholic acid, Sprague-Dawley
rat treated with phenobarbital) [33]; 0.983 (testosterone, Sprague-Dawley rat
treated with phenobarbital) [33]; More (assay conditions [1, 58]) [1, 8, 13,
18, 19, 34–38, 41, 43–46, 58, 80, 81]

K_m-value (mM)

0.0036 (UDPglucuronic acid (+ 1.5 mM morphine), ethanol-induced enzyme) [19]; 0.0054 (UDPglucuronic acid (+ 1.5 mM morphine), non-induced
enzyme) [19]; 0.006 (androsterone, 3alpha-hydroxysteroid UDP-glucuronosyltransferase, Wistar rat, high-activity strain) [3]; 0.0065 (UDPglucuronic
acid (+ 0.75 mM 1-naphthol), ethanol-induced enzyme) [19]; 0.0084 (UDP-
glucuronic acid (+ 0.76 mM 1-naphthol), non-induced enzyme) [19]; 0.009
(androsterone, 3alpha-hydroxysteroid UDP-glucuronosyltransferase,
Sprague-Dawley rat) [3]; 0.012 (testosterone, 17beta-hydroxysteroid
UDP-glucuronosyltransferase, Wistar rat, high-activity strain [3], 1-naphthol,
presence of 5 mM $MgCl_2$ as activator [30]) [3, 30]; 0.013 (testosterone,
17beta-hydroxysteroid UDP-glucuronosyltransferase, Sprague-Dawley rat)

[3]; 0.0137 (bilirubin, conversion to bilirubin monoglucuronide) [6]; 0.015
(1-naphthol, no activator present) [30]; 0.0193 (bilirubin, conversion to bilirubin diglucuronide) [6]; 0.02 (testosterone, 17beta-hydroxysteroid
UDP-glucuronosyltransferase, Wistar rat, low activity strain) [3]; 0.03
(1-naphthol, presence of 50 mM $MgCl_2$ as activator) [30]; 0.045–0.047 (es-
tradiol-17beta) [59, 61]; 0.064 (bilirubin (+ 2.35 mM UDPglucuronate)) [26];
0.067 (1-naphthol, presence of 50 mM $MgCl_2$, 0.1% Triton X-100 as activator) [30]; 0.083 (chenodeoxycholic acid) [78]; 0.120 (androsterone,
3alpha-hydroxysteroid UDP-glucuronosyltransferase, Wistar rat, low-activity
strain) [3]; 0.26 (4-hydroxybiphenyl) [62]; 0.3 (UDPglucuronate) [3]; 0.36
(UDPglucuronate (+ 0.017 mM bilirubin monoglucuronide) at pH 6.5) [5];
0.37 (UDPglucuronate (+ 2 mM 1-naphthol), presence of 50 mM $MgCl_2$)
[30]; 0.44 (UDPglucuronate (+ 0.016 mM bilirubin monoglucuronide) at pH
8.1) [5]; 0.59 (UDPglucuronate (+ 0.2 mM 1-naphthol), no activator present)
[30]; 0.71 (UDPglucuronate (+ 0.2 mM 1-naphthol), presence of 50 mM
$MgCl_2$, 0.1% Triton X-100 as activator) [30]; 0.96–1.0 (UDPglucuronate) [62,
78]; 1.06 (UDPglucuronate (+ 0.125 mM bilirubin)) [26]; 2.0 (4-nitrophenol,
17beta-hydroxysteroid UDP-glucuronosyltransferase, Sprague-Dawley and
Wistar rat) [3]; 11.6 (3'-azido-3'-deoxythymidine, monkey) [13]; 13.4
(3'-azido-3'-deoxythymidine, human) [13]; More (effect of temperature [20],
effect of UDP-N-acetylglucosamine as activator [52], kinetics of phospholi-
pase-treated, UDP-N-acetylglucosamine-activated, detergent-activated and
sonicated enzyme [56]) [20, 22, 52, 53, 56, 61]

pH-optimum
6.5 (conversion of bilirubin monoglucuronide to bilirubin diglucuronide, second optimum at pH 8.1) [5]; 6.7 (chenodeoxycholic acid) [78]; 7.0 (1-naph-
thol) [8]; 7.3–7.8 (4-thionitrophenol, thiophenol) [52]; 7.3–8.7 (estrone) [48];
7.4 (4-nitrophenol) [52]; 7.5 (4-nitrophenol) [34]; 7.6 (1-naphthol [19], estra-
diol-17beta, Tris/HCl buffer [61]) [19, 61]; 7.8 (bilirubin) [55]; 8.0 (bilirubin
[58], estradiol-17beta, Tris/maleate or phosphate buffer [61]) [58, 61];
8.0–8.8 (glucuronidation of bilirubin) [5]; 8.0–9.0 (estrone) [34]; 8.1 (conversion of bilirubin monoglucuronide to bilirubin diglucuronide, second optimum at pH 6.5) [5]

pH-range
6.0–8.0 (substrate chenodeoxycholic acid) [78]

Temperature optimum (°C)
37 [8, 55, 58]; 42 [59]; 55 [48]

Temperature range (°C)
50 (increase of activity up to 50°C, half-life at 50°C: 15 min) [61]

3 ENZYME STRUCTURE

Molecular weight

41500 (rat, bilirubin monoglucuronide forming enzyme, monomeric, radiation inactivation) [27]

73500 (rat, estrone UDP-glucuronyltransferase, dimeric enzyme, radiation inactivation) [27]

109000 (rat, 4-nitrophenol UDP-glucuronyltransferase, dimeric or trimeric enzyme, radiation inactivation) [27]

142000 (rat, testosterone UDP-glucuronyltransferase, trimeric or tetrameric enzyme, radiation inactivation) [27]

159000 (rat, phenolphthalein UDP-glucuronyltransferase, tetrameric enzyme, radiation inactivation) [27]

175000 (rat, formation of bilirubin diglucuronide from bilirubin monoglucuronide, tetrameric enzyme, radiation inactivation) [27]

230000 (rabbit, gel filtration) [34]

316000–318000 (rat, gel filtration, polyacrylamide gradient slab gel electrophoresis) [33]

Subunits

Tetramer (isozyme for conversion of bilirubin monoglucuronide to bilirubin diglucuronide, a + b + c + d, subunits a and b have a low K_m for UDPglucuronide, c and d have a high K_m for UDPglucuronide, a and b can catalyze the conversion of bilirubin to bilirubin monoglucuronide, rat, radiation inactivation) [27]

? (x × 58500, human, sequence of cDNA, anomalous migration of hydrophobic protein on SDS-PAGE [15], x × 50000, rat, 17beta-hydroxysteroid UDP-glucuronyltransferase [24], rat, deglycosylated isozymes GT-1 and GT-2 [18], x × 51000, mouse [28], x × 52000, rat [23], rat, 3alpha-hydroxysteroid UDP-glucuronyltransferase [24], x × 53000, human, isozyme 1 [67], x × 54000, rat, 1-naphthol glucuronyltransferase [4, 21], bilirubin UDP-glucuronosyltransferase [9, 23], rat, isozymes GT-1 and GT-2, identical subunit weights, but different N-terminal amino acid sequences [18], human, isozyme 2 [67], x × 55000, guinea pig [35], x × 56000, rat, morphine glucuronyltransferase [4], rat, 4-nitrophenol UDP-glucuronyltransferase [24, 31], additional values for rat isozymes ranging from 51000 to 59000 [25, 33, 36, 38, 43–45, 64, 66, 80], x × 57000, rabbit [34, 37], x × 52000–57000, Pleurotes platessa [8], all values determined by SDS-PAGE) [4, 8, 9, 15, 18, 21, 23–25, 28, 31, 33–38, 43–45, 64, 66, 67, 80]

Glycoprotein/Lipoprotein

Phospholipoprotein (less than 1 mol phospholipid per mol of enzyme [28], less than 3 mol phospholipid per mol of enzyme [33], composition [35]) [2, 28, 33, 35]; Glycoprotein (mannose-type oligosaccharide chains [18, 21], N-glycosylated [62], no glycoprotein [33]) [9, 18, 21, 62]

4 ISOLATION/PREPARATION

Source organism

Rat (distribution in Wistar and Gunn rat [42, 52], bilirubin UDP-glucuronosyltransferase not in Gunn rat [9, 42, 77], Wistar rat [1, 3, 4, 7, 9, 12, 13, 17, 18, 21, 23, 25–27, 29, 30, 36, 40–43, 45, 46, 52, 56, 58, 68, 80], Sprague-Dawley rat [6, 24, 31, 33, 44, 79], Gunn rat [13, 17, 23, 42, 52, 77], R/A Pfd strain [16, 22], evaluation of heterogeneity [51]) [1–4, 6, 7, 9, 10, 12, 13, 16–18, 21–27, 29–31, 33, 36, 38–46, 50–52, 55, 56, 58, 62, 64, 66, 68, 70, 77, 79, 80]; Human [2, 13, 15, 26, 67, 69, 76, 78, 84]; Cat [5]; Pleurotes platessa (plaice) [8, 63]; Pig [11, 14, 20, 32, 48, 59–61, 82]; Rabbit [19, 34, 37, 71, 81]; Mouse [28, 56, 57]; Guinea pig [35, 47, 50, 53, 54, 56]; Chicken [52]; Rainbow trout [65]

Source tissue

Liver [1–17, 19–29, 31–47, 49–58, 62–64, 67–71, 76–81, 84]; Kidney [4, 10, 13, 18, 42, 48, 52, 65, 66]; Small intestine [4, 30, 59–61, 82]; Duodenum [42]; Lung [4, 42]; Skin [4]; Adrenal [4]; Spleen [4]; Stomach (mouse) [52]

Localization in source

Microsomes [1–82]; Membranes of endoplasmic reticulum (sequencing data indicate that the enzymes are anchored in the membrane by a carboxy-terminal hydrophobic domain, with a bulk of the polypeptide chain located within the cisternal lumen of the endoplasmic reticulum [6, 83], the binding of agylcone substrates such as bilirubin occurs at the amino-terminal half of the polypeptide chain [84], active site at the lumenal face of the endoplasmic reticulum membrane [22]) [6, 22, 51, 55, 83, 84]; Mitochondria [59]; Cytosol [59]

Purification

Rat (overview procedures [2]) [2, 4, 9, 12, 17, 21, 23, 25, 31, 33, 36, 38, 40, 41, 44–46, 62, 64, 66, 68, 80]; Pleurotes platessa [8]; Rabbit [19, 37, 81]; Mouse [28]; Guinea pig [35]; Pig [61]; Human [67]

Crystallization

–

Cloned

(2 bilirubin UDP-glucuronyltransferases, expression in COS-1 cells [76], estriol UDPGT, isozyme pI 7.4 [15], expression in V 79 Chinese hamster lung fibroblasts [84]) [15, 76, 84]

Renatured

(reactivation of lipid-depleted enzyme [35], phospholipid required for reconstitution [67]) [35, 67]

5 STABILITY

pH
   11.1 (unstable above) [14]

Temperature (°C)
   –20 (rapid loss of activity) [4]; 24 (1 d, more than 50% loss of activity) [28];
   30 (15 min, absence of substrates, slight inactivation) [48]; 35 (reversible
   change from active to inactive form, this transition temperature is lowered by
   cholesterol) [11]; 37 (inactivation in 15 min [36, 43], stable up to [61]) [36,
   43, 61]; 40 (inactivation above [55], inactivation in absence of substrates
   [59]) [55, 59]; 44 (irreversible inactivation, stabilization by cholesterol) [11];
   48 (15 min, 50% loss of activity) [48]; 50 (half-life 15 min) [61]; 62 (complete
   inactivation) [61]; 70 (15 min, complete inactivation) [48]

Oxidation

Organic solvent

General stability information
   $MgCl_2$, 5 mM, unstable to, lysophosphatidylcholine restores activity of partially denatured enzyme [4]; Dialysis and vacuum dialysis cause 40% loss of
   activity [36]; Freezing/thawing causes complete loss of activity [46]

Storage
   –20°C, microsomal preparation, more than 4 months [48]; –20°C, several
   months [32]; –20°C, at least 1 month [33]; –20°C, stable [61]; –20°C, 20%
   w/v glycerol, complete loss of activity [36, 43]; 0°C, half-life 18–20 days [36,
   34, 43, 45]; 0°C, 30 days, 4-nitrophenol UDP-glucuronosyltransferase loses
   11% of activity, 2-aminophenol UDP-glucuronosyltransferase loses 19% of
   activity [46]; 0°C, 14 days, 80% loss of activity [61]; 0–4°C, 0.1 mg/ml lysophosphatidylcholine, 1 mM EDTA, 11% loss of activity in 6 days, 20% loss of
   activity in 12 days [4, 80]; 4°C, bilirubin glucuronosyltransferases: isozyme
   F1 3 weeks stable, isozyme F3 inactivation in 1 week [23]; 4°C, at least 6
   days [28]; 4°C, at least 2 days [44]; 4°C, microsomal preparation: increase
   of activity towards 4-nitrophenol, maximum of activation at 3 days storage
   [52]; 4°C, solubilized enzyme stable for 7 days, concentrated form loses
   12% of activity [55]; 22°C, lyophilized microsomes, 15 days [27]

6 CROSSREFERENCES TO STRUCTURE DATABANKS

PIR/MIPS code
   PIR2:B35704 (bovine (fragments)); PIR2:A27878 (human); PIR3:S11309
   (human); PIR3:S17512 (human); PIR2:S00163 (mouse); PIR2:A23520 (rat);
   PIR2:A35343 (rat); PIR2:A40467 (rat); PIR3:S15089 (rat); PIR2:A39092
   (1 precursor bilirubin specific human); PIR2:A42233 (2 rat); PIR2:B39092
   (2 precursor bilirubin specific human (fragment)); PIR2:A28460 (3 precursor

Enzyme Handbook © Springer-Verlag Berlin Heidelberg 1996
Duplication, reproduction and storage in data banks are only
allowed with the prior permission of the publishers

rat); PIR2:A26064 (4 rat); PIR2:A36276 (5 precursor rat); PIR2:A31340 (precursor human); PIR2:A35366 (precursor human); PIR2:A24600 (precursor 3 methylcholanthrene-inducible rat); PIR2:B47113 (UGT2B13 precursor rabbit); PIR2:C47113 (UGT2B14 precursor rabbit); PIR2:A61266 (4-hydroxybiphenyl specific rat (fragment)); PIR2:A24324 (hepatic rat); PIR2:S07390 (hepatic rat); PIR2:PX0008 (hepatic rat (fragment)); PIR2:A47113 (p-nitrophenol rabbit (fragment))

Brookhaven code

---

7 LITERATURE REFERENCES

[1] Bock, K.W., Burchell, B., Dutton, G.J., Hänninen, O., Mulder, G.J., Owens, I.S., Siest, G., Tephly, T.R.: Biochem. Pharmacol.,32,953–955 (1983) (Review)
[2] Burchell, B.: Rev. Biochem. Toxicol.,3,1–32 (1981) (Review)
[3] Green, M.D., Falany, C.N., Kirkpatrick, R.B., Tephly, T.R.: Biochem. J.,230,403–409 (1985)
[4] Bock, K.W., Josting, D., Lilienblum, W., Pfeil, H.: Eur. J. Biochem.,98,19–26 (1979)
[5] Jansen, P.L.M.: Biochim. Biophys. Acta,338,170–182 (1974)
[6] Crawford, J.M., Ransil, B.J., Narciso, J.P., Gollan, J.L.: J. Biol. Chem.,267, 16943–16950 (1992)
[7] Said, M., Noort, D., Magdalou, J., Ziegler, J.C., van der Marel, G.A., van Boom, J.H., Mulder, G.J., Siest, G.: Biochem. Biophys. Res. Commun.,187,140–145 (1992)
[8] Clarke, D.J., George, S.G., Burchell, B.: Biochem. J.,284,417–423 (1992)
[9] Clarke, D.J., Keen, J.N., Burchell, B.: FEBS Lett.,299,183–186 (1992)
[10] Yokoda, H., Fukuda, T., Yuasa, A.: J. Biochem.,110,50–53 (1991)
[11] Rotenberg, M., Zakim, D.: J. Biol. Chem.,266,4159–4161 (1991)
[12] Noort, D., Coughtrie, M.W.H., Burchell, B., van der Marel, G.A., van Boom, J.H., van der Gen, A., Mulder, G.J.: Eur. J. Biochem.,188,309–312 (1990)
[13] Haumont, M., Magdalou, J., Lafaurie, C., Ziegler, J.-M., Siest, G., Colin, J.-N.: Arch. Biochem. Biophys.,281,264–270 (1990)
[14] Dannenberg, A., Wong, T., Zakim, D.: Arch. Biochem. Biophys.,277,312–317 (1990)
[15] Coffman, B.L., Tephly, T.R., Irshaid, Y.M., Green, M.D., Smith, C., Jackson, M.R., Wooster, R., Burchell, B.: Arch. Biochem. Biophys.,281,170–175 (1990)
[16] Vanstapel, F., Hammaker, L., Pua, K., Blanckaert, N.: Biochem. J.,259,659–663 (1989)
[17] Fournel-Gigleux, S., Shepherd, S.R.P., Carre, M.-C., Burchell, B., Siest, G., Caubere, P.: Eur. J. Biochem.,183,653–659 (1989)
[18] Yokota, H., Ohgiya, N., Ishihara, G., Ohta, K., Yuasa, A.: J. Biochem.,106,248–252 (1989)
[19] Hutabarat, R.M., Yost, G.S.: Arch. Biochem. Biophys.,273,16–25 (1989)
[20] Dannenberg, A., Rotenberg, M., Zakim, D.: J. Biol. Chem.,264,238–242 (1989)
[21] Yokota, H., Yuasa, A., Sato, R.: J. Biochem.,104,531–536 (1988)
[22] Vanstapel, F., Blanckaert, N.: Arch. Biochem. Biophys.,263,216–225 (1988)
[23] Odell, G., Mogilevsky, W.S., Siegel, F.L.: Biochem. Biophys. Res. Commun.,154, 1212–1221 (1988)
[24] Falany, C.N., Green, M.D., Swain, E., Tephly, T.R.: Biochem. J.,238,65–73 (1986)

[25] Roy Chowdhury, J., Roy Chowdhury, N., Falany, C.N., Tephly, T.R., Arias, I.M.:
     Biochem. J.,233,827–837 (1986)
[26] Fournel, S., Gergoire, B., Magdalou, J., Carre, M.-C., Lafaurie, C., Siest, G.,
     Caubere, P.: Biochim. Biophys. Acta,883,190–196 (1986)
[27] Peters, W.H.M., Jansen, P.L.M., Nauta, H.: J. Biol. Chem.,259,11701–11706 (1984)
[28] Mackenzie, P.I., Hjelmeland, L.M., Owens, I.S.: Arch. Biochem. Biophys.,231,
     487–497 (1984)
[29] Burchell, B., Blanckaert, N.: Biochem. J.,223,461–465 (1984)
[30] Koster, A.S., Noordhoek, J.: Biochem. Biophys. Acta,761,76–85 (1983)
[31] Falany, C.N., Tephly, T.R.: Arch. Biochem. Biophys.,227,248–258 (1983)
[32] Hochman, Y., Zakim, D.: J. Biol. Chem.,258,4143–4146 (1983)
[33] Matern, H., Matern, S., Gerok, W.: J. Biol. Chem.,257,7422–7429 (1982)
[34] Tukey, R.H., Tephly, T.R.: Arch. Biochem. Biophys.,209,565–578 (1981)
[35] Singh, O.M., Graham, A.B., Wood, G.C.: Eur. J. Biochem.,116,311–316 (1981)
[36] Burchell, B., Weatherill, P.: Methods Enzymol.,77,169–177 (1981)
[37] Tukey, R.H., Tephly, T.R.: Methods Enzymol.,77,177–188 (1981)
[38] Burchell, B.: Methods Enzymol.,77,188–192 (1981)
[39] Bock, K.W., Lilienblum, W., Pfeil, H.: FEBS Lett.,121,269–272 (1980)
[40] Weatherill, P.J., Burchell, B.: Biochem. J.,189,377–380 (1980)
[41] Burchell, B.: FEBS Lett.,111,131–135 (1980)
[42] Leakey, J.E.A.: Biochem. J.,175,1119–1124 (1978)
[43] Burchell, B.: Biochem. J.,173,749–757 (1978)
[44] Gorski, J.P., Kasper, C.B.: J. Biol. Chem.,252,1336–1343 (1977)
[45] Burchell, B.: FEBS Lett.,78,101–104 (1977)
[46] Burchell, B.: Biochem. J.,161,543–549 (1977)
[47] Zakim, D., Vessey, D.A.: Biochem. J.,157,667–673 (1976)
[48] Rao, M.L., Rao, G.S., Breuer, H.: Biochim. Biophys. Acta,452,89–100 (1976)
[49] Zakim, D., Vessey, D.A.: Biochem. Soc. Trans.,2,1165–1167 (1974)
[50] Graham, A.B., Pechey, D.T., Wood, G.C., Woodcock, B.G.: Biochem. Soc. Trans.,2,
     1167–1172 (1974)
[51] Mulder, G.J.: Biochem. Soc. Trans.,2,1172–1176 (1974)
[52] Illing, H.P.A., Dutton, G.J.: Biochem. J.,131,139–147 (1973)
[53] Vessey, D.A., Goldenberg, J., Zakim, D.: Biochim. Biophys. Acta,309,58–66 (1973)
[54] Zakim, D., Goldenberg, J., Vessey, D.A.: Biochemistry,12,4068–4074 (1973)
[55] Gregory, D.H., Strickland, R.D.: Biochim. Biophys. Acta,327,36–45 (1973)
[56] Winsnes, A.: Biochim. Biophys. Acta,284,394–405 (1972)
[57] Winsnes, A.: Biochim. Biophys. Acta,289,88–96 (1972)
[58] Heirwegh, K.P.M., van de Vijver, M., Fevery, J.: Biochem. J.,129,605–618 (1972)
[59] Schumacher, R., Rao, G.S., Rao, L.M., Breuer, H.: Hoppe-Seyler's Z. Physiol. Chem.,
     353,1784–1788 (1972)
[60] Rao, G.S., Schumacher, R., Rao, M.L., Breuer, H.: Hoppe-Seyler's Z. Physiol. Chem.,
     353,1789–1797 (1972)
[61] Götze, W., Grube, E., Rao, G.S., Rao, M.L., Breuer, H.: Hoppe-Seyler's Z. Physiol.
     Chem.,352,1223–1230 (1971)
[62] Styczynski, P., Green, M., Puig, J., Coffman, B., Tephly, T.: Mol. Pharmacol.,40,
     80–84 (1991)
[63] Clarke, D.J., Burchell, B., George, S.: Mar. Environ. Res.,24,105–109 (1988)
[64] Bock, K.W., Schirmer, G., Green, M.D., Tephly, T.R.: Biochem. Pharmacol.,37,
     1439–1443 (1988)

[65] Pesonen, M., Andersson, T.: Biochem. Pharmacol.,36,823–829 (1987)
[66] Coughtrie, M.W.H., Burchell, B., Bend, J.R.: Biochem. Pharmacol.,36,245–251 (1987)
[67] Irshaid, Y.M., Tephly, T.R.: Mol. Pharmacol.,31,27–34 (1987)
[68] Puig, J.F., Tephly, T.R.: Mol. Pharmacol.,30,558–565 (1986)
[69] Matern, S., Matern, H., Farthmann, E.H., Gerok, W.: J. Clin. Invest.,74,402–410 (1984)
[70] Aitio, A., Parkki, M.G.: Xenobiotica,11,97–102 (1981)
[71] Tukey, R., Tephly, T.: Life Sci.,27,2471–2476 (1980)
[72] Moulder, G.J.: Annu. Rev. Pharmacol. Toxicol.,32,25–49 (1992) (Review)
[73] Burchell, B., Coughtrie, M.W.H.: Pharmacol. Ther.,43,261–289 (1989) (Review)
[74] Tephly, T.R.: Chem. Res. Toxicol.,3,509–516 (1990) (Review)
[75] Dutton, G.J. in "Glucuronidation of Drugs and other Compounds", CRC Press, Boca Raton, Fl. (1980)
[76] Ritter, J.K., Crawford, J.M., Owens, I.S.: J. Biol. Chem.,266,1043–1047 (1991)
[77] Roy-Chowhury, J., Huang, T., Kesari, K., Lederstein, M., Arias, I.M., Roy-Chowhury, N.: J. Biol. Chem.,266,18294–18298 (1991)
[78] Matern, H., Matern, S., Schelzig, C., Gerok, W.: FEBS Lett.,118,251–254 (1980)
[79] Hauser, S.C., Ziurys, J.C., Gollan, J.L.: J. Biol. Chem.,259,4527–4533 (1984)
[80] Bock, K.W., Josting, D., Lilienblum, W., Pfeil, H.: Eur. J. Biochem.,98,19–26 (1979)
[81] Tukey, R.H., Billings, R.E., Tephly, T.R.: Biochem. J.,171,659–663 (1978)
[82] Grube, E., Götze, W., Rao, G.S., Rao, M.L.: Hoppe-Seyler's Z. Physiol. Chem.,352, 1215–1222 (1971)
[83] Jansen, P.L.M., Mulder, G.J., Burchell, B., Bock, K.W.: Hepatology (Baltimore), 15,532–544 (1992)
[84] Wooster, R., Sutherland, L., Ebner, T., Clarke, D., Da Cruz, E., Silva, O., Burchell, B.: Biochem. J.,278,465–469 (1991)

1 NOMENCLATURE

EC number
2.4.1.18

Systematic name
1,4-alpha-D-Glucan:1,4-alpha-D-glucan 6-alpha-D-(1,4-alpha-D-glucano)-transferase

Recommended name
1,4-alpha-Glucan branching enzyme

Synonyms
Branching enzyme
Amylo-(1,4–1,6)-transglycosylase
Q-enzyme
Glycosyltransferase, alpha-glucan-branching
alpha-Glucan-branching glycosyltransferase
Amylose isomerase
Branching factor, enzymatic
Branching glycosyltransferase
Enzyme Q
Glucosan transglycosylase
Glycogen branching enzyme
Plant branching enzyme
alpha-1,4-Glucan:alpha-1,4-glucan-6-glycosyltransferase [2]
Starch branching enzyme [5]
alpha-1,4-Glucan:alpha-1,4-glucan 6-glycosyltransferase [17]
More (the recommended name requires a qualification depending on the product, glycogen or amylopectin, e.g. glycogen branching enzyme, amylo-pectin branching enzyme. The latter has frequently been termed Q-enzyme)

CAS Reg. No.
9001-97-2

---

2 REACTION AND SPECIFICITY

Catalysed reaction
Transfers a segment of a 1,4-alpha-D-glucan chain to a primary hydroxyl group in a similar glucan chain

Reaction type
Hexosyl group transfer

Enzyme Handbook © Springer-Verlag Berlin Heidelberg 1996
Duplication, reproduction and storage in data banks are only
allowed with the prior permission of the publishers

Natural substrates

Substrate spectrum
1 Amylose [5, 10–12, 17–19]
2 Pea starch [5]
3 Waxy maize starch [5]
4 Amylopectin [5, 8, 11, 13, 17–19]
5 Glycogen (from E. coli and rabbit liver) [17]
6 beta-Limit dextrin [17]

Product spectrum
1 Amylose containing alpha-1,6-glucosidic linkages (characterization of products, smallest chains transferred contain 5–7 glucose units [17]) [10, 17, 19]
2 ?
3 ?
4 Amylopectin with additional alpha-1,6-glucosidic linkages [13]
5 ?
6 ?

Inhibitor(s)
Oxidized glutathione [5]; $HgCl_2$ [7]; Citrate [7]; Urea (up to 2 M, reversible inhibition, denaturation above) [20], Maltoheptaose [20]; Maltohexaose [20]; Maltooctaose [20]; Maltononanose [20]; Mg^{2+} [21]; Mn^{2+} [21]; Ca^{2+} [21]

Cofactor(s)/prosthetic group(s)/activating agents

Metal compounds/salts
Citrate (0,15 M [20], activation [20, 21], inhibition [7]) [20, 21]; Potassium alpha-D-glucose 1-phosphate (0.05 M [20], activation [20, 21]) [20, 21]; SO_4^{2-} (activation) [21]; Cl^- (activation) [21]; PO_4^{3-} (activation) [21]; Borate (activation) [21]; Acetate (activation) [21]

Turnover number (min^{-1})
60000 (amylose) [1]; More [2]

Specific activity (U/mg)
More (assay method [3, 21]) [1–3, 10, 13, 14, 16, 17, 19, 21]

K_m-value (mM)
More (amylose: 0.018 mg/ml [7], 0.02 mg/ml [1], 0.29 mg/ml [12]) [1, 3, 7, 8, 12]

pH-optimum
6.0–6.2 [21]; 6.5–8.0 [7]; 6.8–7.4 (citrate buffer) [20]; 7.2–7.6 (glucose 1-phosphate buffer) [20]; 7.5 [3]; 8.0 [13]

pH-range

Temperature optimum (°C)
15–20 (isozyme BE IIb) [3]; 25 (isozyme BE IIa) [3]; 27 [13]; 30 (Oryza sativa) [12]; 30–35 [20]; 33 (isozyme BE I) [3]; 36 (Zea mays) [12]

Temperature range (°C)

3 ENZYME STRUCTURE

Molecular weight
103000 (Solanum tuberosum, native PAGE) [1]
92000–103000 (rabbit, sucrose density centrifugation) [14, 20]
90000 (Zea mays) [7]
85000 (Oryza sativa, isozyme QE II [4], Solanum tuberosum, sucrose density centrifugation [18]) [4, 18]
80000 (Oryza sativa, isozyme QE I [4], Neurospora crassa, native PAGE [13]) [4, 13]
70000–90000 (Zea mays) [8]
64000 (Solanum tuberosum) [5]
60000 (rabbit, gel filtration) [16]

Subunits
Monomer (1 × 103000, Solanum tuberosum, SDS-PAGE [1], 1 × 85000, Solanum tuberosum, SDS-PAGE [18], 1 × 70000–90000, Zea mays, SDS-PAGE [8], 1 × 80000, Neurospora crassa, SDS-PAGE [13], 1 × 71000–77000, rabbit, SDS-PAGE [14, 16]) [1, 8, 13, 14, 16, 18]

Glycoprotein/Lipoprotein
No glycoprotein [14]

4 ISOLATION/PREPARATION

Source organism
Solanum tuberosum (potato) [1, 5, 18, 19]; Saccharomyces cerevisiae [2]; Zea mays (corn) [3, 7, 8, 10]; Oryza sativa (rice) [4, 12]; Gossypium sp. (cotton) [6]; Ricinum communis (castor bean) [9]; Neurospora crassa [13]; Sorghum bicolor [11]; Rabbit [14, 16, 20, 21]; E. coli [15, 17]

Source tissue
Tubers [1, 5, 18, 19]; Endosperm [4, 7, 9, 12]; Kernels (developing [8]) [8, 10]; Leaf [6, 10]; Mycelia [13]; Liver [14]; Skeletal muscle [15, 16, 20, 21]

Localization in source
Soluble [10, 18, 19]; Particle-bound [18]

Purification
Solanum tuberosum [1, 5, 18, 19]; Oryza sativa (2 isozymes) [4]; Gossypium sp. [6]; Zea mays (2 isozymes [8]) [8, 10]; Sorghum bicolor (4 isozymes) [11]; Neurospora crassa [13]; Rabbit [14, 16, 20, 21]; E. coli [17]

Crystallization
–

Cloned
–

Renatured
–

5 STABILITY

pH
5.0 (irreversible inactivation below) [1]; 7.3–8.1 [13]

Temperature (°C)
35 (stable below) [13]

Oxidation

Organic solvent

General stability information
Urea, 2 M irreversibly inactivates [20]; Unstable in diluted solutions [21]

Storage
–85°C or 4°C, several months [17]; –80°C, presence of SH-group reducing agents [5]; –20°C, 3 months [1]; –20°C, 50 mM glycylglycine buffer, pH 8.0, 5 mM mercaptoethanol, 1 mM EDTA, 0.02% NaN_3, 25% glycerol [13]; 4°C, 10 mM Tris/HCl buffer, pH 7.0, 15 mM mercaptoethanol, several months [14]; 0–4°C, 50 mM Tris/HCl buffer, pH 7.5, at least 1 month [16]

6 CROSSREFERENCES TO STRUCTURE DATABANKS

PIR/MIPS code
PIR3:S23856 (Bacillus caldolyticus); PIR3:S18599 (Bacillus stearothermophilus); PIR3:S31839 (Bacillus stearothermophilus (fragment)); PIR3:S36624 (Bacillus subtilis); PIR3:S40048 (Bacillus subtilis); PIR2:B41328 (Butyrivibrio fibrisolvens); PIR2:S28422 (cassava (fragments)); PIR1:NQECA (Escherichia coli); PIR2:JT0968 (maize (fragment)); PIR3:S34730 (potato); PIR3:S38732 (potato); PIR3:S38733 (potato); PIR3:S34218 (Streptomyces coelicolor); PIR2:JQ0550 (Synechococcus sp. (PCC 7942)); PIR2:S50448

(yeast (Saccharomyces cerevisiae)); PIR2:JX0243 (I precursor rice); PIR2:S34037 (sbe1 precursor rice); PIR2:S34039 (sbe2 rice (fragment)); PIR3:PQ0696 (Rice (fragment))

Brookhaven code

---

7 LITERATURE REFERENCES

[1] Blennow, A., Johansson, G.: Phytochemistry,30,437–444 (1991)
[2] Swinton, S.J., Woods, L.F.J.: Food Biotechnol.,3,197–202 (1989)
[3] Takeda, Y., Guan, H.P., Preiss, J.: Carbohydr. Res.,240,253–263 (1993)
[4] Nakamura, Y., Takeichi, T., Kawaguchi, K., Yamanouchi, H.: Physiol. Plant.,84, 329–335 (1992)
[5] Praznik, W., Rammesmayer, G., Spies, T.: Carbohydr. Res.,227,171–182 (1992)
[6] Chang, C.W.: Microchem. J.,28,363–373 (1983)
[7] Baba, T., Arai, Y., Ono, T., Munakata, A., Yamaguchi, H., Itoh, T.: Carbohydr. Res., 107,215–230 (1982)
[8] Boyer, C.D., Preiss, J.: Carbohydr. Res.,61,321–334 (1978)
[9] Goldner, W., Beevers, H.: Phytochemistry,28,1809–1812 (1989)
[10] Dang, P.L., Boyer, C.D.: Phytochemistry,27,1255–1259 (1988)
[11] Boyer, C.D.: Phytochemistry,24,15–18 (1985)
[12] Smyth, D.A.: Plant Sci.,57,1–8 (1988)
[13] Matsumoto, A., Kamata, T., Matsuda, K.: J. Biochem.,94,451–458 (1983)
[14] Zimmermann, C.P, Gold, A.M.: Biochemistry,22,3387–3392 (1983)
[15] Holmes, E., Boyer, C., Preiss, J.: J. Bacteriol.,151,1444–1453 (1982)
[16] Caudwell, F.B., Cohen, P.: Eur. J. Biochem.,109,391–394 (1980)
[17] Boyer, C., Preiss, J.: Biochemistry,16,3693–3699 (1977)
[18] Borovsky, D., Smith, E.E., Whelan, W.J.: Eur. J. Biochem.,59,615–625 (1975)
[19] Drummond, G.S., Smith, E.E., Whelan, W.J.: Eur. J. Biochem.,26,168–176 (1972)
[20] Gibson, W.B., Illingworth Brown, B., Brown, D.H.: Biochemistry,10,4253–4262 (1971)
[21] Illingworth Brown, B., Brown, D.H.: Methods Enzymol.,8,395–403 (1966)

1 NOMENCLATURE

EC number
2.4.1.19

Systematic name
1,4-alpha-D-Glucan 4-alpha-D-(1,4-alpha-D-glucano)-transferase (cyclizing)

Recommended name
Cyclomaltodextrin glucanotransferase

Synonyms
Glycosyltransferase, cyclodextrin
alpha-Cyclodextrin glucanotransferase
alpha-Cyclodextrin glycosyltransferase
beta-Cyclodextrin glucanotransferase
beta-Cyclodextrin glycosyltransferase
gamma-Cyclodextrin glycosyltransferase
Cyclodextrin glycosyltransferase
Cyclomaltodextrin glucotransferase
Cyclomaltodextrin glycosyltransferase
Konchizaimu
Bacillus macerans amylase
Cyclodextrin glucanotransferase
alpha-1,4-Glucan 4-glycosyltransferase, cyclizing [1]
BMA [1]
CGTase [2, 5]
Neutral-cyclodextrin glycosyltransferase [19]
More (cf. EC 2.4.1.25)

CAS Reg. No.
9030-09-5

---

2 REACTION AND SPECIFICITY

Catalysed reaction
Cyclizes part of a 1,4-alpha-D-glucan chain by formation of a 1,4-alpha-D-glucosidic bond

Reaction type
Hexosyl group transfer

Natural substrates
Starch [12]

Substrate spectrum

1 Starch (and derivatives: soluble starch [14, 21, 24], corn starch [24], wheat starch [24], waxy maize starch [24], glucidex 2B [24], maltohep-taose [24], maltohexaose [24], potato starch [14, 17, 19, 20, 24], amylo-pectin (65–70% converted to cyclodextrin [19]) [12, 19], maltooligo-saccharides [2], amylose (85–90% converted to cyclodextrin [19]) [19, 23], glycogen (55–60% converted to cyclodextrin [19]) [19, 23, 24], amylopectin beta-limit dextrin (45–50% converted to cyclodextrin) [19], maltotriose (20–25% converted to cyclodextrin [19]) [19, 24], maltose (10–15% converted to cyclodextrin [19]) [19, 23, 24]) [1–27]
2 Cyclomaltohexose + D-lactose (transglycosylation) [10]
3 alpha-Cyclodextrin + ascorbic acid [23]
4 p-Nitrophenyl-(glucose)$_{3-8}$ + ? [25]

Product spectrum

1 Cyclodextrins (Schardinger dextrins of various size: 6,7,8 etc. glucose units, alpha-, beta- and gamma-cyclodextrins in the proportion 1:67:1.6 [4], Bacillus macerans IFO 3490 enzyme produces alpha-cyclodextrin as the major hydrolysis product [11], Bacillus stearothermophilus NO2 enzyme produces alpha- and beta-cyclodextrins [11], gamma-cyclodextrin + beta-cyclodextrin, the amount of beta-cyclodextrin increases gradually with time from the start of the reaction [14], addition of ethanol enhances formation of: gamma-cyclodextrins [14], beta-cyclodextrins [17], enzyme forms mainly beta-cyclodextrins with a small amount of alpha-cyclodex-trins and gamma cyclodextrins [17], Triton X-100 enhances specificity and yield of beta-cyclodextrins [17], Bacillus sp. No. 5 enzyme mainly produces cycloheptaamylose [18], Bacillus macerans enzyme mainly produces cyclohexaamylose [18], Schardinger beta-dextrin is preferen-tially produced from starch, and alpha- or gamma-dextrin is gradually formed after prolonged incubation [19], ratio of alpha-:beta-:gamma-cy-clodextrin produced after 20 h at 50°C is 8.1:8.9:1.0 [20], ratio of alpha-, beta- and gamma-cyclodextrin produced is 1:7:2 in the initial phase of the reaction and 3:3:1 at the equilibrium [24]) [1–27]
2 ?
3 ?
4 p-Nitrophenylglucosides (p-nitrophenyl-(glucose)$_1$: main product when chain length of substrate is 4 glucose or less, p-nitrophenyl-(glucose)$_2$: when substrate chain length is 5 or more glucose residues) [25]

Inhibitor(s)

Tetranitromethane [13]; Zn^{2+} [14, 22]; Fe^{2+} [14, 22]; Cu^{2+} [14]; Mo^{6+} [14]; Hg^{2+} (HgCl$_2$ [17], HgSO$_4$ [20], relatively stable to Hg^{2+} [14]) [17, 20]; Cyclo-dextrin (slight) [17]; Urea [20]; Guanidinium chloride [20]; SDS [20]; Co^{2+} [22]; Mg^{2+} [22]; Al$_2$(SO$_4$)$_3$ [20]; CrO$_3$ [20]; CuSO$_4$ [20]; (NH$_4$)$_6$Mo$_7$O$_{24}$ (weak) [20]; p-Nitrophenyl-alpha-D-glucose [25]; p-Nitrophenyl-beta-D-glucoside

[25]; Amygdalin [25]; Palatinose [25]; Dodecyl-beta-D-maltoside [25];
Deoxynojirimycin (noncompetitive) [25]; Acarbose (uncompetitive) [25];
Heptyl thioglucoside [25]; Helicin [25]; Cellobiose [25]; Maltitol [25]; Methyl-alpha-D-glucoside [25]; Methyl-beta-D-glucoside [25]; Sucrose (weak)
[25]; Glucose (competitive) [25]; Maltose (competitive) [25]; Salicin (competitive) [25]; 3-O-Methylglucose [25]; Mannose (weak) [25]

Cofactor(s)/prosthetic group(s)/activating agents
Triton X-100 (enhances specificity and yield of beta-cyclodextrins) [17]

Metal compounds/salts
Ca^{2+} (Bacillus coagulans enzyme contains 2 mol of Ca^{2+} per mol of enzyme,
Bacillus macerans enzyme contains 0.88 mol per mol of enzyme) [20]

Turnover number (min^{-1})
99.6 (p-nitrophenyl-(glucose)$_3$) [25]; 50 (p-nitrophenyl-(glucose)$_4$, p-nit-
rophenyl-(glucose)$_5$, p-nitrophenyl-(glucose)$_6$, p-nitrophenyl-(glucose)$_7$) [25];
15 (p-nitrophenyl-(glucose)$_8$) [25]

Specific activity (U/mg)
281.0 [19]; 904 [20]; More [1, 14, 17, 18, 21, 22, 26]

K_m-value (mM)
0.08 (beta-cyclodextrin, cyclodextrin ring opening, Bacillus macerans) [20];
0.25 (gamma-cyclodextrin, cyclodextrin ring opening, Bacillus macerans)
[20]; 0.51 (beta-cyclodextrin, cyclodextrin ring opening, Bacillus coagulans)
[20]; 0.73 (alpha-cyclodextrin, cyclodextrin ring opening, Bacillus coagulans) [20]; 0.83 (beta-cyclodextrin, glycosyl residue transfer action to sucrose) [19]; 0.88 (gamma-cyclodextrin, cyclodextrin ring opening, Bacillus
coagulans) [20]; 2.32 (alpha-cyclodextrin, cyclodextrin-ring opening, Bacillus macerans) [20]; 10 (alpha-cyclodextrin, glycosyl residue transfer action
to sucrose) [19]; 16 (alpha-cyclodextrin (+ ascorbic acid)) [23]; More (K_m:
3.33 mg/ml soluble starch [1], 5.7 mg/ml starch [8], 1.77 g/l starch, Bacillus
coagulans [20], 2.5 g/l, Bacillus macerans [20]) [1, 8, 20, 24, 25]

pH-optimum
5.0 (starch-dextrinizing activity, 2 optima: 5 and 9 [22], L-ascorbic acid-
2-O-alpha-D-glucoside formation [23]) [22, 23]; 5.0–6.0 [17]; 5–5.7 [18];
5.4–5.8 [8]; 5.5 (immobilized [26]) [21, 24, 26]; 5.9 (soluble) [26]; 6.0 [20];
6.1–6.2 [1]; 6.5–8.0 (cyclodextrin-forming activity) [22]; 6.5–8.5 [4]; 7 [19];
7–10 [14]; 9 (starch-dextrinizing activity, 2 optima: 5 and 9) [22]

pH-range
4–11 (4: about 20% of activity maximum, 11: about 70% of activity maximum, starch-dextrinizing activity) [22]; 5.0–6.7 (more than 80% of activity
maximum at pH 5.0 and 6.7) [3]; 5–8.5 (5: about 35% of activity maximum,
8.5: about 70% of activity maximum) [19]; 5.5–9 (about 30% of activity maximum at pH 5.5 and 9, cyclodextrin-forming activity) [22]

Temperature optimum (°C)
40–60 (immobilized enzyme) [26]; 50 (enzyme immobilized on Amberlite
[5], soluble [15]) [5, 15, 19]; 55 (starch-dextrinizing activity [14], soluble [5],
immobilized [15]) [5, 14, 15, 18]; 60 (cyclodextrin-forming activity [14, 22],
soluble [26]) [8, 14, 17, 21, 22, 26]; 70 (starch-dextrinizing activity [22],
L-ascorbic acid-2-O-alpha-D-glucoside formation [23]) [22, 23]; 90–95 [3]

Temperature range (°C)
35–65 (about 60% of activity maximum at 35°C and 65°C) [19]; 40–70
(40°C: about 55% of activity maximum, 70°C: about 65% of activity maximum) [21]

---

3 ENZYME STRUCTURE

Molecular weight
58000 (Bacillus sp. ATCC 21783, size exclusion chromatography) [27]
65000 (Bacillus coagulans, sedimentation equilibrium) [20]
67000 (Bacillus macerans) [8]
68000 (Bacillus autolyticus, gel chromatography [17], Bacillus macerans,
sedimentation equilibrium [20]) [17, 20]
74000 (Bacillus sp. AL-6, gel filtration) [14]
103000 (Bacillus circulans, gel filtration) [21]
139300 (Bacillus macerans, calculation from sedimentation and diffusion
data) [1]

Subunits
? ($x \times 72000$ [6, 7], Bacillus sp. [6], Bacillus sp. IT25 [7], SDS-PAGE [6, 7],
$x \times 85000$–88000, Bacillus sp. ATCC 21783, SDS-PAGE [19], $x \times 68000$,
Bacillus sp. HA3–3–2, SDS-PAGE [22]) [6, 7, 19, 22]
Monomer (1×90000, Bacillus autolyticus, SDS-PAGE [17], 1×67000, Bacil-
lus macerans [8], 1×74000, Bacillus sp. AL-6, SDS-PAGE [14], 1×65000,
Bacillus coagulans, SDS-PAGE [20], 1×67000, Bacillus macerans,
SDS-PAGE [20], 1×78000, Bacillus circulans, SDS-PAGE [24], 1×70500,
Bacillus sp. ATCC 21783, SDS-PAGE [27]) [8, 14, 17, 20, 24, 27]

Glycoprotein/Lipoprotein
–

---

4 ISOLATION/PREPARATION

Source organism
Klebsiella pneumonia (M5 al) [12]; Bacillus autolyticus (11149) [17]; Bacillus
circulans (E 192 [13, 24], C31 [21]) [13, 21, 24, 25]; Bacillus stearothermophilus (construction of mutant CGTase for better understanding of the amyl-
olytive and cyclization mechanism [16], NO2 [11, 16]) [2, 10, 11, 16, 23];

Thermoanaerobacter sp. [3]; Bacillus lentus [4]; Bacillus coagulans [20];
Bacillus sp. (HA3–3–2 [22], IT25 [7], strain No.5 [18], ATCC 21783 [15, 19,
27], AL-6 [14]) [6, 7, 14, 15, 18, 19, 22, 27]; Bacillus macerans (IFO 3490
[11]) [1, 5, 8, 9, 11, 18, 20, 26]

Source tissue
Culture medium [12, 14, 18, 22]

Localization in source
Extracellular [7, 8, 12]

Purification
Bacillus circulans (E 192, 2 isoenzymes [24], C31 [21]) [21, 24]; Bacillus autolyticus [17]; Bacillus coagulans [20]; Bacillus macerans [1, 8, 9]; Bacillus
sp. (IT25 [7], AL-6 [14], No. 5, 2 isoenzymes: Fr. 1 and Fr. 2 [18], HA3–3–2
[22], ATCC 21783 [19, 27]) [6, 7, 14, 18, 19, 22, 27]; Klebsiella pneumoniae
(M5 al) [12]

Crystallization
[7, 9]

Cloned
(gene transferred to a Bacillus host, enabling large-scale production in
acceptable yields [3], cloning and construction of chimeric genes to analyze the cyclization characteristics [11]) [3, 11]

Renatured
–

---

5 STABILITY

pH
5.0–9.0 (40°C, 12 h, stable) [17]; 5–8 (40°C, 3 h, stable) [14]; 5.5–9.0
(stable [21], 40°C, 1 h, stable [20]) [20, 21]; 6.0–8.0 (55°C, 20 min, immobilized enzyme stable [15], 60°C, 30 min, stable [19]) [15, 19]; 6–11 (stable in
the range) [22]; 7.0–10.0 (Bacillus sp. No. 5 enzyme stable) [18]; 8.0–10.0
(Bacillus macerans enzyme stable) [18]

Temperature (°C)
More (glycerol, sorbitol and sucrose effect thermostability [2], 1,4-dioxane
and n-octane increase thermostability [2], Ca^{2+} increases thermal stability
(of Bacillus coagulans enzyme, Bacillus macerans enzyme is not effected
[20]) [4, 15, 19, 20], thermal stability of CGTase immobilized on chitosan is
increased from 50°C to 55°C [5]) [2, 4, 5, 15, 19, 20]; 40 (pH 7.5, 30 min,
stable [14], pH 6.0, 1 h, stable below [17], pH 6, 20 min, 25% loss of activity
[24]) [14, 17, 24]; 45 (rapid inactivation above) [24]; 50 (pH 7.6, 20 min,
stable up to [21], pH 6.0, 30 min, stable [8], pH 6, 20 min, 90% loss of activity

[24]) [8, 21, 24]; 55 (pH 7.0, 10 min, Bacillus sp. No. 5 enzymes Fr. 1 and
Fr. 2 stable) [18]; 60 (pH 7, 30 min, stable up to [19], pH 7.0, 10 min, Bacillus macerans enzyme stable [18], pH 6.0, 30 min, 80% loss of activity [8],
pH 6, 20 min, 99% loss of activity [24]) [8, 18, 19, 24]; 65 (pH 8.0, 20 min,
stable up to) [15]; 70 (10 min, presence of $CaCl_2$, stable) [20]; 100 (stable
above in presence of starch) [3]

Oxidation

Organic solvent

General stability information

Prolonged digestion with trypsin has no effect [8]; No inactivation by treatment with 8 M urea, 2 M guanidium chloride or 0.2% SDS, at 40°C, 10 min,
and dilution with 0.01 M sodium acetate buffer, pH 6.0 to one hundred times
[20]; Substrate, product and/or calcium ions, optimum concentration 5 mM,
protects against heat denaturation [24]; Conformational stability of immobilized enzyme is markedly enhanced in a wide temperature and pH range
[26]

Storage

4°C, pH 5.0–10.5, 1 month [20]

---

6 CROSSREFERENCES TO STRUCTURE DATABANKS

PIR/MIPS code

PIR1:ALBSGC (precursor Bacillus circulans); PIR1:ALBSMX (precursor
Bacillus licheniformis); PIR1:ALBSGR (precursor Bacillus macerans);
PIR1:ALBSXR (precursor Bacillus macerans); PIR1:ALBSX1 (precursor Ba-
cillus sp. (strain 1–1)); PIR1:ALBSG1 (precursor Bacillus sp. (strain 1011));
PIR1:ALBSG7 (precursor Bacillus sp. (strain 17–1)); PIR1:ALBSG6 (precur-
sor Bacillus sp. (strain 6.6.3)); PIR1:ALBSG3 (precursor Bacillus sp. (strain
no. 38–2)); PIR1:ALBSXF (precursor Bacillus stearothermophilus);
PIR1:ALKBG (precursor Klebsiella pneumoniae)

Brookhaven code

1CDG ((Bacillus Circulans, strain 251)); 1CGT ((Bacillus Circulans, strain
8)); 1CGU ((Bacillus Circulans, strain 8)); 1CGV ((Bacillus Circulans) strain:
251); 1CGW ((Bacillus Circulans) strain 251); 1CGX ((Bacillus Circulans)
strain 251); 1CGY ((Bacillus Circulans) strain 251); 1CXG ((Bacillus Circulans, strain 251)); 1CYG ((Bacillus Stearothermophilus))

7 LITERATURE REFERENCES

[1] DePinto, J.A., Campbell, L.L.: Biochemistry,7,114–120 (1968)
[2] Ahn, J.H., Hwang, J.B., Kim, S.H.: Sanop Misaengmul Hakhoechi,19,368–371 (1991)
[3] Norman, B.E., Joergensen, S.T.: Denpun Kagaku,39,101–108 (1992)
[4] Sabioni, J.G., Park, Y.K.: Starch Staerke,44,225–229 (1992)
[5] Lee, S.H., Shin, H.D., Lee, Y.H.: J. Microbiol. Biotechnol.,1,54–62 (1991)
[6] Prema, P., Sreedharan V.P., Raja, K.C.M., Ramakrishna, S.V.in "Minutes Int. Symp. Cyclodextrins",5th Ed. (Duchene, D., Ed.) ,46–49, Sante, Paris Fr. (1990)
[7] Aoki, H., Yao, D., Misawa, M.: Prog. Biotechnol.,3 (Ind. Polysaccharides) ,81–93 (1987)
[8] Stavn, A., Granum, P.E.: Carbohydr. Res.,75,243–250 (1979)
[9] Kobayashi, S., Kainuma, K., Suzuki, S.: Carbohydr. Res.,61,229–238 (1978)
[10] Shibuya, T., Miwa, Y., Nakano, M., Yamauchi, T., Chaen, H., Sakai, S., Kurimoto, M.: Biosci. Biotechnol. Biochem.,57,56–60 (1993)
[11] Fujiwara, S., Kakihara, H., Woo, K.B., Lejeune, A., Kanemoto, M., Sakaguchi, K., Imanaka, T.: Appl. Environ. Microbiol.,58,4016–4025 (1992)
[12] Bender, H.: Arch. Microbiol.,111,271–282 (1977)
[13] Villette, J.R., Helbecque, N., Albani, J.R., Sicard, P.J., Bouquelet, J.-L.: Biotechnol. Appl. Biochem.,17,205–216 (1993)
[14] Fujita, Y., Tsubouchi, H., Inagi, Y., Tomita, K., Ozaki, A., Nakanishi, K.: J. Ferment. Bioeng.,70,150–154 (1990)
[15] Kato, T., Horikoshi, K.: Biotechnol. Bioeng.,26,595–598 (1984)
[16] Fujiwara, S., Kakihara, H., Sakaguchi, K., Imanaka, T.: J. Bacteriol.,174,7478–7481 (1992)
[17] Tomita, K., Kaneda, M., Kawamura, K., Nakanishi, K.: J. Ferment. Bioeng.,75,89–92 (1993)
[18] Kitahata, S., Tsuyama, N., Okada, S.: Agric. Biol. Chem.,38,387–393 (1974)
[19] Nakamura, N., Horikoshi, K.: Agric. Biol. Chem.,40,1785–1791 (1976)
[20] Akimaru, K., Yagi, T., Yamamoto, S.: J. Ferment. Bioeng.,71,322–328 (1991)
[21] Pongsawasdi, P., Yagisawa, M.: Agric. Biol. Chem.,52,1099–1103 (1988)
[22] Nomoto, M., Chen, C.-C., Sheu, D.-C.: Agric. Biol. Chem.,50,2701–2707 (1986)
[23] Tanaka, M., Muto, N., Yamamoto, I.: Biochim. Biophys. Acta,1078,127–132 (1991)
[24] Bovetto, L.J., Backer, D.P., Villette, J.R., Sicard, P.J., Bouquelet, S.J.-L.: Biotechnol. Appl. Biochem.,15,48–58 (1992)
[25] Bovetto, L.J., Villette, J.R., Fontaine, I.F., Sicard, P.J., Bouquelet, S.J.-L.: Biotechnol. Appl. Biochem.,15,59–68 (1992)
[26] Ivony, K., Szajani, B., Seres, G.: J. Appl. Biochem.,5,158–164 (1983)
[27] Mäkelä, M., Mattsson, P., Schinina, M.E., Korpela, T.: Biotechnol. Appl. Biochem., 10,414–427 (1988)

1 NOMENCLATURE

EC number
2.4.1.20

Systematic name
Cellobiose:orthophosphate alpha-D-glucosyltransferase

Recommended name
Cellobiose phosphorylase

Synonyms
Phosphorylase, cellobiose

CAS Reg. No.
9030-20-0

---

2 REACTION AND SPECIFICITY

Catalysed reaction
Cellobiose + phosphate →
→ alpha-D-glucose 1-phosphate + D-glucose (mechanism [6, 7])

Reaction type
Hexosyl group transfer

Natural substrates

Substrate spectrum
1 Cellobiose + phosphate (r, beta-D- better substrate than alpha-D-cellobi-ose [6], no substrates are sucrose, laminaribiose, kojibiose, cellodextrin [5, 8], salicin [1], lactose, maltose, gentiobiose [1, 5, 8]) [1, 3, 5–8]
2 Cellobiose + arsenate [1]
3 Celtrobiose + phosphate (i.e. 4-O-beta-D-glucopyranosyl-D-altrose, poor substrate) [3]
4 alpha-D-Glucose 1-phosphate + D-glucose (r, non-specific, glucosyl acceptors are D-glucose, 2-deoxyglucose (poor [6]) or 6-deoxyglucose (51% as effective as D-glucose [6]) [3, 5, 6], D-glucosamine (9% as effec-tive as D-glucose [6]) [1, 3, 5, 6], D-mannose (poor [6]) [2, 3, 5, 6], D-altrose (poor [3]), L-galactose (poor [3]), L-fucose, D-arabinose (not L-) [3, 5], D-xylose (not Ruminococcus flavefaciens [2], not L-) [1–3, 5, 6], no acceptors are D-fructose, D-allose, D-lyxose, D-ribose [3, 5, 6], 2-deoxyri-bose, D-glucitol, D-gluconate, 3-O-methyl-D-glucose [3, 5], D/L-arabinose, L-xylose, D-galactose, L-idose, D-glucuronic acid [6], alpha- or beta-meth-yl-D-glucose [3, 5], myo-inositol, L-sorbose [3, 6], N-acetyl-D-glucosamine

[3, 5, 6], alpha- or beta-methyl-D-glucoside, 1,5-anhydro-D-glucitol, D-glucono-delta-lactone, D-mannitol, D-glucitol [6], no glucosyl donor: beta-D-glucose 1-phosphate [5]) [1–3, 5, 6]

Product spectrum
1 alpha-D-Glucose 1-phosphate + D-glucose [1, 3, 5]
2 ?
3 ?
4 Cellobiose + phosphate [1–3, 5]

Inhibitor(s)
F$^-$ (phosphorolysis) [1]; Phlorizin (phosphorolysis) [1]; D-Glucose (phosphorolysis [1, 3], substrate inhibition [6], product inhibition [7]) [1, 3, 6, 7]; 6-Deoxyglucose (substrate inhibition) [6]; D-Glucosamine (substrate inhibition) [6]; Glucose 1-phosphate (product inhibition) [7]; Nojirimycin (i.e. D-glycopiperidinose, competitive to cellobiose) [5, 8]; alpha-Oxogluconate (weak) [5, 8]; Glucono-delta-lactone (weak) [5, 8]; Methyl-alpha-D-glucoside (weak) [5, 8]; 6-Phosphogluconate (weak) [5, 8]; p-Nitrophenyl-beta-glucopyranoside (weak) [5, 8]; N-Ethylmaleimide [5, 8]; PCMB [5, 8]; p-Substituted mercuribenzoate (0.02 mM, partially reversible by cysteine) [1]

Cofactor(s)/prosthetic group(s)/activating agents
More (no pyridoxal 5-phosphate required) [5, 8]

Metal compounds/salts
Mg^{2+} (requirement) [5, 8]

Turnover number (min^{-1})

Specific activity (U/mg)
27.4 [5, 8]; 33.3 [3]; 40.8 [7]

K$_m$-value (mM)
More (kinetic study (disaccharide synthesis [6], phosphorolysis [7])) [1, 6, 7]; 0.77 (phosphate) [5, 8]; 1.25 (cellobiose) [5, 8]; 2.1 (disaccharide synthesis, glucose 1-phosphate [3], D-glucose [6]) [3]; 2.9 (phosphate) [3]; 9.2 (disaccharide synthesis, 6-deoxyglucose) [2, 3]; 9.5 (disaccharide synthesis, D-glucosamine) [3]; 13 (disaccharide synthesis, D-glucosamine) [6]; 24 (disaccharide synthesis, 6-deoxy-D-glucose) [6]; 35 (disaccharide synthesis, D-xylose) [3]; 73 (disaccharide synthesis, 2-deoxy-D-glucose) [3]; 84 (disaccharide synthesis, D-xylose) [6]; 85 (disaccharide synthesis, mannose) [3]; 115 (disaccharide synthesis, mannose) [6]; 160 (disaccharide synthesis, L-fucose) [3]; 168 (disaccharide synthesis, 2-deoxy-D-glucose) [6]

pH-optimum
More (pI: 6.0) [7]; 6.5 (disaccharide synthesis, 2-deoxyglucose or D-xylose as acceptors) [3]; 7.0 [1, 7]; 7.5 (disaccharide synthesis, D-glucosamine) [3]; 7.6 [5, 8]

pH-range
4.6–8.1 (detectable activity in this range) [1]; 4.8–8.2 (about half-maximal activity at pH 4.8 and 8.2) [3]

Temperature optimum (°C)
20 (assay at) [5]; 37 (assay at) [3, 6, 7]; 39 (assay at) [4]

Temperature range (°C)

---

3 ENZYME STRUCTURE

Molecular weight
180000 (Cellvibrio gilvus, gel filtration) [7]

Subunits
Dimer (2 × 85000, Cellvibrio gilvus, SDS-PAGE) [7]

Glycoprotein/Lipoprotein
–

---

4 ISOLATION/PREPARATION

Source organism
Clostridium thermocellum [1–3]; Ruminococcus flavefaciens [1, 2, 4]; Cellvibrio gilvus [1, 5–8]; Cellulomonas sp. [9]; Fomes annosus [10]

Source tissue
Cell [1–9]

Localization in source
Cytoplasm [8]

Purification
Clostridium thermocellum (partial) [1, 3]; Ruminococcus flavefaciens (partial) [4]; Cellvibrio gilvus [5, 7, 8]

Crystallization
–

Cloned

Renatured
–

5 STABILITY

pH
   7.6 (below and above, rapid decrease of activity) [5]

Temperature (°C)
   40 (inactivation above) [5, 8]; 60 (complete inactivation within 10 min) [5]

Oxidation

Organic solvent
   Dithiothreitol stabilizes [5, 8]; 2-Mercaptoethanol stabilizes [5, 8]

General stability information

Storage
   –20°C, appreciable loss of activity within 2 weeks, DTT or 2-mercaptoethanol stabilizes [5]

---

6 CROSSREFERENCES TO STRUCTURE DATABANKS

PIR/MIPS code

Brookhaven code

---

7 LITERATURE REFERENCES

[1] Doudoroff, M. in "The Enzymes",2nd Ed. (Boyer, P.D., Lardy, H., Myrbäck, K., Eds.) 5,229–236 Academic Press, New York (1961), (Review)
[2] Mieyal, J.J., Abeles, R.H. in "Enzymes",3rd Ed. (Boyer, P.D., Ed.) 7,515–532 Academic Press, New York (1972), (Review)
[3] Alexander, J.K.: J. Biol. Chem.,243,2899–2904 (1968)
[4] Ayers, W.: J. Biol. Chem.,234,2819–2822 (1959)
[5] Sasaki, T.: Methods Enzymol.,160,468–472 (1988) (Review)
[6] Kitaoka, M., Sasaki, T., Taniguchi, H.: J. Biochem.,112,40–44 (1992)
[7] Kitaoka, M., Sasaki, T., Taniguchi, H.: Biosci. Biotechnol. Biochem.,56,652–655 (1992)
[8] Sasaki, T., Tanaka, T., Nakagawa, S., Kainuma, K.: Biochem. J.,209,803–807 (1983)
[9] Schimz, K.-L., Broll, B., John, B.: Arch. Microbiol.,135,241–245 (1983)
[10] Hutterman, A., Volger, C.: Nature New Biol.,245,64ff. (1973)

1 NOMENCLATURE

EC number
  2.4.1.21

Systematic name
  ADPglucose:1,4-alpha-D-glucan 4-alpha-D-glucosyltransferase

Recommended name
  Starch synthase

Synonyms
  ADPglucose-starch glucosyltransferase
  Glucosyltransferase, adenosine diphosphoglucose-starch
  Adenosine diphosphate glucose-starch glucosyltransferase
  Adenosine diphosphoglucose-starch glucosyltransferase
  ADP-glucose starch synthase
  ADP-glucose synthase
  ADP-glucose transglucosylase
  ADP-glucose-starch glucosyltransferase
  ADPG starch synthetase
  ADPG-starch glucosyltransferase
  Starch synthetase
  More (the recommended name varies according to the source of the enzyme and the nature of its synthetic product, e.g. starch synthase, bacterial glycogen synthase. Similar to EC 2.4.1.11 but the preferred or mandatory nucleoside diphosphate sugar substrate is ADPglucose. The entry covers starch and glycogen synthases utilizing ADPglucose)

CAS Reg. No.
  9030-10-8

---

2 REACTION AND SPECIFICITY

Catalysed reaction
  ADPglucose + (1,4-alpha-D-glucosyl)$_n$ →
  → ADP + (1,4-alpha-D-glucosyl)$_{n+1}$

Reaction type
  Hexosyl group transfer

Natural substrates
  ADPglucose + ADPglucose (starch biosynthesis) [7, 12]

Substrate spectrum
1 ADPglucose + alpha-1,4-polyglucan (r [20], citrate independent enzyme requires a glucan primer e.g. rabbit liver glycogen [2–5, 7, 8, 10, 11, 20, 26, 28], amylose [2, 20, 26], amylopectin [2–4, 8–10, 12, 13, 26, 29], oyster glycogen [3–5, 8, 10, 26], bovine liver glycogen [5], starch [3, 6, 23, 29], maltose [20], maltotriose [26], beta-limit dextrin [20], E. coli glycogen [26], phytoglycogen [29], maltotetrose [29], maltopentose [29], specific for ADPglucose [24]) [2–13, 19, 20, 23–26, 28, 29]
2 ADPglucose + ADPglucose (unprimed activity in presence of citrate) [7, 9, 10, 12, 13, 19]
3 dADPglucose + alpha-1,4-polyglucan (8% of the activity with ADPglucose) [29]
4 More (glucose cannot function as acceptor [20], no activity with UDPglucose [1], reduced activity with UDPglucose or GDPglucose [13]) [1, 13, 20]

Product spectrum
1 ADP + alpha-1,4-glucosyl-polyglucan [2, 6, 12]
2 ADP + alpha-1,4-glucosyl-polyglucan [12]
3 dADP + alpha-1,4-glucosyl-polyglucan
4 ?

Inhibitor(s)
Cu^{2+} [3]; ATP [3, 29]; ADP [3, 28, 29]; Calcium calmodulin [16]; NH_4^+ (above 20 mM) [18]; Cs^+ (above 20 mM) [18]; Na^+ (above 20 mM) [18]; Li^+ (above 20 mM) [18]; 1,5-Gluconolactone [20]; Ethanol [27]; Methanol [27]; $MnCl_2$ [29]; AMP [29]; Adenosine [29]; dADPglucose [29]; p-Chloromercuribenzoate [29]; $HgCl_2$ [29]; $ZnSO_4$ [29]; $MgCl_2$ [29]

Cofactor(s)/prosthetic group(s)/activating agents
Glutathione (activation) [28]; Bovine plasma albumin (activation [28], activation of unprimed reaction [24]) [24, 28]; Isopropanol (above 70% v/v, activation) [27]; n-Propanol (85% v/v, activation) [27]; beta-Lactoglobulin (activation of unprimed reaction) [24]; Ovomucoid (activation of unprimed reaction) [24]; Myoglobin (activation of unprimed reaction) [24]; Hemoglobin (activation of unprimed reaction) [24]; 1,4-alpha-Glucan branching enzyme (EC 2.4.1.18, stimulation of unprimed reaction) [24]

Metal compounds/salts
Citrate (activation of unprimed synthase) [4, 7, 8, 12, 13, 22, 24]; K^+ (slight activation [3], 50–100 mM, activation [18], activation [28]) [3, 18, 28]; Mg^{2+} (slight activation) [3]; Rb^+ (50–100 mM, activation) [18]; NH_4^+ (below 20 mM, slight activation, inhibition above) [18]; Cs^+ (below 20 mM, slight activation, inhibition above) [18]; Na^+ (below 20 mM, slight activation, inhibition above) [18]; Li^+ (below 20 mM, slight activation, inhibition above) [18]

Turnover number (min^{-1})

Specific activity (U/mg)
15.87 [1]; 2.3 (primed synthase) [7]; 1.1 (unprimed synthase) [7]; More [2, 8, 12, 21, 22, 25, 26]

K_m-value (mM)
0.033 (ADPglucose) [7]; 0.035 (ADPglucose) [13, 20]; 0.05–0.1 (ADPglucose) [8]; 0.05–0.14 (ADPglucose, granule-bound isozymes) [9]; 0.077 (ADPglucose) [22]; 0.1–0.12 (ADPglucose, soluble isozymes) [9]; 0.11–0.22 (ADPglucose) [12]; 0.29 (ADPglucose, synthase I, absence of citrate) [1]; 0.48 (ADPglucose, synthase II, absence of citrate) [1]; 0.51 (ADPglucose, synthase II, presence of citrate) [1]; 0.81 (ADPglucose) [2]; 0.97 (ADPglucose, synthase I, presence of citrate) [1]; 2.7 (UDPglucose) [13]; More (amylopectin: 0.19–0.79 mg/ml, value depends on isozyme and presence of citrate [1], 0.002–1.5 mg/ml, value depends on isozyme and presence of citrate [9], rabbit liver glycogen: 0.298 mg/ml [20], soluble potato amylose: 0.833 mg/ml [20], influence of citrate on values for primed and unprimed reaction [22]) [1, 9, 20, 22, 28, 29]

pH-optimum
6–9 [28]; 7.0–9.5 [20]; 7.5–8.5 [26]; 8–9 [2]; 8.3 [29]; 8.5 [3]

pH-range
6–9 [3]; 6.5–9.7 (less than half-maximal activity above and below) [29]

Temperature optimum (°C)
24–30 [29]; 30 (various isozymes) [26]; 37 (unprimed synthase activity of fraction III) [26]; 50 [3]

Temperature range (°C)
22–44 [26]

---

3 ENZYME STRUCTURE

Molecular weight
110000 (Oryza sativa, synthase I, gel filtration) [21]
92700 (E. coli, sucrose density centrifugation) [20]
69000 (Oryza sativa, synthase II, gel filtration) [21]

Subunits
Decamer (10 × 12000, Oryza sativa, synthase I, SDS-PAGE) [21]
Hexamer (6 × 12000, Oryza sativa, synthase II, SDS-PAGE) [21]
Dimer (2 × 49000, E. coli, SDS-PAGE) [20]
Oligomer (Oryza sativa enzyme active in at least 2 oligomeric forms, bands in SDS-PAGE of MW 11500, 20000, 35000, 50000, 68000, and in gel filtration of MW 22000 and 67000) [25]

? (x × 60000, synthase I, x × 77000, synthase II, Pisum sativum, SDS-PAGE
[1], x × 60000, Oryza sativa, SDS-PAGE [14]) [1, 14]

Glycoprotein/Lipoprotein
–

---

4 ISOLATION/PREPARATION

Source organism
Pisum sativum (pea) [1, 17]; Solanum tuberosum (potato) [2, 16, 18, 27];
Glycine max (soybean) [3, 6]; Ricinus communis (castor bean) [4]; Zea
mays (corn, ssp. mexicana [10]) [5, 7, 9, 10, 15, 16, 18, 23]; Sorghum bicolor [8]; Zea diploperennis (teosinte) [10]; Hordeum vulgare (barley) [12];
E. coli [13, 20, 24]; Oryza sativa (rice) [14, 21, 25]; Sweet corn [19, 29];
Spinacia oleracea (spinach) [22, 26]; Vitis vinifera (grape) [23]; Arthrobacter
sp. NRRL B1973 [28]; Enterobacter hafniae [11]; Aeromonas hydrophila
[11]; Aeromonas formicans [11]; Aeromonas liquefaciens [11]; Edwardsiella
tarda [11]; Klebsiella pneumoniae [11]; Salmonella enteritidis [11]; Shigella
dysenteriae [11]

Source tissue
Seedlings [1]; Tubers [2, 16, 18, 27]; Cell suspension culture [4, 6]; Endosperm [4, 12, 14]; Leaves [5, 14, 15, 18, 22, 23, 25, 26]; Developing seeds
[8, 21]; Seeds [7, 9, 10, 16, 18, 21, 23, 29]; Cotyledons [17]

Localization in source
Soluble [1, 2, 5, 8, 9, 12, 14, 17, 21, 22, 25, 26, 28, 29]; Granule-bound [3,
6, 9, 15, 16, 18, 28]

Purification
Pisum sativum [1, 17]; Solanum tuberosum [2, 18]; Ricinus communis (partial) [4]; Zea mays (6 isozymes [9], 2 isozymes [23]) [5, 7, 9, 15, 18, 23];
Hordeum vulgare [12]; Oryza sativa (2 isozymes) [21]; Sweet corn (partial)
[19, 29]; E. coli (2 isozymes) [20]; Spinacia oleracea (multiple forms [26])
[22, 26]; Vitis vinifera [23]; Arthrobacter sp. NRRL B1973 [28]

Crystallization
–

Cloned
–

Renatured
–

4

5 STABILITY

pH

Temperature (°C)
15 (stable at) [1]; 25 (stable at) [1]; 35 (inactivation) [1]; 40 (10 min, 78% loss of activity) [29]; 50 (stable up to [3], 1 min, 75% loss of activity [29]) [3, 29]; 62 (inactivation at) [29]

Oxidation

Organic solvent
More (activation by isopropanol (above 70% v/v) or n-propanol (85% v/v)) [27]

General stability information

Storage
–85°C, at least 1 year [20]; –70°C, 50 mM Tris/acetate buffer, pH 7.5, 2.5 mM DTT, 10 mM EDTA, 5% w/v sucrose [7]; –15°C, fraction IV, 6 months [29]; –10°C, fraction V, unstable [29]; 0–4°C, as ammonium sulfate precipitate [25]

6 CROSSREFERENCES TO STRUCTURE DATABANKS

PIR/MIPS code
PIR1:YUPOY (precursor potato); PIR2:JQ2322 (precursor rice); PIR3:S36627 (Bacillus subtilis); PIR3:S40051 (Bacillus subtilis); PIR3:S43341 (cassava)

Brookhaven code

7 LITERATURE REFERENCES

[1] Denyer, K., Smith, A.M.: Planta,186,609–617 (1992)
[2] Baba, T., Noro, M., Hiroto, M., Arai, Y.: Phytochemistry,29,719–723 (1990)
[3] Miyamoto, J., Ishigami, T., Hayashi, T., Nakajima, T., Ichishima, E., Matsuda, K.: Agric. Biol. Chem.,53,1987–1988 (1989)
[4] Goldner, W., Beevers, H.: Phytochemistry,28,1809–1812 (1989)
[5] Dang, P.L., Boyer, C.D.: Phytochemistry,27,1255–1259 (1988)
[6] Miyamoto, J., Nakajima, T., Matsuda, K.: Agric. Biol. Chem.,51,1697–1699 (1987)
[7] Pollock, C., Preiss, J.: Arch. Biochem. Biophys.,204,578–588 (1980)
[8] Boyer, C.D.: Phytochemistry,24,15–18 (1985)
[9] MacDonald, F.D., Preiss, J.: Plant Physiol.,78,849–852 (1985)
[10] Boyer, C.D., Fisher, M.B.: Phytochemistry,23,733–737 (1984)
[11] Yung, S.-G., Paule, M., Beggs, R., Greenberg, E., Preiss, J.: Arch. Microbiol.,138, 1–8 (1984)

[12] Kreis, M.: Planta,148,412–416 (1980)
[13] Holmes, E., Preiss, J.: Arch. Biochem. Biophys.,196,436–448 (1979)
[14] Taira, T., Uematsu, M., Nakano, Y., Morikawa, T.: Biochem. Genet.,29,301–311 (1991)
[15] Dang, P.L., Boyer, C.D.: Biochem. Genet.,27,521–532 (1989)
[16] Preusser, E., Chudy, M., Grundel, M., Khalil, F.A.: Acta Physiol. Plant.,10,133–142 (1988)
[17] Matters, G.L., Boyer, C.D.: Biochem. Genet.,20,833–848 (1982)
[18] Preusser, E., Khalil, F.A., Goering, H.: Biochem. Physiol. Pflanz.,176,744–752 (1981)
[19] Schiefer, S., Lee, E.Y.C., Whelan, W.J.: Carbohydr. Res.,61,239–252 (1978)
[20] Fox, J., Kawaguchi, K., Greenberg, E., Preiss, J.: Biochemistry,15,849–856 (1976)
[21] Pisigan, R.A., del Rosario, E.J.: Phytochemistry,15,71–73 (1976)
[22] Hawker, J.S., Ozbun, J.L., Ozaki, H., Greenberg, E., Preiss, J.: Arch. Biochem. Biophys.,160,530–551 (1974)
[23] Hawker, J.S., Downton, W.J.: Phytochemistry,13,893–900 (1974)
[24] Fox, J., Kennedy, L.D., Hawker, J.S., Ozbun, J.L., Greenberg, E., Lammel, C., Preiss, J.: Ann. N.Y. Acad. Sci.,210,90–103 (1973)
[25] Antonio, A.A., del Rosario, E.J., Juliano, B.: Phytochemistry,12,1929–1932 (1973)
[26] Ozbun, J.L., Hawker, J.S., Preiss, J.: Biochem. J.,126,953–963 (1972)
[27] Judewicz, N.D., Lavintman, N., Cardini, C.E.: Phytochemistry,11,2213–2215 (1972)
[28] Greenberg, E., Preiss, J.: J. Biol. Chem.,240,2341–2348 (1965)
[29] Frydman, R.B., Cardini, C.E.: Biochim. Biophys. Acta,96,294–303 (1965)

1 NOMENCLATURE

EC number
2.4.1.22

Systematic name
UDPgalactose:D-glucose 4-beta-D-galactotransferase

Recommended name
Lactose synthase

Synonyms
UDPgalactose-glucose galactosyltransferase
N-Acetyllactosamine synthase
Galactosyltransferase, uridine diphosphogalactose-glucose
Lactose synthetase
UDP-galactose-glucose galactosyltransferase
More (cf. EC 2.4.1.38 and EC 2.4.1.90)

CAS Reg. No.
9030-11-9

2 REACTION AND SPECIFICITY

Catalysed reaction
UDPgalactose + D-glucose →
→ UDP + lactose (mechanism [1])

Reaction type
Hexosyl group transfer

Natural substrates
UDPgalactose + glucose (biosynthesis of lactose) [1, 4]

Substrate spectrum
1 UDPgalactose + D-glucose (presence of alpha-lactalbumin required)
  [1–4, 6, 8–13]
2 UDPgalactose + xylose [1]
3 UDPgalactose + maltose [1]
4 UDPgalactose + 2-deoxyglucose [1]
5 UDPgalactose + alpha-methylglucose [1]
6 dUDPgalactose + glucose [4]

7 More (the enzyme is a complex of two proteins A and B. In the absence of the B protein (alpha-lactalbumin) the enzyme catalyzes the transfer of galactose from UDPgalactose to N-acetylglucosamine (EC 2.4.1.38, EC 2.4.1.90)) [1]

Product spectrum
1 Lactose + UDP [1–3]
2 ?
3 ?
4 ?
5 ?
6 ?
7 ?

Inhibitor(s)
Iodoacetate [1]; N-Acetylimidazole [1]; 2-Hydroxy-5-nitrobenzyl bromide [1]; Dimethyl-(2-hydroxy-5-nitrobenzyl)-sulfonium bromide [1]; 2-Nitrophenyl-sulfenyl chloride [1]; Formylkynurenine [1]; Maleic anhydride [1]; 2,4,6-Trinitrobenzene sulfonate [1]; Mg^{2+} (above 4 mM) [4]; Ca^{2+} (above 4 mM) [4]; Tunicamycin [5]; Vincristine [5]

Cofactor(s)/prosthetic group(s)/activating agents
alpha-Lactalbumin (i.e. B protein, in the absence of alpha-lactalbumin the reaction of EC 2.4.1.38/EC 2.4.1.90 is catalyzed) [1]

Metal compounds/salts
Mn^{2+} (required, partially replaceable by Mg^{2+}, Ca^{2+} [1], not replaceable by Mg^{2+}, Ca^{2+} [4], K_m: 0.2 mM [6]) [1, 4, 6]

Turnover number (min^{-1})

Specific activity (U/mg)
More (assay methods [1]) [1, 6]

K_m-value (mM)
0.0023 (alpha-lactalbumin) [8]; 0.006 (alpha-lactalbumin, rat serum) [10]; 0.007 (alpha-lactalbumin, rat liver, Golgi apparatus) [10]; 0.009 (UDPgalactose, rat serum) [10]; 0.012 (alpha-lactalbumin, bovine milk) [10]; 0.023 (UDPgalactose, rat liver, Golgi apparatus) [10]; 0.041 (UDPgalactose, recombinant enzyme) [9]; 0.049 (UDPgalactose) [9]; 0.06 (UDPgalactose, bovine milk) [10]; 1.7 (glucose) [8]; 1.8 (glucose, rat serum) [10]; 2.0 (glucose, bovine milk) [10]; 2.3 (glucose, rat liver, Golgi apparatus) [10]; 2.9 (glucose) [12]; 3 (glucose (+ 0.1 mM UDPgalactose), presence of 0.2 mg alpha-lactalbumin/ml) [2]; 3.4 (glucose) [6]; 4 (glucose (+ 0.05 or 0.2 mM UDPgalactose), presence of 0.2 mg alpha-lactalbumin/ml) [2]; 4.9 (glucose)

[9]; 5 (glucose (+ 0.4 mM UDPgalactose), presence of 0.2 mg alpha-lactal-bumin/ml) [2]; 11 (glucose (+ 0.1 mM UDPgalactose), presence of 0.2 mg alpha-lactalbumin/ml) [2]; 12.0 (glucose, bovine milk [11], human HeLa cells, recombinant enzyme [9]) [9, 11]; 14 (glucose (+ 0.2 mM UDPgalac-tose), presence of 0.05 mg alpha-lactalbumin/ml) [2]; 15.4 (glucose (+ 0.4 mM UDPgalactose), presence of 0.05 mg alpha-lactalbumin/ml) [2]; 21.0 (glucose, bovine eyes, cornea) [11]; 33 (glucose (+ 0.05 mM UDPga-lactose), presence of 0.05 mg alpha-lactalbumin/ml) [2]; More (kinetic studies) [1, 4]

pH-optimum

pH-range

Temperature optimum (°C)

Temperature range (°C)

---

3 ENZYME STRUCTURE

Molecular weight

80000 (human, A protein, gel filtration) [6]
70000–75000 (bovine, A protein, gel filtration) [2]
45300–46200 (bovine, A protein, calculation from sedimentation and diffu-sion coefficients) [7]
40000–44000 (mammalian A protein, sucrose density centrifugation, gel chromatography) [4]
14600 (bovine, B protein, i.e. alpha-lactalbumin, calculation from sedimen-tation and diffusion coefficients) [7]
14214 (guinea pig, alpha-lactalbumin, i.e. B protein, amino acid sequence analysis) [1]
14179 (bovine, alpha-lactalbumin, i.e. B protein, amino acid sequence anal-ysis) [1]
14071 (human, alpha-lactalbumin, i.e. B protein, amino acid sequence anal-ysis) [1]
More (conformation) [1]

Subunits

Dimer (2 × 49000, human, A protein, SDS-PAGE [6], 1 × 45300–46200 + 1 × 14600, bovine, complex consisting of transferase, i.e. A protein and alpha-lactalbumin, i.e. B protein [7]) [6, 7]

Glycoprotein/Lipoprotein

Glycoprotein (A protein contains 11% carbohydrate [6]) [1, 6]

4 ISOLATION/PREPARATION

Source organism
Mammals (female) [1, 4]; Bovine (cow) [1–3, 7, 10, 11]; Guinea pig [1, 3]; Rat [5, 10]; Human [1, 6, 9, 12, 13]; Mouse [8]

Source tissue
Mammary gland (during lactation period) [1, 3–5, 8]; Milk [2, 7, 10]; Serum [6, 10, 12]; Eyes (cornea) [11]; Liver [10]; HeLa cells [9]; Amniotic fluid [13]

Localization in source
Soluble (alpha-lactalbumin, B protein) [1, 8, 10–12]; Membranes of Golgi vesicles [1, 5, 8, 10]

Purification
Bovine (A protein) [2, 11]; Human (A protein) [6, 12]; Rat (partial) [10]

Crystallization
(alpha-lactalbumin) [1]

Cloned
–

Renatured
–

5 STABILITY

pH

Temperature (°C)

Oxidation

Organic solvent

General stability information

Storage
4°C, A protein: 10 mM Tris/HCl buffer, pH 7.5, 10 mM $MnCl_2$ [6]

6 CROSSREFERENCES TO STRUCTURE DATABANKS

PIR/MIPS code
PIR3:S37717 (human); PIR2:A24229 (human (fragment)); PIR2:A24251 (human (fragment))

Brookhaven code
1HML (Human (Homo Sapiens))

7 LITERATURE REFERENCES

[1] Hill, R.L., Brew, K. in "Adv. Enzymol. Relat. Areas Mol. Biol",43,411–490 (1975) (Review)
[2] Fitzgerald, D.K., Brodbeck, U., Kiyosawa, I., Mawal, R., Colvin, B., Ebner, K.E.: J. Biol. Chem.,245,2103–2108 (1970)
[3] Watkins, W.M., Hassid, W.Z.: J. Biol. Chem.,237,1432–1440 (1962)
[4] Ebner, K.E. in "The Enzymes",3rd Ed. (Boyer, P.D., ed.) 9,363–377, Academic Press, N.Y. (1973) (Review)
[5] West, D.W.: Biochem. Soc. Trans.,13,694–695 (1985)
[6] Fujita-Yamaguchi, Y., Yoshida, A.: J. Biol. Chem.,256,2701–2706 (1981)
[7] Ivatt, R.J., Rosemeyer, M.A.: Eur. J. Biochem.,64,233–242 (1976)
[8] Jones, E.A.: Biochem. J.,126,67–78 (1972)
[9] Krezdorn, C.H., Watzele, G., Kleene, R.B., Ivanov, S.X., Berger, E.G.: Eur. J. Biochem.,212,113–120 (1993)
[10] Paquet, M.R., Moscarello, M.A.: Biochem. J.,218,745–751 (1984)
[11] Christner, J.E., Distler, J.J., Jourdian, G.W.: Arch. Biochem. Biophys.,192,548–558 (1979)
[12] Bella, A., Whitehead, J.S., Kim, Y.S.: Biochem. J.,167,621–628 (1977)
[13] Nelson, J.D., Jato-Rodriguez, J.J., Mookerjea, S.: Can. J. Biochem.,52,42–50 (1974)

1 NOMENCLATURE

EC number
2.4.1.23

Systematic name
UDPgalactose:sphingosine 1-beta-galactotransferase

Recommended name
Sphingosine beta-galactosyltransferase

Synonyms
Galactosyltransferase, uridine diphosphogalactose-sphingosine beta-
Galactosyl-sphingosine transferase [1]
Psychosine-uridine diphosphate galactosyltransferase
UDPgalactose:sphingosine O-galactosyl transferase
Uridine diphosphogalactose-sphingosine beta-galactosyltransferase
Psychosine-UDP galactosyltransferase

CAS Reg. No.
9032-90-0

---

2 REACTION AND SPECIFICITY

Catalysed reaction
UDPgalactose + sphingosine →
→ UDP + psychosine

Reaction type
Hexosyl group transfer

Natural substrates

Substrate spectrum
1 UDPgalactose + DL-erythro-trans-sphingosine (maximal activity) [1]
2 UDPgalactose + N-acetyl-DL-threo-trans-sphingosine [1]
3 More (basic structure for the acceptor seems to be a 1-hydroxy-2-amino
hydrocarbon of about 18 carbon atoms) [1]

Product spectrum
1 UDP + psychosine (i.e. galactosyl-sphingosine) [1]
2 ?
3 ?

Inhibitor(s)

Cofactor(s)/prosthetic group(s)/activating agents
   Tween-20 (4 mg/ml, highest activity with DL-erythro-trans-sphingosine,
   solubilizing effect) [1]

Metal compounds/salts
   Mg^{2+} (0.002–0.008 M: stimulation) [1]; Mn^{2+} (0.002–0.003 M: slight stimulation) [1]

Turnover number (min^{-1})

Specific activity (U/mg)

K_m-value (mM)
   0.25 (UDPgalactose) [1]

pH-optimum
   8.4 [1]

pH-range
   7–9 (70% of maximal activity at pH 7, 85% of maximal activity at pH 9) [1]

Temperature optimum (°C)
   37 (assay at) [1]

Temperature range (°C)

3 ENZYME STRUCTURE

Molecular weight

Subunits

Glycoprotein/Lipoprotein
   –

4 ISOLATION/PREPARATION

Source organism
   Guinea pig [1]; Rat [1, 2]

Source tissue
   Brain [1, 2]

Localization in source
   Microsomes [1, 2]

Purification
   Guinea pig (partial) [1]; Rat (partial) [1]

Crystallization
–

Cloned
–

Renatured
–

---

5 STABILITY

pH

Temperature (°C)

Oxidation

Organic solvent

General stability information

Storage
Lyophilized powder is indefinitely stable in the deep-freeze [1]

---

6 CROSSREFERENCES TO STRUCTURE DATABANKS

PIR/MIPS code

Brookhaven code

---

7 LITERATURE REFERENCES

[1] Cleland, W.W., Kennedy, E.P.: J. Biol. Chem.,235,45–51 (1960)
[2] Neskovic, N.M., Sarlieve, L.L., Mandel, P.: J. Neurochem.,20,1419–1430 (1973)

1 NOMENCLATURE

EC number
2.4.1.24

Systematic name
1,4-alpha-D-Glucan:1,4-alpha-D-glucan(D-glucose) 6-alpha-D-glucosyltransferase

Recommended name
1,4-alpha-Glucan 6-alpha-glucosyltransferase

Synonyms
Oligoglucan-branching glycosyltransferase
1,4-alpha-D-Glucan 6-alpha-D-glucosyltransferase
T-Enzyme
Glucosyltransferase, 1,4-alpha-glucan 6-alpha-
D-Glucosyltransferase [1]

CAS Reg. No.
9030-12-0

---

2 REACTION AND SPECIFICITY

Catalysed reaction
Transfers an alpha-D-glucosyl residue in a 1,4-alpha-D-glucan to the primary hydroxyl group of glucose, free or combined in a 1,4-alpha-D-glucan

Reaction type
Hexosyl group transfer

Natural substrates

Substrate spectrum
1 Maltose (r) [1]
2 Maltotriose [1]
3 Isomaltose [1]
4 Panose [1]

Product spectrum
1 Glucose [1]
2 ?
3 ?
4 ?

Enzyme Handbook © Springer-Verlag Berlin Heidelberg 1996
Duplication, reproduction and storage in data banks are only
allowed with the prior permission of the publishers

Inhibitor(s)

Cofactor(s)/prosthetic group(s)/activating agents

Metal compounds/salts

Turnover number (min^{-1})

Specific activity (U/mg)

K_m-value (mM)
   0.18 (maltotriose) [1]; 0.67 (maltose) [1]; 2.84 (panose) [1]; 4.54 (isomaltose) [1]

pH-optimum
   4.5 (maltose) [1]

pH-range

Temperature optimum (°C)

Temperature range (°C)

---

3 ENZYME STRUCTURE

Molecular weight
   101000 (Aspergillus niger, gel chromatography) [1]

Subunits
   Monomer (1 × 109000, Aspergillus niger, SDS-PAGE) [1]

Glycoprotein/Lipoprotein
   Glycoprotein (16% D-mannose, 0.1% D-glucose, 2.95% 2-acetamido-
   2-deoxy-D-glucose) [1]

---

4 ISOLATION/PREPARATION

Source organism
   Aspergillus niger [1, 2]

Source tissue
   Culture medium [1]

Localization in source
   Extracellular [1]

Purification
   Aspergillus niger [1]

Crystallization
   —

Cloned

–

Renatured

–

5 STABILITY

pH

Temperature (°C)

Oxidation

Organic solvent

General stability information

Storage

6 CROSSREFERENCES TO STRUCTURE DATABANKS

PIR/MIPS code

Brookhaven code

7 LITERATURE REFERENCES

[1] Kobrehel, D., Deponte, R.: Enzyme Microb. Technol.,4,185–190 (1982)
[2] Barker, S.A., Carrington, T.R.: J. Chem. Soc.,3588–3593 (1953)

1 NOMENCLATURE

EC number
2.4.1.25

Systematic name
1,4-alpha-D-Glucan:1,4-alpha-D-glucan 4-alpha-D-glycosyltransferase

Recommended name
4-alpha-Glucanotransferase

Synonyms
Disproportionating enzyme
Dextrin glycosyltransferase
D-enzyme (plant enzyme is termed D-enzyme)
Debranching enzyme (EC 3.2.1.33 and EC 2.4.1.25 [12–14], the two different
enzyme activities reside on the same polypeptide chain [12, 14]) [12–14]
Maltodextrin glycosyltransferase [3]
4-alpha-Glucanotransferase [3]
Amylomaltase [3, 4]
Glycosyltransferase, dextrin
Dextrin transglycosylase
EC 2.4.1.3 (formerly)
Glucosyltransferase, maltose 4-
Maltose 4-glucosyltransferase

CAS Reg. No.
9032-09-1

---

2 REACTION AND SPECIFICITY

Catalysed reaction
Transfers a segment of a 1,4-alpha-D-glucan to a new position in an acceptor, which may be glucose or a 1,4-alpha-D-glucan (mechanism [4])

Reaction type
Hexosyl group transfer

Natural substrates
More (enzyme may be of importance in the metabolism of starch in the
bacterium [1], role in chloroplast starch degradation [7]) [1, 7]

Enzyme Handbook © Springer-Verlag Berlin Heidelberg 1996
Duplication, reproduction and storage in data banks are only
allowed with the prior permission of the publishers

Substrate spectrum

1 Maltotriose + maltotriose [6, 7]
2 Maltopentaose + maltotriose [10]
3 Maltoheptaose + maltotriose [10]
4 Maltononaose + maltotriose [10]
5 More (transfers glucanosyl segments from maltopentaose, higher molecular weight maltohomologues or starch to maltopentaose and other maltohomologues to yield new oligosaccharides [1], malto-oligosaccharides (G_3-G_7) [2, 7], products from all substrates except G4 are those resulting from maltosyl transfer as the predominant reaction, G_{n-2} and G_{n+2} produced from G_n, maltose not formed in any case [2], transfers 1,4-alpha-glucanosyl chains, resulting in disproportionation of 1,4-alpha-glucans [3], converts soluble starch, amylopectin and amylose [3], addition of low-molecular-mass maltooligosaccharides, which act as glucanosyl acceptors, enhances reaction and results in formation of a series of linear maltohomologues from 2 to more than nine glucose units in size [3], use of either of the malto-oligosaccharides, maltotetraose, maltopentaose, maltohexaose or maltoheptaose as sole substrate yields linear maltohomologues [3], maltose and maltotriose are not disproportionated [3], maltose and maltotriose are good acceptors for glucanosyl transfer [3], the chain length of glucanosyl segments transferred ranges from 2 to probably far more than six glucose residues [3], glucose is not an acceptor in transferase reaction and not a reaction product [3], Pseudomonas stutzeri produces D-enzyme and amylomaltase with different action pattern, D-enzyme: with maltotriose as initial substrate, glucose, maltopentaose, maltoheptaose, maltononaose and maltoundecaose are formed as major products, amylomaltase: acts on maltotriose, maltotetraose and maltopentaose to form a series of homologous 1,4-alpha-glucans, no chain-lengthening reaction with maltohexaose [10], efficient acceptors: D-mannose, D-glucosamine, N-acetyl-D-glucosamine, D-xylose, D-allose, isomaltose, cellobiose [9], transfer of glycosyl residues only to the 4-hydroxyl groups of D-mannose, N-acetyl-D-glucosamine, D-allose and D-xylose, producing oligosaccharides terminated by 4-O-alpha-D-glucopyranosyl-D-mannose, 4-O-alpha-D-glucopyranosyl-N-acetyl-D-glucosamine, 4-O-alpha-glucopyranosyl-D-allose and 4-O-alpha-D-glucopyranosyl-D-xylose at the reducing ends [9], enzyme exhibits both glucosyl-transfer and 4-alpha-glucanosyl-transfer specificity [4], 4-alpha-glucanosyl chains containing up to at least 9 glucosyl units can be transferred [4], maltose serves only as acceptor substrate, no function as donor substrate [4], when maltodextrin serves as donor, the portion of the molecule transferred by the enzyme is that containing the nonreducing-end group [4], catalyzes transfer of maltooligosaccharides from one 1,4-alpha-D-glucan molecule to another or to glucose [5], maltooligosaccharides are effective donor molecules, but short-chain amylose and amylopectin may also

function as donors [5], disproportionates maltotriose and higher maltodextrins by transferring maltosyl or maltodextrinyl groups between maltodextrins resulting in the production of glucose and different maltodextrins,
but not maltose [6], acts on: maltooligosaccharides, maltotertraitol, malto-
pentaitol, maltosylsucrose [8], not: maltitol, maltotriitol, glucosylsucrose,
isomaltose, panose, isopanose, isomaltosylmaltose [8]) [1–10]

Product spectrum
1 Maltopentaose + glucose [6, 7]
2 Maltoheptaose + glucose [10]
3 Maltononaose + glucose [10]
4 Maltoundecaose + glucose [10]
5 ?

Inhibitor(s)
Hg^{2+} [2]; Ag^+ [2]; Bay e 4609 [6]; Acarbose [6]; Glycerol [7]

Cofactor(s)/prosthetic group(s)/activating agents

Metal compounds/salts

Turnover number (min^{-1})

Specific activity (U/mg)
More (radioisotope method for assay of amylomaltase and D-enzyme [11]) [3,
8, 10–12]; 1.13 [6]; 3.24 [7]; 30 (commercial maltose substrate) [4]; 47.5 [5]

K_m-value (mM)
3.3 (maltotriose) [7]; 7.1 (maltotetraose, amylomaltase) [10]; 7.3 (maltotriose) [2]; 8.3 (maltose, D-enzyme) [10]

pH-optimum
5 [1]; 6.5 [6, 8]; 7–8 [3]; 7.5–8 [7]; 7.6 (D-enzyme) [10]; 7.7 (amylomaltase) [10]

pH-range
3–8 (pH 3: about 60% of activity maximum, pH 8: about 20% of activity
maximum) [1]; 6.7–9.5 (about 50% of activity maximum at pH 6.7 and 9.5)
[7]; 7.7–9.0 (pH 7.7: activity maximum, pH 9.0: about 80% of activity maximum, amylomaltase) [10]

Temperature optimum (°C)
30 (D-enzyme [10], assay at [1]) [1, 10]; 35 [8]; 37 (amylomaltase [10],
assay at [5, 6]) [5, 6, 10]; 60 (assay at) [3]; 70 [3]

Temperature range (°C)
55–80 (about 90% of activity maximum at 55°C and 80°C) [3]

3 ENZYME STRUCTURE

Molecular weight
  56000 (Thermotoga maritima, size exclusion chromatography) [3]
  70000–80000 (Bacillus amyloliquefaciens, sucrose gradient centrifugation) [1]
  74000 (Pseudomonas stutzeri, amylomaltase) [10]
  98000 (Pisum sativum, gel filtration) [7]
  115000 (Pseudomonas stutzeri, D-enzyme) [10]
  159000–170000 (rabbit, debranching enzyme, EC 3.2.1.33 and EC 2.4.1.25,
  the two different enzyme activities reside on the same polypeptide chain,
  gel chromatography in presence of 6 M guanidine-HCl, analytical gel chromatography under non-denaturing conditions) [14]
  164000–266000 (rabbit, debranching enzyme: EC 3.2.1.33 and EC 2.4.1.25,
  the two different enzyme activities reside on the same polypeptide chain,
  high-speed sedimentation equilibrium) [12]
  More (dependent on buffer concentration the enzyme exists in interconvert-
  ible low-molecular-weight form, MW 71000, and high-molecular-weight form,
  MW 370000) [4]

Subunits
  Monomer (1 × 53000, Thermotoga maritima, SDS-PAGE) [3]
  Dimer (2 × 50000, Pisum sativum, gel filtration) [7]
  ? (x × 60000, Solanum tuberosum, SDS-PAGE [5], x × 93000, E. coli,
  SDS-PAGE [8]) [5, 8]

Glycoprotein/Lipoprotein
  –

4 ISOLATION/PREPARATION

Source organism
  Solanum tuberosum [5]; Arabidopsis thaliana [6]; Pseudomonas stutzeri
  (produces D-enzyme and amylomaltase with different action pattern) [10];
  Bacillus amyloliquefaciens [1]; Sweet potato (beta-amylase-deficient variety)
  [2]; Thermotoga maritima (gene cloned and expressed in E. coli) [3];
  E. coli (inducible [4], IFO 3806 [8, 9]) [4, 8, 9]; Pisum sativum [7]; Rabbit
  (debranching enzyme, EC 3.2.1.33 and EC 2.4.1.25, the two different
  enzyme activities reside on the same polypeptide chain) [12, 14]; Saccharomyces cerevisiae (debranching enzyme, EC 3.2.1.33 and EC 2.4.1.25)
  [13]; Streptococcus mutans (6715–49) [11]

Source tissue
  Cell [3, 4, 11]; Leaf [6, 7]; Tuber [5]; Stipules [7]; Muscle (skeletal [14],
  debranching enzyme, EC 3.2.1.33 and EC 2.4.1.25) [12, 14]

Localization in source
Chloroplast [7]

Purification
Arabidopsis thaliana (2 forms: D1 and D2) [6]; Pseudomonas stutzeri (D-enzyme and amylomaltase with different action pattern) [10]; Bacillus amyloliquefaciens (partial) [1]; Thermotoga maritima (gene cloned and expressed in E. coli) [3]; E. coli (IFO 3806 [8]) [4, 8]; Solanum tuberosum [5]; Pisum sativum [7]; Rabbit (debranching enzyme, EC 3.2.1.33 and EC 2.4.1.25, the two different enzyme activities reside on the same polypeptide chain) [12]; Saccharomyces cerevisiae (debranching enzyme, EC 3.2.1.33 and EC 2.4.1.25) [13]

Crystallization
−

Cloned
(Thermotoga maritima MSB8 chromosomal gene cloned and expressed in E. coli [3], potato tuber enzyme cloned and expressed in E. coli [5]) [3, 5]

Renatured
−

5 STABILITY

pH

Temperature (°C)
37 (30 min, stable) [3]; 45 (30 min, pH 7.0, stable up to) [8]; 50 (15 min, 50% loss of activity [1], 30 min, stable [3], rapid inactivation above, amylomaltase [10]) [1, 3, 10]; 60 (10 min, loss of activity [6], 15 min, complete loss of activity [1], 30 min, stable [3]) [1, 3, 6]; 80 (half-life: 3 h) [3]

Oxidation

Organic solvent

General stability information
Purified enzyme unstable to freezing [7]

Storage
4°C, 20 mM bis-Tris-propane buffer, pH 6.5, 10% glycerol, stable for 6 months [6]; 4°C, stable for several months [7]

6 CROSSREFERENCES TO STRUCTURE DATABANKS

PIR/MIPS code

PIR2:S03774 (Escherichia coli); PIR2:A45049 (potato); PIR2:A40203 (/amylo-1,6-glucosidase (EC 3.2.1.33) human); PIR2:B40203 (/amylo-1,6-glucosidase (EC 3.2.1.33) pig (fragments))

Brookhaven code

7 LITERATURE REFERENCES

[1] Pazur, J.H., Okada, S.: J. Biol. Chem.,243,4732–4738 (1968)
[2] Toshihiko, S., Shinji, S., Shigeo, F., Tomonori, N.: Carbohydr. Res.,212,201–212 (1991)
[3] Liebl, W., Feil, R., Gabelsberger, J., Kellermann, J., Schleifer, K.-H.: Eur. J. Biochem.,207,81–88 (1992)
[4] Palmer, T.N., Ryman, B., Whelan, W.J.: Eur. J. Biochem.,69,105–115 (1976)
[5] Takaha, T., Yanase, M., Okada, S., Smith, S.M.: J. Biol. Chem.,268,1391–1396 (1993)
[6] Lin, T.-P., Preiss, J.: Plant Physiol.,86,260–265 (1988)
[7] Kakefuda, G., Duke, S.H.: Plant Physiol.,91,136–143 (1989)
[8] Kitahata, S., Murakami, H., Okada, S.: Agric. Biol. Chem.,53,2653–2659 (1989)
[9] Kitahata, S., Murakami, H., Sone, Y., Misaki, A.: Agric. Biol. Chem.,53,2661–2666 (1989)
[10] Schmidt, J., John, M.: Biochim. Biophys. Acta,566,100–114 (1979)
[11] Medda, S., Smith, E.E.: Anal. Biochem.,138,354–359 (1984)
[12] Taylor, C., Cox, A.J., Kernohan, J.C., Cohen, P.: Eur. J. Biochem.,51,105–115 (1975)
[13] Tabata, S., Hizukuri, S.: Eur. J. Biochem.,206,345–348 (1992)
[14] White, R.C., Nelson, T.E.: Biochim. Biophys. Acta,400,154–161 (1975)

1 NOMENCLATURE

EC number
2.4.1.26

Systematic name
UDPglucose:DNA alpha-D-glucosyltransferase

Recommended name
DNA alpha-glucosyltransferase

Synonyms
Glucosyltransferase, uridine diphosphoglucose-deoxyribonucleate alpha-
UDPglucose-DNA alpha-glucosyltransferase
Uridine diphosphoglucose-deoxyribonucleate alpha-glucosyltransferase
T2-HMC-alpha-glucosyl transferase [1]
T4-HMC-alpha-glucosyl transferase [1]
T6-HMC-alpha-glucosyl transferase [1]

CAS Reg. No.
9030-13-1

---

2 REACTION AND SPECIFICITY

Catalysed reaction
Transfers an alpha-D-glucosyl residue from UDPglucose to a hydroxymethylcytosine residue in DNA

Reaction type
Hexosyl group transfer

Natural substrates
UDPglucose + 5-hydroxymethylcytosine containing DNA [1]

Substrate spectrum
1 UDPglucose + 5-hydroxymethylcytosine containing DNA (extent of
glucosylation varies between 50% and 80% of the amount of hydroxymethylcytosine residues calculated to be present, T4 DNA with no unsubstituted hydroxymethylcytosine groups does not serve as acceptor, T2
DNA: no activity with enzyme from T2-infected cells, small activity with
enzyme from T4- and T6-infected cells [1]) [1–3]

Product spectrum
1 UDP + monoglucosyl-hydroxymethylcytosine containing DNA [1]

Inhibitor(s)
  $MgCl_2$ [1]; Phosphate buffer [1]

Cofactor(s)/prosthetic group(s)/activating agents
  Sulfhydryl reagents (required) [1]

Metal compounds/salts

Turnover number (min^{-1})

Specific activity (U/mg)
  More [1, 2]

K_m-value (mM)

pH-optimum
  7.5 (assay at) [1]

pH-range

Temperature optimum (°C)
  30 (assay at) [1, 2]

Temperature range (°C)

3 ENZYME STRUCTURE

Molecular weight

Subunits
  ? (x × 46651, phage-T4, E. coli infected with, nucleotide sequence) [3]

Glycoprotein/Lipoprotein
  –

4 ISOLATION/PREPARATION

Source organism
  Bacteriophage-T2 (E. coli infected with) [1]; Bacteriophage-T4 (E. coli infected with) [1–3]; Bacteriophage-T6 (E. coli infected with) [1]

Source tissue
  Cell [1]

Localization in source

Purification
  Bacteriophage-T2 (partial, E. coli infected with) [1]; Bacteriophage-T4 (partial, E. coli infected with) [1, 2]; Bacteriophage-T6 (partial, E. coli infected with) [1]

Crystallization

–

Cloned

[2]

Renatured

–

---

5 STABILITY

pH

Temperature (°C)

Oxidation

Organic solvent

General stability information

Storage

---

6 CROSSREFERENCES TO STRUCTURE DATABANKS

PIR/MIPS code

Brookhaven code

---

7 LITERATURE REFERENCES

[1] Kornberg, S.R., Zimmerman, S.B., Kornberg, A.: J. Biol. Chem.,236,1487–1493
    (1961)
[2] Tomaschewski, J., Gram, H., Crabb, J.W., Rüger, W.: Nucleic Acids Res.,13,
    7551–7568 (1985)
[3] Gram, H., Rüger, W.: EMBO J.,4,257–264 (1985)

1 NOMENCLATURE

EC number
2.4.1.27

Systematic name
UDPglucose:DNA beta-D-glucosyltransferase

Recommended name
DNA beta-glucosyltransferase

Synonyms
T4-HMC-beta-glucosyl transferase [1]
T4-beta-glucosyl transferase [1]
T4 phage beta-glucosyltransferase [3]
Glucosyltransferase, uridine diphosphoglucose-deoxyribonucleate beta-
UDP glucose-DNA beta-glucosyltransferase
Uridine diphosphoglucose-deoxyribonucleate beta-glucosyltransferase

CAS Reg. No.
9030-14-2

2 REACTION AND SPECIFICITY

Catalysed reaction
Transfers a beta-D-glucosyl residue from UDPglucose to a hydroxymethylcytosine residue in DNA

Reaction type
Hexosyl group transfer

Natural substrates
UDPglucose + 5-hydroxymethylcytosine containing DNA (one of two
enzymes in glucosylation of T4 phage DNA) [3]

Substrate spectrum
1 UDPglucose + 5-hydroxymethylcytosine containing DNA (amount of glucose transferred to the DNA of T2 and T6 phages is close to the amount
of unglucosylated hydroxymethylcytosine in these DNAs, no reaction with
T4 DNA (absence of unglucosylated hydroxymethylcytosine residues),
extensive reaction with synthetic hydroxymethylcytosine-DNA [1]) [1–3]

Product spectrum
1 UDP + monoglucosyl-hydroxymethyl containing DNA [1–3]

Enzyme Handbook © Springer-Verlag Berlin Heidelberg 1996
Duplication, reproduction and storage in data banks are only
allowed with the prior permission of the publishers

Inhibitor(s)

Cofactor(s)/prosthetic group(s)/activating agents

Metal compounds/salts
    Phosphate buffer (stimulates) [1]; $MgCl_2$ (stimulates) [1]

Turnover number (min^{-1})

Specific activity (U/mg)
    More [1]

K_m-value (mM)

pH-optimum
    7.8 (assay at) [1]

pH-range

Temperature optimum (°C)
    30 (assay at) [1, 2]

Temperature range (°C)

3 ENZYME STRUCTURE

Molecular weight

Subunits

Glycoprotein/Lipoprotein
    −

4 ISOLATION/PREPARATION

Source organism
    Bacteriophage-T4 (E. coli infected with) [1–3]

Source tissue
    Cell [1, 2]

Localization in source

Purification
    Bacteriophage-T4 (E. coli infected with) [1, 2]

Crystallization
    [3]

Cloned
    [2]

Renatured

–

---

5 STABILITY

pH

Temperature (°C)

Oxidation

Organic solvent

General stability information

Storage

---

6 CROSSREFERENCES TO STRUCTURE DATABANKS

PIR/MIPS code
    PIR1:XUBPB4 (phage T4)

Brookhaven code
    1BGT (Bacteriophage t4 recombinant form expressed in (Escherichia coli));
    1BGU (Bacteriophage t4 recombinant form expressed in (Escherichia coli))

---

7 LITERATURE REFERENCES

[1] Kornberg, S.R., Zimmerman, S.B., Kornberg, A.: J. Biol. Chem.,236,1487–1493 (1961)
[2] Tomaschewski, J., Gram, H., Crabb, J.W., Rüger, W.: Nucleic Acids Res.,13,
    7551–7568 (1985)
[3] Freemont, P.S., Rüger, W.: J. Mol. Biol.,203,525–526 (1988)

1 NOMENCLATURE

EC number
2.4.1.28

Systematic name
UDPglucose:D-glucosyl-DNA beta-D-glucosyltransferase

Recommended name
Glucosyl-DNA beta-glucosyltransferase

Synonyms
T6-glucosyl-HMC-beta-glucosyl transferase [1]
T6-beta-glucosyl transferase [1]
Glucosyltransferase, uridine diphosphoglucose-glucosyldeoxyribonucleate beta-
Uridine diphosphoglucose-glucosyldeoxyribonucleate beta-glucosyltrans-ferase

CAS Reg. No.
9030-15-3

---

2 REACTION AND SPECIFICITY

Catalysed reaction
Transfers a beta-D-glucosyl residue from UDPglucose to a glucosylhydroxymethylcytosine residue in DNA

Reaction type
Hexosyl group transfer

Natural substrates
UDPglucose + glucosylhydroxymethylcytosine residue in DNA [1]

Substrate spectrum
1 UDPglucose + glucosylhydroxymethylcytosine residue in DNA (no trans-fer of glucose to synthetic 5-hydroxymethylcytosine-DNA or to T6 DNA, enzyme glucosylates glucosylhydroxymethylcytosine residues of either alpha or beta configuration) [1]

Product spectrum
1 UDP + diglucosylhydroxymethylcytosine containing DNA [1]

Inhibitor(s)

Cofactor(s)/prosthetic group(s)/activating agents
    Sulfhydryl reagents (required) [1]

Metal compounds/salts
    $MgCl_2$ (relatively inert in absence of, stimulates) [1]

Turnover number (min^{-1})

Specific activity (U/mg)
    More [1]

K_m-value (mM)

pH-optimum
    7.5 (assay at) [1]

pH-range

Temperature optimum (°C)
    30 (assay at) [1]

Temperature range (°C)

3 ENZYME STRUCTURE

Molecular weight

Subunits

Glycoprotein/Lipoprotein
    –

4 ISOLATION/PREPARATION

Source organism
    Bacteriophage-T6 (E. coli infected with) [1]

Source tissue
    Cell [1]

Localization in source

Purification
    Bacteriophage-T6 (E. coli infected with, partial) [1]

Crystallization
    –

Cloned

−

Renatured

−

---

5 STABILITY

pH

Temperature (°C)

Oxidation

Organic solvent

General stability information

Storage

---

6 CROSSREFERENCES TO STRUCTURE DATABANKS

PIR/MIPS code

Brookhaven code

---

7 LITERATURE REFERENCES

[1] Kornberg, S.R., Zimmerman, S.B., Kornberg, A.: J. Biol. Chem.,236,1487–1493 (1961)

1 NOMENCLATURE

EC number
2.4.1.29

Systematic name
GDPglucose:1,4-beta-D-glucan 4-beta-D-glucosyltransferase

Recommended name
Cellulose synthase (GDP-forming)

Synonyms
Glucosyltransferase, guanosine diphosphoglucose-1,4-beta-glucan
Cellulose synthase (guanosine diphosphate-forming)
Cellulose synthetase
Guanosine diphosphoglucose-1,4-beta-glucan glucosyltransferase
Guanosine diphosphoglucose-cellulose glucosyltransferase
More (a similar enzyme EC 2.4.1.12 utilizes UDPglucose)

CAS Reg. No.
9027-18-3

2 REACTION AND SPECIFICITY

Catalysed reaction
GDPglucose + (1,4-beta-D-glucosyl)$_n$ $\rightarrow$
$\rightarrow$ GDP + (1,4-beta-D-glucosyl)$_{n+1}$

Reaction type
Hexosyl group transfer

Natural substrates
GDPglucose + (1,4-beta-D-glucosyl)$_n$ (involved in the synthesis of cellulose)
[1]

Substrate spectrum
1 GDPglucose + (1,4-beta-D-glucosyl)$_n$ [1, 2]

Product spectrum
1 GDP + (1,4-beta-D-glucosyl)$_{n+1}$ [1, 2]

Inhibitor(s)

Cofactor(s)/prosthetic group(s)/activating agents

Metal compounds/salts

Turnover number (min^{-1})

Specific activity (U/mg)

K_m-value (mM)

pH-optimum

pH-range

Temperature optimum (°C)

Temperature range (°C)

3 ENZYME STRUCTURE

Molecular weight

Subunits

Glycoprotein/Lipoprotein

–

4 ISOLATION/PREPARATION

Source organism
  Phaseolus aureus [1, 2]; Lupinus albus [1]

Source tissue
  Seedlings [1]

Localization in source

Purification

Crystallization
  –

Cloned
  –

Renatured
  –

5 STABILITY

pH

Temperature (°C)

Oxidation

Organic solvent

General stability information

Storage

6 CROSSREFERENCES TO STRUCTURE DATABANKS

PIR/MIPS code

Brookhaven code

7 LITERATURE REFERENCES

[1] Flowers, H.M., Batra, K.K., Kemp, J., Hassid, W.Z.: J. Biol. Chem.,244,4969–4974 (1969)
[2] Chambers, J., Elbein, A.D.: Arch. Biochem. Biophys.,138,620–631 (1970)

1 NOMENCLATURE

EC number
2.4.1.30

Systematic name
1,3-beta-D-Oligoglucan:orthophosphate alpha-D-glucosyltransferase

Recommended name
1,3-beta-Oligoglucan phosphorylase

Synonyms
Phosphorylase, 1,3-beta-oligoglucan
beta-1,3-Oligoglucan phosphorylase
beta-1,3-Oligoglucan:orthophosphate glucosyltransferase II

CAS Reg. No.
37257-28-6

---

2 REACTION AND SPECIFICITY

Catalysed reaction
$(1,3$-beta-D-Glucosyl$)_{n-1}$ + alpha-D-glucose 1-phosphate $\rightarrow$
$\rightarrow (1,3$-beta-D-glucosyl$)_n$ + phosphate

Reaction type
Hexosyl group transfer

Natural substrates

Substrate spectrum
1 $(1,3$-beta-D-Glucosyl$)_{n-1}$ + alpha-D-glucose 1-phosphate (r) [1, 2]
2 Glucose + alpha-D-glucose 1-phosphate [1]
3 Laminaribiose + alpha-D-glucose 1-phosphate [1, 2]
4 Laminaritriose + alpha-D-glucose 1-phosphate [1]
5 Laminaritetraose + alpha-D-glucose 1-phosphate [1]
6 Laminaripentaose + alpha-D-glucose 1-phosphate [1]
7 Laminaribiosyl-beta-1,4-glucose + alpha-D-glucose 1-phosphate [1]
8 Laminaritriosyl-p-hydroquinone + alpha-D-glucose 1-phosphate [1]
9 Laminaritriosyl-saligenin + alpha-D-glucose 1-phosphate [1]
10 Arbutin + alpha-D-glucose 1-phosphate [1]
11 Salicin + alpha-D-glucose 1-phosphate [1]
12 Laminaritetraosyl-saligenin + alpha-D-glucose 1-phosphate [1]
13 Laminaribiosyl-p-hydroquinone + alpha-D-glucose 1-phosphate [1]
14 beta-Phenylglucoside + alpha-D-glucose 1-phosphate [1]

15 beta-Methylglucoside + alpha-D-glucose 1-phosphate [1]
16 Cellobiose + alpha-D-glucose 1-phosphate [1]
17 Laminaribiosyl-saligenin + alpha-D-glucose 1-phosphate [1]
18 Laminaridextrins + alpha-D-glucose 1-phosphate [1]
19 Laminariheptaose + alpha-D-glucose 1-phosphate [1]
20 Gentiobiose + alpha-D-glucose 1-phosphate [1]
21 Laminaripentaose + phosphate [2]
22 Laminaritetraose + phosphate [2]
23 Laminaribiose + phosphate [2]
24 More (does not act on laminarin and paramylon [1], EC 2.4.1.30 and EC 2.4.1.31 catalyze the same reaction but with different quantitative specificity, EC 2.4.1.30 phosphorolyzes laminaritriose and higher homologues at a greater rate than laminaribiose, while the opposite behavior is observed with EC 2.4.1.31 i.e. laminaribiose phosphorylase [2]) [1, 2]

Product spectrum
1 $(1,3\text{-beta-D-Glucosyl})_n$ + phosphate [1, 2]
2 ?
3 Laminaritriose + phosphate [1, 2]
4 ?
5 ?
6 ?
7 ?
8 ?
9 ?
10 ?
11 ?
12 ?
13 ?
14 ?
15 ?
16 ?
17 ?
18 ?
19 ?
20 ?
21 ?
22 ?
23 ?
24 ?

Inhibitor(s)
    4-Hydroxyquinone (inhibits in absence of 2-mercaptoethanol) [1]; UMP
    (phosphorolysis) [1]

Cofactor(s)/prosthetic group(s)/activating agents
    More (absolute requirement for sulfhydryl groups) [1, 2]

Metal compounds/salts

Turnover number (min^{-1})

Specific activity (U/mg)
    0.3 (laminaribiose) [1]; 0.5 (glucose) [1]

K$_m$-value (mM)
    1.8 (glucose 1-phosphate (+ laminaribiose)) [1]; 2 (laminaripentaose) [1];
    3 (laminaritetraose) [1]; 4 (laminaribiose) [1]; 4.5 (laminaritriose) [1];
    40 (glucose) [1]

pH-optimum
    7–7.5 (glucose or laminaribiose as acceptor) [1]

pH-range

Temperature optimum (°C)
    30 (assay at) [1]

Temperature range (°C)

3 ENZYME STRUCTURE

Molecular weight

Subunits

Glycoprotein/Lipoprotein
    –

4 ISOLATION/PREPARATION

Source organism
    Euglena gracilis [1, 2]

Source tissue
    Cells [1]

Localization in source

Purification

Crystallization
–

Cloned
–

Renatured
–

---

5 STABILITY

pH

Temperature (°C)

Oxidation

Organic solvent

General stability information

Storage

---

6 CROSSREFERENCES TO STRUCTURE DATABANKS

PIR/MIPS code

Brookhaven code

---

7 LITERATURE REFERENCES

[1] Marechal, L.R.: Biochim. Biophys. Acta,146,417–430 (1967)
[2] Marechal, L.R.: Biochim. Biophys. Acta,146,431–442 (1967)

1 NOMENCLATURE

EC number
2.4.1.31

Systematic name
3-beta-D-Glucosyl-D-glucose:orthophosphate alpha-D-glucosyltransferase

Recommended name
Laminaribiose phosphorylase

Synonyms
Phosphorylase, laminaribiose

CAS Reg. No.
37257-29-7

---

2 REACTION AND SPECIFICITY

Catalysed reaction
3-beta-D-Glucosyl-D-glucose + phosphate →
→ D-glucose + alpha-D-glucose 1-phosphate (ordered bi-bi mechanism [4])

Reaction type
Hexosyl group transfer

Natural substrates

Substrate spectrum
1 3-beta-D-Glucosyl-D-glucose + phosphate (i.e. laminaribiose) [1–4]
2 alpha-D-Glucose 1-phosphate + glucose (r [1, 4], at equilibrium synthe-sis of laminaribiose is favored [1], initial rate of synthetic reaction is 4 times faster than that of phosphorolytic reaction [4], specific for alpha-D-glucose 1-phosphate as glucosyl donor [1, 3]) [1–4]
3 alpha-D-Glucose 1-phosphate + phenyl beta-D-glucoside [1, 3, 4]
4 alpha-D-Glucose 1-phosphate + methyl beta-D-glucoside [1, 3, 4]
5 alpha-D-Glucose 1-phosphate + laminaribiose [1, 3, 4]
6 alpha-D-Glucose 1-phosphate + laminaritriose [1, 3]
7 alpha-D-Glucose 1-phosphate + cellobiose [1, 3, 4]
8 alpha-D-Glucose 1-phosphate + arbutin [3]
9 alpha-D-Glucose 1-phosphate + salicin [3, 4]
10 alpha-D-Glucose 1-phosphate + laminaritetraose [3]
11 alpha-D-Glucose 1-phosphate + deoxyglucose [3]
12 alpha-D-Glucose 1-phosphate + laminaribiosyl-beta-1,4-glucose [3]
13 alpha-D-Glucose 1-phosphate + m-nitrophenyl-beta-D-glucoside [4]

---

14 alpha-D-Glucose 1-phosphate + p-nitrophenyl-beta-D-glucoside [4]
15 alpha-D-Glucose 1-phosphate + 1,5-anhydro-D-glucitol [4]
16 Laminaritriose + phosphate [2–4]
17 Laminaritetraose + phosphate [2–4]
18 Laminaripentaose + phosphate [2–4]
19 More (arsenolysis of laminaribiose is observed, but not that of glucose
   1-phosphate [3], in presence of arsenate, laminaribiose is completely
   converted into glucose [1], EC 1.4.1.31 and EC 1.4.1.30 catalyze the
   same reaction but with different quantitative specificity, EC 1.4.1.30
   phosphorolyzes laminaritriose and higher homologues at a greater
   rate than laminaribiose while the opposite behavior is observed with
   EC 1.4.1.31 [2]) [1–3]

Product spectrum
1 D-Glucose + alpha-D-glucose 1-phosphate [1–4]
2 Laminaribiose + laminaritriose + laminaritetraose + laminaripentaose
  (amount of individual products depends on substrate concentration and
  reaction time [3]) [1, 3]
3 ?
4 ?
5 ?
6 ?
7 Phosphate + oligosaccharides (in which one, two or more 1,3-linked glucose residues are attached to the nonreducing residue of cellobiose) [1]
8 ?
9 ?
10 ?
11 ?
12 ?
13 ?
14 ?
15 ?
16 Laminaribiose + glucose 1-phosphate [3]
17 Laminaritriose + glucose 1-phosphate [3]
18 Laminaritetraose + glucose 1-phosphate [3]
19 ?

Inhibitor(s)
Imidazole (weak) [3]; Glucose (substrate inhibition at high concentration) [4]

Cofactor(s)/prosthetic group(s)/activating agents

Metal compounds/salts
Ca^{2+} (slight activation) [1]

Turnover number (min^{-1})

Specific activity (U/mg)
  24.0 (glucose + glucose 1-phosphate) [3]; More [4]

K$_m$-value (mM)
  2.1 (glucose 1-phosphate (+ glucose)) [3]; 2.5 (phosphate (+ laminaribiose))
  [3]; 5 (laminaribiose (+ phosphate)) [3]; 6 (laminaritriose (+ phosphate)) [3];
  19 (glucose (+ glucose 1-phosphate)) [3]

pH-optimum
  6.3–6.9 [3]; 6.5 (assay at) [3]; 7.2 [4]

pH-range

Temperature optimum (°C)
  30 (assay at) [4]; 37 (assay at) [1, 3]

Temperature range (°C)

---

3 ENZYME STRUCTURE

Molecular weight
  200000 (Euglena gracilis, gel filtration) [4]

Subunits
  Dimer (2 × 120000, Euglena gracilis, SDS-PAGE) [4]

Glycoprotein/Lipoprotein
  –

---

4 ISOLATION/PREPARATION

Source organism
  Euglena gracilis (var. bacillaris, IAM E-2 and IAM E-3, Z. IAM E-6 [4]) [2–4];
  Astasia ocellata [1]

Source tissue
  Cells [1]

Localization in source

Purification
  Euglena gracilis (3 isoforms: F0, FI, FII [4]) [3, 4]

Crystallization
  –

Cloned
–

Renatured
–

5 STABILITY

pH

Temperature (°C)
 21 (5, 9, 24 and 37% loss of activity after 2, 6, 24 and 48 h, respectively) [1];
 35 (isoenzyme F1, stable below) [4]; 40 (isoenzymes F0 and F2, stable below) [4]

Oxidation

Organic solvent

General stability information
 Stable to repeated freezing and thawing [3]

Storage
 –14°C, 3 months stable [3]

6 CROSSREFERENCES TO STRUCTURE DATABANKS

PIR/MIPS code
 PIR1:XUBPA4 (phage T4)

Brookhaven code

7 LITERATURE REFERENCES

[1] Manners, D.J., Taylor, D.C.: Arch. Biochem. Biophys.,121,443–451 (1967)
[2] Marechal, L.R.: Biochim. Biophys. Acta,146,431–442 (1967)
[3] Goldemberg, S.H., Marechal, L.R., De Souza, B.C.: J. Biol. Chem.,241,45–50 (1966)
[4] Kitaoka, M., Sasaki, T., Taniguchi, H.: Arch. Biochem. Biophys.,304,508–514 (1993)

4

1 NOMENCLATURE

EC number
2.4.1.32

Systematic name
GDPmannose:glucomannan 1,4-beta-D-mannosyltransferase

Recommended name
Glucomannan 4-beta-mannosyltransferase

Synonyms
GDP-Man-beta-mannan manosyltransferase [2]
Glucomannan-synthase [3]
Mannosyltransferase, glucomannan 4-beta-
Glucomannan 4-beta-mannosyltransferase

CAS Reg. No.
37257-30-0

2 REACTION AND SPECIFICITY

Catalysed reaction
GDPmannose + (glucomannan)$_n$ →
→ GDP + (glucomannan)$_{n+1}$

Reaction type
Hexosyl group transfer

Natural substrates
GDPmannose + (glucomannan)$_n$ [3]

Substrate spectrum
1 GDPmannose + (glucomannan)$_n$ (alkali insoluble) [1–3]

Product spectrum
1 GDP + (glucomannan)$_{n+1}$ (mannose attached by beta-1,4-linkages
[1, 3], homopolymer of beta-1,4-linked mannose with a few branches [2])
[1–3]

Inhibitor(s)
GDP-D-glucose [1–3]; Phosphate buffer (slight) [1]; Glycerol (10%, slight)
[3]

Cofactor(s)/prosthetic group(s)/activating agents

Metal compounds/salts
  Mg^{2+} (required, optimum concentration: 10 mM) [1]

Turnover number (min^{-1})

Specific activity (U/mg)

K_m-value (mM)
  0.085 (GDP-D-mannose) [3]; 0.1 (GDP-D-mannose) [1]

pH-optimum
  7.5 (Tris buffer) [1]

pH-range
  6.5–9 (6.5: 55% of activity maximum, 9: 10% of activity maximum) [1]

Temperature optimum (°C)
  37 (assay at) [1]

Temperature range (°C)

---

3 ENZYME STRUCTURE

Molecular weight

Subunits

Glycoprotein/Lipoprotein
  –

---

4 ISOLATION/PREPARATION

Source organism
  Phaseolus aureus [1]; Acer pseudoplantanus [2]; Pinus sylvestris [3]

Source tissue
  Seedlings [1]; Cell suspension [2]; Cambial cells [3]; Xylem cells (differen-
  tiated and differentiating) [3]

Localization in source
  Membrane [2, 3]

Purification

Crystallization
  –

Cloned

–

Renatured

–

5 STABILITY

pH

Temperature (°C)

Oxidation

Organic solvent

General stability information

Storage

6 CROSSREFERENCES TO STRUCTURE DATABANKS

PIR/MIPS code

Brookhaven code

7 LITERATURE REFERENCES

[1] Elbein, A.D.: J. Biol. Chem.,244,1608–1616 (1969)
[2] Smith, M.M., Axelos, M., Peaud-Lenoel, C.: Biochimie,58,1195–1211 (1976)
[3] Dalessandro, G., Piro, G., Northcote, D.H.: Planta,169,564–574 (1986)

1 NOMENCLATURE

EC number
2.4.1.33

Systematic name
GDP-D-mannuronate:alginate D-mannuronyltransferase

Recommended name
Alginate synthase

Synonyms
Synthase, alginate
Mannuronosyl transferase [1]

CAS Reg. No.
37257-31-1

2 REACTION AND SPECIFICITY

Catalysed reaction
GDP-D-mannuronate + (alginate)$_n$ $\rightarrow$
$\rightarrow$ GDP + (alginate)$_{n+1}$

Reaction type
Hexosyl group transfer

Natural substrates

Substrate spectrum
1 GDP-D-mannuronate + (alginate)$_n$ [1]
2 alpha-D-Glucuronate 1-phosphate + (alginate)$_n$ (25% of activity with GDP-D-mannuronate) [1]
3 alpha-D-Mannuronate 1-phosphate + (alginate)$_n$ (25% of activity with GDP-D-mannuronate) [1]

Product spectrum
1 GDP + (alginate)$_{n+1}$
2 ?
3 ?

Inhibitor(s)
$MnCl_2$ (at 0.25 mM: 15% increase of activity, above 0.75 mM: 80% inhibition) [1]; ATP (5% inhibition) [1]

Cofactor(s)/prosthetic group(s)/activating agents
  GDP (0.3 mM: 15% stimulation) [1]; 5'-GMP (slight stimulation) [1]; More (GTP, UTP, CTP, TTP, a mixture of NAD^+ and $NADP^+$ have no significant effect) [1]

Metal compounds/salts
  $MnCl_2$ (no essential requirement, at 0.25 mM: 15% increase of activity, inhibition above 0.75 mM) [1]

Turnover number (min^{-1})

Specific activity (U/mg)

K_m-value (mM)

pH-optimum
  6.5–7.0 [1]

pH-range
  5.6–8.5 (65% of maximal activity at pH 5.6 and pH 8.5) [1]

Temperature optimum (°C)
  36 [1]

Temperature range (°C)

---

3 ENZYME STRUCTURE

Molecular weight

Subunits

Glycoprotein/Lipoprotein
  –

---

4 ISOLATION/PREPARATION

Source organism
  Fucus gardneri (marine brown alga) [1]

Source tissue
  Thallus [1]

Localization in source

Purification
  Fucus gardneri (partial) [1]

Crystallization
  –

2

Cloned

–

Renatured

–

5 STABILITY

pH

Temperature (°C)
10 (15 min, 23% remaining activity) [1]; 36 (15 min, 72% remaining activity) [1];
45 (15 min, 60% remaining activity) [1]

Oxidation

Organic solvent

General stability information

Storage
–15°C, stable for 3 days, 80% remaining activity [1]; 0–4°C, unstable [1]

6 CROSSREFERENCES TO STRUCTURE DATABANKS

PIR/MIPS code

Brookhaven code

7 LITERATURE REFERENCES

[1] Lin, T.-Y., Hassid, W.Z.: J. Biol. Chem.,241,5284–5297 (1966)

1 NOMENCLATURE

EC number

2.4.1.34

Systematic name

UDPglucose:1,3-beta-D-glucan 3-beta-D-glucosyltransferase

Recommended name

1,3-beta-Glucan synthase

Synonyms

1,3-beta-D-Glucan-UDP glucosyltransferase
UDPglucose-1,3-beta-D-glucan glucosyltransferase
Callose synthetase
Paramylon synthetase [1]
UDP-glucose-beta-glucan glucosyltransferase [4]
GS-II [20]
Glucosyltransferase, uridine diphosphoglucose-1,3-beta-glucan
(1,3)-beta-Glucan (callose) synthase
beta-1,3-Glucan synthase
beta-1,3-Glucan synthetase
1,3-beta-D-Glucan synthetase
1,3-beta-D-Glucan synthase
1,3-beta-Glucan-uridine diphosphoglucosyltransferase
Callose synthase
UDP-glucose-1,3-beta-glucan glucosyltransferase
UDP-glucose:(1,3)beta-glucan synthase
Uridine diphosphoglucose-1,3-beta-glucan glucosyltransferase

CAS Reg. No.

9037-30-3

2 REACTION AND SPECIFICITY

Catalysed reaction

UDPglucose + (1,3-beta-D-glucosyl)$_n$ $\rightarrow$
$\rightarrow$ UDP + (1,3-beta-D-glucosyl)$_{n+1}$

Reaction type

Hexosyl group transfer

Enzyme Handbook © Springer-Verlag Berlin Heidelberg 1996
Duplication, reproduction and storage in data banks are only
allowed with the prior permission of the publishers

Natural substrates

UDPglucose + (1,3-beta-D-glucosyl)$_n$ (synthesis of callose in higher plants, enzyme is latent in intact and undamaged cells, it is activated only under perturbed conditions, possible involvement in wound-healing process and in defense against pathogens, enzyme seems to play a vital role in a number of specialized developmental processes in plants, namely pollen maturation and sieve pore formation and gravitropism [10], production of cell wall polysaccharide found in yeast cells [11]) [10, 11]

Substrate spectrum

1 UDPglucose + (1,3-beta-D-glucosyl)$_n$ (specific for UDPglucose as glucosyl donor [10]) [1–22]

Product spectrum

1 UDP + (1,3-beta-D-glucosyl)$_{n+1}$ (chain length 60–80 [11]) [1–22]

Inhibitor(s)

UDP [1, 10, 11, 17, 22]; Glycylglycine buffer (0.75 M) [1]; NaCl [1]; Na$_2$SO$_4$ [1]; KCl [1]; Phosphoenolpyruvate [1]; p-Hydroxymercuribenzoate [1]; Natural inhibitor in green Euglena cells [1]; N-Ethylmaleimide (0.16 mM, partial, A. ambisexualis enzyme not [5], weak [22]) [5, 22]; Showdomycin (0.1 mM, partial, A. ambisexualis enzyme not) [5]; Sirufluor (fluorochrome from aniline blue) [7]; EDTA (fully reversible by addition of Ca^{2+} [10]) [10, 22]; EGTA (fully reversible by addition of Ca^{2+} [10]) [10, 22]; 8-Hydroxyquinoline [22]; ATP (enhances activity [12]) [10]; GTP (enhances activity [12]) [10]; CTP [10]; UTP [10]; UMP [10]; p-Chloromercuribenzoate [22]; Glucono-delta-lactone [22]; Co^{2+} [22]; Mn^{2+} (at high concentration, stimulation at low concentration) [22]; Ca^{2+} (at high concentration, stimulation at low concentration) [22]; Octylglucoside [10]; Triton X-100 [10]; Nonidet P-40 [10]; 1,2-Bis(O-aminophenoxy)ethane-N,N,N',N'-tetraacetate [10]; TDPglucose [12]; Unsaturated fatty acids (trienoic acids most effective) [16]; Lysophosphatidylcholine (inhibition in presence of digitonin, stimulation in absence of digitonin within a certain concentration range) [16]; Platelet-activating factor (inhibition in presence of digitonin, stimulation in absence of digitonin within a certain concentration range) [16]; Acylcarnitine (inhibition in presence of digitonin, stimulation in absence of digitonin within a certain concentration range) [16]; Echinocandin B (inhibition in presence of digitonin, stimulation in absence of digitonin within a certain concentration range) [16]; Congo red (non competitive) [14]

Cofactor(s)/prosthetic group(s)/activating agents

beta-Furfuryl-beta-glucoside (stimulation) [17]; GTP (and its analogs stimulate, GTPgammaS is the most potent stimulator [12], inhibits [10]) [5, 12]; Laminaribiose (stimulates) [1]; D-Glucosides (stimulate [1, 7], up to 12-fold [7]) [1, 7]; Salicin (stimulates) [1]; Cellobiose (stimulates [1, 2, 6, 7, 17, 18], no effect [10, 22]) [1, 2, 6, 7, 17, 18]; Cellobiosylglucose (stimulates) [1]; Spermine (stimulates) [2, 17]; Polyamines (stimulate) [3]; Polyols (stimulate) [3, 17]; beta-linked Glucosides (stimulate) [4]; Bovine serum albumin (enhances activity) [5]; Glycerol (stimulates Saccharomyces cerevisiae enzyme) [5]; Laminaribiose (stimulates [7, 13], no reaction as primer [13]) [7, 13]; Gentiobiose (stimulates) [7, 13]; Glucose (stimulates) [7, 18]; beta-Methylglucoside (stimulates) [7]; Hydroquinone-beta-glucoside (stimulates) [7]; Maltose (stimulates) [7, 18]; Digitonin (stimulates) [10]; Sucrose (stimulates) [18]; 3-[(Cholamidopropyl)dimethyl(ammonio)]propanesulfonate (stimulates) [10]; ATP (enhances activity [12], inhibits [10]) [12]; Lysophosphatidylcholine (inhibition in presence of digitonin, stimulation in absence of digitonin within a certain concentration range) [16]; Platelet-activating factor (inhibition in presence of digitonin, stimulation in absence of digitonin within a certain concentration range) [16]; Acylcarnitine (inhibition in presence of digitonin, stimulation in absence of digitonin within a certain concentration range) [16]; Echinocandin B (inhibition in presence of digitonin, stimulation in absence of digitonin within a certain concentration range) [16]; More (enzyme shows activity without addition of a primer [1], activation by substrate [4], maximal activity in presence of 0.75 mM Ca^{2+}, 0.5 mM EGTA and 5 mM cellobiose at pH 7.5 and 30°C [8], reaction requires addition of glycerol, bovine serum albumin and ATP or GTP for maximal activity [11], membrane-bound stimulator, probably a glycoprotein may represent a natural effector which modulates enzyme activity during membrane flow leading to the delivery of active enzymes at the cell surface [15]) [1, 4, 8, 11, 15]

Metal compounds/salts

More (no requirement for metal cations) [11]; Ca^{2+} (activates [2, 6], stimulates (at low concentration, inhibits at high concentrations [22]) [3, 7, 8, 20, 22], activation is half-maximal at about 0.05 mM [7], required [10, 17], optimum concentration 2–5 mM [10]) [2, 3, 6–8, 10, 17, 20, 22]; Mg^{2+} (stimulates [4, 6, 7, 22], 15 mM $MgCl_2$ stimulates at 0.01 mM UDPglucose, decreases at 1 mM UDPglucose [18], activates by increasing their affinity to Ca^{2+} [17]) [4, 6, 7, 17, 18, 22]; Mn^{2+} (stimulates (at low concentrations, inhibits at high concentrations [22]) [7, 22], less effective than Ca^{2+} [7], less effective than Mg^{2+} [22]) [7, 22]; Sr^{2+} (stimulates, less effective than Ca^{2+}) [7]

Turnover number (min^{-1})

Specific activity (U/mg)
    More [9, 10]

K$_m$-value (mM)
    0.43 (UDPglucose) [6]; 0.6 (UDPglucose) [1]; 0.67 (UDPglucose, Arachis
    hypogaea [10], Hansenula anomala [5]) [5, 10]; 0.8 (UDPglucose, Schizo-
    phyllum commune) [5]; 0.86 (UDPglucose, Cryptococcus laurentii) [5]; 1.8
    (UDPglucose, Wangiella dermatitidis) [5]; 2.9 (UDPglucose, Neurospora
    crassa) [5]; 3.8 (UDPglucose, Saccharomyces cerevisiae) [5]; 7.1 (UDPglu-
    cose, Achlya ambisexualis) [5]; More (in absence and presence of GTP
    [12]) [12, 13, 17, 18, 20, 22]

pH-optimum
    6.7 [22]; 7.2–7.6 [6]; 7.4 [10]; 7.5 (glycylglycine buffer [1]) [1, 18]; 7.5–8
    (Saccharomyces cerevisiae) [5]; 8.0 [11, 20]

pH-range
    6.5–9 (6.5 and 9: about 30% of activity maximum) [10]; 6.8–8.8 (6.8 and 8.8:
    about 50% of activity maximum) [11]; 7–7.8 (7: about 60% of activity maxi-
    mum, 7.8: about 70% of activity maximum) [1]; 7.5–8.5 (7.5: activity maxi-
    mum, 8.5: about 50% of activity maximum) [18]

Temperature optimum (°C)
    23 [1]; 24 [22]; 25 (assay at) [6]; 30 (assay at) [7, 12]

Temperature range (°C)
    17–37 (17°C: 80% of activity maximum, 30°C: 90% of activity maximum,
    37°C: 80% of activity maximum) [1]; 0–45 (0°C: 16% of activity maximum,
    37°C: 44% of activity maximum, 45°C: 6% of activity maximum) [22]

---

3 ENZYME STRUCTURE

Molecular weight

Subunits
    ? (x × 48000, Arachis hypogaea, SDS-PAGE) [10]
    More (one or both of the 150000 and 57000 MW polypeptide represent the
    UDPglucose binding subunit of glucan synthase [8]) [8, 9, 21]

Glycoprotein/Lipoprotein
    –

4 ISOLATION/PREPARATION

Source organism
Daucus carota [8]; Saprolegnia monoica [9, 14, 15, 19]; Euglena gracilis [1]; Brassica oleracea (cauliflower) [2]; Beta vulgaris (sugar beet) [3, 7]; Gossypium hirsutum [4, 6]; Saccharomyces cerevisiae [5, 11, 22]; Arachis hypogaea [10]; Candida albicans [12]; Achlya ambisexualis [5]; Hansenula anomala [5]; Neurospora crassa [5, 13]; Cryptococcus laurentii [5]; Schizophyllum commune [5]; Wangiella dermatitidis [5]; Glycine max [16]; Spinach [17]; Citrus aurantifolia (Christm. Swing., mexican lime) [18]; Oryza sativa [20]; Apium graveolens [21]

Source tissue
Dark grown cells [1]; Cotton fibers [4, 6]; Petiole tissue [7]; Cotyledons (germinating) [10]; Budding and filamentous cultures [12]; Suspension cultured cells [16]; Leaf [17]; Bark [18]

Localization in source
Plasma membrane (integral transmembrane protein [3]) [2, 3, 7, 10, 17, 19–21]; Membrane (bound) [8]; Microsomes [16, 19]

Purification
Brassica oleracea [2]; Saprolegnia monoica [9, 19]; Arachis hypogaea [10]; Apium graveolens [21]

Crystallization
–

Cloned
–

Renatured
–

---

5 STABILITY

pH

Temperature (°C)
28 (2 h, 90% loss of activity) [2]

Oxidation

Organic solvent

General stability information
Inactivation at 30°C is greatly accelerated by the presence of 1–2 mM EDTA [5]; Guanosine nucleotides prevent inactivation at 30°C [5]

Storage
Frozen, 60% loss of activity after 3 days [22]; –20°C, 50 mM Tris-HCl, pH
7.4, 1% digitonin, 0.01% beta-mercaptoethanol, 20% sucrose or glycerol,
stable for at least 1 week, losing less than 10% of initial activity [10]; –14°C,
2 weeks, 20% loss of activity [1]; –80°C, several months [5]; 4°C, solubilized
enzyme stable for 4 h [8]; –80°C, solubilized enzyme stable for 8 h [8]

6 CROSSREFERENCES TO STRUCTURE DATABANKS

PIR/MIPS code
PIR2:S50235 (chain FKS1 yeast (Saccharomyces cerevisiae)); PIR2:S50240
(chain FKS2 yeast (Saccharomyces cerevisiae))

Brookhaven code

7 LITERATURE REFERENCES

[1] Marechal, L.R., Goldemberg, S.H.: J. Biol. Chem.,239,3163–3167 (1964)
[2] Fredrikson, K., Kjellbom, P., Larsson, C.: Physiol. Plant.,81,289–294 (1991)
[3] Fredrikson, K., Larsson, C.: Physiol. Plant.,77,196–201 (1989)
[4] Delmer, D.P., Heiniger, U., Kulow, C.: Plant Physiol.,59,713–718 (1977)
[5] Cabib, E., Kang, M.S.: Methods Enzymol.,138,637–642 (1987)
[6] Li, L., Brown, R.M.: Plant Physiol.,101,1143–1148 (1993)
[7] Morrow, D.L., Lucas, W.J.: Plant Physiol.,81,171–176 (1986)
[8] Lawson, S.G., Mason, T.L., Sabin, R.D., Sloan, M.E., Drake, R.R., Haley, B.E.,
Wasserman, B.P.: Plant Physiol.,90,101–108 (1989)
[9] Bulone, V., Girard, V., Fevre, M.: Plant Physiol.,94,1748–1755 (1990)
[10] Kamat, U., Garg, R., Sharma, C.B.: Arch. Biochem. Biophys.,298,731–739 (1992)
[11] Shematek, E.M., Braatz, J.A., Cabib, E.: J. Biol. Chem.,255,888–894 (1980)
[12] Orlean, P.A.B.: Eur. J. Biochem.,127,397–403 (1982)
[13] Quigley, D.R., Selitrennikoff, C.P.: Curr. Microbiol.,15,181–184 (1987)
[14] Nodet, P., Girard, V., Fevre, M.: FEMS Microbiol. Lett.,69,225–228 (1990)
[15] Girard, V., Fevre, M.: Plant Sci.,76,193–200 (1991)
[16] Kauss, H., Jeblick, W.: Plant Physiol.,80,7–13 (1986)
[17] Fredrikson, K., Larsson, C.: Biochem. Soc. Trans.,20,710–713 (1992) (Review)
[18] Beltran, J.P., Carbonell, J.: Phytochemistry,17,1531–1532 (1978)
[19] Girard, V., Bulone, V., Fevre, M.: Plant Sci.,82,145–153 (1992)
[20] Kuribayashi, I., Kimura, S., Morita, T., Igaue, I.: Biosci. Biotechnol. Biochem.,
56,388–393 (1992)
[21] Slay, R.M., Watada, A.E., Frost, D.J., Wasserman, B.P.: Plant Sci.,86,125–136 (1992)
[22] Lopez-Romero, E., Ruiz-Herrera, J.: Antonie Leeuwenhoek,44,329–339 (1978)

1 NOMENCLATURE

EC number
2.4.1.35

Systematic name
UDPglucose:phenol beta-D-glucosyltransferase

Recommended name
Phenol beta-glucosyltransferase

Synonyms
UDPglucosyltransferase
Glucosyltransferase, uridine diphospho-
Phenol-beta-D-glucosyltransferase
UDP glucosyltransferase
UDP-glucose glucosyltransferase
Uridine diphosphoglucosyltransferase

CAS Reg. No.
9046-69-9

---

2 REACTION AND SPECIFICITY

Catalysed reaction
UDPglucose + a phenol →
→ UDP + an aryl beta-D-glucoside

Reaction type
Hexosyl group transfer

Natural substrates

Substrate spectrum
1 UDPglucose + 2-aminophenol (specific for UDPglucose, not replaceable by ADPglucose, CDPglucose, or TDPglucose) [1]
2 UDPglucose + 4-nitrophenol [1–3]
3 UDPglucose + phenol [3, 6]
4 UDPglucose + benzyl alcohol [3]
5 UDPglucose + 2-phenylethanol [3]
6 UDPglucose + 4-hydroxyphenylethanol [3]
7 UDPglucose + 4-hydroxybenzoic acid methyl ester [4]
8 UDPglucose + butyl alcohol [6]
9 More (overview: acceptors for D-glucose) [6]

Product spectrum

1 UDP + 2-aminophenyl beta-D-glucoside [1]
2 UDP + 4-nitrophenyl beta-D-glucoside
3 UDP + phenyl beta-D-glucoside
4 UDP + benzyl beta-D-glucoside
5 UDP + 2-phenylethyl beta-D-glucoside
6 UDP + 4-hydroxyphenylethyl beta-D-glucoside
7 UDP + ?
8 UDP + butyl beta-D-glucoside
9 ?

Inhibitor(s)

Hg^{2+} [3], Ag^{+} [3]; Cu^{2+} [3]; UDP [3]; p-Chloromercuribenzene sulfonate [6]; Mn^{2+} [6]; Co^{2+} [6]; Fe^{2+} [6]; EDTA (at 10 mM, partially) [6]; Zn^{2+} (10 mM, strong) [6]

Cofactor(s)/prosthetic group(s)/activating agents

Triton X-100 (0.8%, activation) [2]

Metal compounds/salts

Mg^{2+} (1.5–15 mM results in 40% activation [1], 0.17 mM activate [2], slight stimulation [6]) [1–3, 6]

Turnover number (min^{-1})

Specific activity (U/mg)

K_m-value (mM)

0.07 (4-nitrophenol) [3]; 0.17 (UDPglucose) [3]

pH-optimum

7.0 [4]; 7.5 [5]; 8.0–8.5 (Arion ater, UDPglucose + 4-nitrophenol) [1]; 9.3 (Arion ater, UDPglucose + 2-aminophenol) [1]

pH-range

5.3–10.4 (less than half-maximal activity above and below) [3]

Temperature optimum (°C)

15 [5]; 37 [1]; 50 [3]

Temperature range (°C)

3 ENZYME STRUCTURE

Molecular weight

56000 (Carica papaya, gel filtration) [3]
62000 (Phaseolus aureus, gel filtration) [6]

Subunits
  Dimer (2 × 28000, Carica papaya, SDS-PAGE) [3]

Glycoprotein/Lipoprotein
  –

---

4 ISOLATION/PREPARATION

Source organism
  Helix pomatia [1]; Arion ater [1]; Musca domestica (phenobarbital induced)
  [2]; Carica papaya (papaya) [3]; Beta vulgaris (sugar beet) [4]; Chrysochro-
  mulina chiton (marine unicellular alga) [5]; Phaseolus aureus (mung bean)
  [6]

Source tissue
  Alimentary tract (crop and foregut, anterior, stomach, digestive gland) [1];
  Fruit [3]; Germinating seeds [6]; Whole plants (young state) [4]; More (not in
  skin) [1]

Localization in source
  Microsomes [4]; Endoplasmic reticulum [5]; Particulate [1]

Purification
  Carica papaya [3]; Beta vulgaris (partial) [4]; Phaseolus aureus (partial) [6]

Crystallization
  –

Cloned
  –

Renatured
  –

---

5 STABILITY

pH
  8 (half-life 5 days) [3]

Temperature (°C)
  More (Arrhenius plot) [3]

Oxidation

Organic solvent

General stability information
  Dithiothreitol stabilizes [6]

Storage

−15°C, 2 days stable [1]

6 CROSSREFERENCES TO STRUCTURE DATABANKS

PIR/MIPS code

Brookhaven code

7 LITERATURE REFERENCES

[1] Dutton, G.J.: Arch. Biochem. Biophys.,116,399–405 (1966)
[2] Morello, A., Repetto, Y.: Biochem. J.,177,809–812 (1979)
[3] Keil, U., Schreier, P.: Phytochemistry,28,2281–2284 (1989)
[4] Stölzel, G., Pommer, U., Hartung, J., Gräser, H.: J. Chromatogr.,280,331–342 (1983)
[5] Senanayake, N.D., Northcote, D.H.: Phytochemistry,18,741–748 (1979)
[6] Storm, D.L., Hassid, W.Z.: Plant Physiol.,54,840–845 (1974)

1 NOMENCLATURE

EC number
   2.4.1.36

Systematic name
   GDPglucose:D-glucose-6-phosphate 1-alpha-D-glucosyltransferase

Recommended name
   alpha,alpha-Trehalose-phosphate synthase (GDP-forming)

Synonyms
   GDPglucose-glucose-phosphate glucosytransferase
   GDP glucose-glucosephosphate glucosyltransferase
   Guanosine diphosphoglucose-glucose phosphate glucosytransferase
   Trehalose phosphate synthase (GDP-forming)
   Glucosyltransferase, guanosine diphosphoglucose-glucose phosphate
   More (cf. EC 2.4.1.15)

CAS Reg. No.
   37257-32-2

---

2 REACTION AND SPECIFICITY

Catalysed reaction
   GDPglucose + glucose 6-phosphate →
   → GDP + alpha,alpha-trehalose 6-phosphate

Reaction type
   Hexosyl group transfer

Natural substrates

Substrate spectrum
   1 GDPglucose + glucose 6-phosphate [1]

Product spectrum
   1 GDP + alpha,alpha-trehalose 6-phosphate [1]

Inhibitor(s)

Cofactor(s)/prosthetic group(s)/activating agents

Metal compounds/salts

Turnover number (min⁻¹)

Specific activity (U/mg)

K_m-value (mM)

pH-optimum

pH-range

Temperature optimum (°C)

Temperature range (°C)

3 ENZYME STRUCTURE

Molecular weight

Subunits

Glycoprotein/Lipoprotein

–

4 ISOLATION/PREPARATION

Source organism
  Streptomyces hygroscopicus [1]

Source tissue

Localization in source
  Soluble [1]

Purification

Crystallization

–

Cloned

–

Renatured

–

5 STABILITY

pH

Temperature (°C)

Oxidation

Organic solvent

General stability information

Storage

6 CROSSREFERENCES TO STRUCTURE DATABANKS

PIR/MIPS code

Brookhaven code

7 LITERATURE REFERENCES

[1] Elbein, A.D.: J. Biol. Chem.,242,403–406 (1967)

Enzyme Handbook © Springer-Verlag Berlin Heidelberg 1996
Duplication, reproduction and storage in data banks are only
allowed with the prior permission of the publishers

1 NOMENCLATURE

EC number
2.4.1.37

Systematic name
UDPgalactose:glycoprotein-alpha-L-fucosyl-(1,2)-D-galactose 3-alpha-D-galactosyltransferase

Recommended name
Fucosylglycoprotein 3-alpha-galactosyltransferase

Synonyms
UDPgalactose:O-alpha-L-fucosyl(1→2)D-galactose alpha-D-galactosyltrans-ferase [3]
Galactosyltransferase, [blood-group substance] alpha-
Blood-group substance B-dependent galactosyltransferase
Histo-blood substance B-dependent galactosyltransferase
Histo-blood group B transferase
[Blood group substance] alpha-galactosyltransferase

CAS Reg. No.
37257-33-3

---

2 REACTION AND SPECIFICITY

Catalysed reaction
UDPgalactose + glycoprotein alpha-L-fucosyl-(1,2)-D-galactose →
→ UDP + glycoprotein alpha-D-galactosyl-(1,3)-[alpha-L-fucosyl-(1,2)]-D-galactose

Reaction type
Hexosyl group transfer

Natural substrates
UDPgalactose + glycoprotein alpha-L-fucosyl-(1,2)-D-galactose (acts on blood group substance) [1]

Substrate spectrum
1 UDPgalactose + glycoprotein alpha-L-fucosyl-(1,2)-D-galactose (enzyme transfers D-galactose in alpha-linkage to oligosaccharides, glycolipids and glycoproteins with terminal non-reducing H-active structures and confers blood group B activity on group O erythrocytes [2], converts blood group O red blood cells to B-cells [4]) [1–4]
2 UDPgalactose + O-alpha-L-fucosyl(1→2)galactose [2]

3 UDPgalactose + 2'-fucosyllactose [2, 3]
4 UDPgalactose + lacto-N-fucopentaose I [2]
5 UDPgalactose + H-active glycoprotein [2]
6 UDPgalactose + N-acetyllactosamine [3]
7 More (2 distinct alpha-3-D-galactosyltransferases: one which is more tightly membrane-bound, resembles the human B-gene-specific transferase in its acceptor specificity, and the second, which is a more soluble enzyme transfers D-galactose to the same positional linkage in unsubstituted beta-D-galactosyl residues) [3]

Product spectrum
1 UDP + glycoprotein alpha-D-galactosyl-(1,3)-[alpha-L-fucosyl-(1,2)]-D-galactose
2 ?
3 ?
4 ?
5 UDP + B-active substance [2]
6 ?
7 ?

Inhibitor(s)
UDP (competitive with respect to UDPgalactose) [2]; UDP-N-acetyl-galactosamine (weak, competitive with respect to UDPgalactose) [2]

Cofactor(s)/prosthetic group(s)/activating agents

Metal compounds/salts
Mn^{2+} (required, maximal activity at 15–30 mM [4], divalent cation required, Mn^{2+} most effective, optimum concentration 20 mM [2, 3]) [2–4]; Co^{2+} (slight activation) [2]

Turnover number (min^{-1})

Specific activity (U/mg)
1.92 [2]; More [4]

K_m-value (mM)
0.01 (UDPgalactose) [2]; 0.036 (UDPgalactose (+ 2'-fucosyllactose)) [3]; 0.050 (UDPgalactose (+ N-acetyllactosamine)) [3]; 0.08 (2'-fucosyllactose) [3]; 0.5 (2'-fucosyllactose) [2]; 0.8 (N-acetyllactosamine) [3]; 2.2 (O-alpha-L-fucosyl(1→2)galactose) [2]; 2.5 (lacto-N-fucopentaose I) [2]

pH-optimum
6.4–6.8 (N-acetyllactosamine) [3]; 6.5 [2]; 6.8 (2'-fucosyllactose) [3]; 7.0–7.5 [4]

pH-range

Temperature optimum (°C)

Temperature range (°C)

3 ENZYME STRUCTURE

Molecular weight
  80000 (human, gel filtration, 0.2 M NaCl added to the buffer) [4]

Subunits
  Dimer (2 × 40000, human, SDS-PAGE) [4]

Glycoprotein/Lipoprotein
  –

4 ISOLATION/PREPARATION

Source organism
  Rabbit (2 distinct alpha-3-D-galactosyltransferases: one which is more tightly membrane-bound, resembles the human B-gene-specific transferase in its acceptor specificity, and the second, which is a more soluble enzyme transfers D-galactose to the same positional linkage in unsubstituted beta-D-galactosyl residues) [3]; Human (blood group B donors) [1, 2, 4]; Baboon (blood group B donors) [1]

Source tissue
  Stomach [1]; Submaxillary gland [1]; Serum [2]; Stomach mucosa [3]; Blood plasma [4]

Localization in source
  Membrane (the enzyme catalyzing the transfer of galactose to 2'-fucosyllactose is much more tightly attached to membrane than the transferase which utilizes N-acetyllactosamine) [3]

Purification
  Human (one-step procedure involving absorption onto group O erythrocyte membranes followed by elution with the low molecular weight H-active trisaccharide 2'-fucosyllactose [2]) [2, 4]

Crystallization
  –

Cloned
  –

Renatured

–

---

5 STABILITY

pH

Temperature (°C)
55 (5 min, 20% loss of activity, 20 min, 60% loss of activity) [2]

Oxidation

Organic solvent

General stability information
EDTA, 1 mM and glycerol, 5% v/v stabilize [4]

Storage
4°C, 25% loss of activity after 2 weeks, crude extract [4]

---

6 CROSSREFERENCES TO STRUCTURE DATABANKS

PIR/MIPS code

Brookhaven code

---

7 LITERATURE REFERENCES

[1] Race, C., Zideman, D., Watkins, W.M.: Biochem. J.,107,733–735 (1968)
[2] Carne, L.R., Watkins, W.M.: Biochem. Biophys. Res. Commun.,77,700–707 (1977)
[3] Betteridge, A., Watkins, W.M.: Eur. J. Biochem.,132,29–35 (1983)
[4] Nagai, M., Dave, V., Muensch, H., Yoshida, A.: J. Biol. Chem.,253,380–381 (1978)

1 NOMENCLATURE

EC number
2.4.1.38

Systematic name
UDPgalactose:N-acetyl-beta-D-glucosaminylglycopeptide beta-1,4-galactosyltransferase

Recommended name
beta-N-Acetylglucosaminylglycopeptide beta-1,4-galactosyltransferase

Synonyms
UDPgalactose-glycoprotein galactosyltransferase
Glycoprotein 4-beta-galactosyltransferase
beta-N-Acetyl-D-glucosaminide beta-1,4-galactosyltransferase
Galactosyltransferase, uridine diphosphogalactose-glycoprotein
beta1–4-Galactosyltransferase
Galactosyltransferase, thyroid glycoprotein beta-
Glycoprotein beta-galactosyltransferase
Thyroid galactosyltransferase
UDP-galactose-glycoprotein galactosyltransferase
Uridine diphosphogalactose-glycoprotein galactosyltransferase
More (cf. EC 2.4.1.22, not distinguishable from EC 2.4.1.90)

CAS Reg. No.
37237-43-7

2 REACTION AND SPECIFICITY

Catalysed reaction
UDPgalactose + N-acetyl-beta-D-glucosaminyl-glycopeptide →
→ UDP + beta-D-galactosyl-1,4-N-acetyl-beta-D-glucosaminylglycopeptide
(the acceptor sugar N-acetylglucosamine may be the free monosaccharide
or the non-reducing terminal monosaccharide of a carbohydrate side-chain
of a glycoprotein or glycolipid. Not distinguishable from EC 2.4.1.90 which
has identical substrate specificities. EC 2.4.1.38/90 is identical to the A protein of EC 2.4.1.22. The sequences of cDNA isolated from mammary and F9
cell lines are identical, thus indicating that EC 2.4.1.38 and EC 2.4.1.90 are
non-distinguishable [1]. In the presence of alpha-lactalbumin the reaction of
EC 2.4.1.22 is catalyzed.)

Reaction type
Hexosyl group transfer

Natural substrates

Substrate spectrum

Product spectrum

Inhibitor(s)

Cofactor(s)/prosthetic group(s)/activating agents

Metal compounds/salts

Turnover number (min^{-1})

Specific activity (U/mg)

K_m-value (mM)

pH-optimum

pH-range

Temperature optimum (°C)

Temperature range (°C)

3 ENZYME STRUCTURE

Molecular weight

Subunits

Glycoprotein/Lipoprotein
—

4 ISOLATION/PREPARATION

Source organism

Source tissue

Localization in source

Purification

Crystallization
—

Cloned
—

Renatured
—

5 STABILITY

pH

Temperature (°C)

Oxidation

Organic solvent

General stability information

Storage

6 CROSSREFERENCES TO STRUCTURE DATABANKS

PIR/MIPS code

Brookhaven code

7 LITERATURE REFERENCES

[1] Nakazawa, K., Furukawa, K., Kobata, A., Narimatsu, H.: Eur. J. Biochem.,196, 363–368 (1991)

1 NOMENCLATURE

EC number
2.4.1.39

Systematic name
UDP-N-acetyl-D-glucosamine:estradiol-17alpha-3-D-glucuronoside
17alpha-N-acetylglucosaminyltransferase

Recommended name
Steroid N-acetylglucosaminyltransferase

Synonyms
Acetylglucosaminyltransferase, hydroxy steroid
Steroid acetylglucosaminyltransferase
Uridine diphosphoacetylglucosamine-steroid acetylglucosaminyltransferase

CAS Reg. No.
9033-56-1

2 REACTION AND SPECIFICITY

Catalysed reaction
UDP-N-acetyl-D-glucosamine + estradiol-17alpha 3-D-glucuronoside →
→ UDP + 17alpha-(N-acetyl-D-glucosaminyl)-estradiol 3-D-glucuronoside

Reaction type
Hexosyl group transfer

Natural substrates

Substrate spectrum
1 17alpha-Estradiol-3-glucuronoside + UDP-N-acetylglucosamine [1]
2 16-Epiestriol-3-glucuronoside + UDP-N-acetylglucosamine [1]
3 16,17-Epiestriol-3-glucuronoside + UDP-N-acetylglucosamine [1]

Product spectrum
1 UDP + 17alpha-(N-acetyl-D-glucosaminyl)-estradiol 3-D-glucuronoside [1]
2 UDP + 16-(N-acetyl-D-glucosaminyl)-epiestriol 3-D-glucuronoside [1]
3 UDP + 16,17-(N-acetyl-D-glucosaminyl)-epiestriol 3-D-glucuronoside [1]

Inhibitor(s)
 Ethanol [1]; Butanol-1 [1]; Eugenol (i.e. 2-methoxy-4-(2-propenyl)phenol) [1];
 Estrone [1]; Diethylstilbestrol [1]; Phospholipase C [2]; Trypsin [2]; Chymo-
 trypsin [2]

Cofactor(s)/prosthetic group(s)/activating agents

Metal compounds/salts

Turnover number (min^{-1})

Specific activity (U/mg)

K_m-value (mM)
 0.0684 (UDP-N-acetylglucosamine) [1]; 0.168 (17alpha-estradiol-3-glucu-
 ronoside) [1]

pH-optimum
 8.0 [1]

pH-range

Temperature optimum (°C)

Temperature range (°C)

---

3 ENZYME STRUCTURE

Molecular weight

Subunits

Glycoprotein/Lipoprotein
 –

---

4 ISOLATION/PREPARATION

Source organism
 Rabbit [1, 2]

Source tissue
 Liver [1, 2]; Kidney [1]; Small intestine [1]; Large intestine [1]

Localization in source
 Microsomes [1]

Purification

Crystallization
 –

Cloned

–

Renatured

–

5 STABILITY

pH

Temperature (°C)

Oxidation

Organic solvent

General stability information

Storage
   –10°C, half-life 25 d [1]

6 CROSSREFERENCES TO STRUCTURE DATABANKS

PIR/MIPS code

Brookhaven code

7 LITERATURE REFERENCES

[1] Collins, D.C., Jirku, H., Layne, D.S.: J. Biol. Chem.,243,2928–2933 (1968)
[2] Labow, R.S., Williamson, D.G., Layne, D.S.: Biochemistry,12,1548–1551 (1973)

1 NOMENCLATURE

EC number
2.4.1.40

Systematic name
UDP-N-acetyl-D-galactosamine:alpha-L-fucosyl-(1,2)-D-galactose
3-N-acetyl-D-galactosaminyltransferase

Recommended name
Fucosylgalactose alpha-N-acetylgalactosaminyltransferase

Synonyms
A-transferase
Histo-blood group A glycosyltransferase (Fucalpha1→2Galalpha1→3-N-ace-
tylgalactosaminyltransferase) [4]
UDP-GalNAc:Fucalpha1→2Galalpha1→3-N-acetylgalactosaminyltransf-
erase [4]
alpha-3-N-Acetylgalactosaminyltransferase [5, 6]
Acetylgalactosaminyltransferase, [blood-group substance] alpha-
Acetylgalactosaminyltransferase, fucosylgalactose
Blood-group substance alpha-acetyltransferase
Blood-group substance A-dependent acetylgalactosaminyltransferase
Fucosylgalactose acetylgalactosaminyltransferase
Histo-blood group A acetylgalactosaminyltransferase
Histo-blood group A transferase
UDP-N-acetyl-D-galactosamine:alpha-L-fucosyl-1,2-D-galactose 3-N-acetyl-
D-galactosaminyltransferase

CAS Reg. No.
9067-69-0

---

2 REACTION AND SPECIFICITY

Catalysed reaction
UDP-N-acetyl-D-galactosamine + alpha-L-fucosyl-1,2-D-galactose →
→ UDP + N-acetyl-alpha-D-galactosaminyl-1,3-[alpha-L-fucosyl-1,2]-D-galac-
tose

Reaction type
Hexosyl group transfer

Natural substrates
More (transfers N-acetylgalactosamine from UDP-N-acetylgalactosamine to
H-active structures to form A determinants) [6]

Substrate spectrum

1 UDP-N-acetyl-D-galactosamine + alpha-L-fucosyl-1,2-D-galactose [1–6]
2 UDP-N-acetyl-D-galactosamine + 2'-fucosyllactose [1, 3–5]
3 UDP-N-acetyl-D-galactosamine + lacto-N-fucopentaose I (i.e. Fuc-alpha(1–2)Galbeta(1–3)GlcNAcbeta(1–3)Galbeta(1–4)Glc) [1, 5, 6]
4 UDP-N-acetyl-D-galactosamine + asialo-porcine submaxillary mucin of A-negative blood-type [2]
5 More (effective acceptor substrates contain a subterminal beta-galactosyl residue substituted at the O-2 position with fucose [6], enzyme also catalyzes the transfer of galactose in alpha-linkage to 2'-fucosyllactose, the transfer rate of galactose is much lower than that of N-acetylgalactosamine [5]) [5, 6]

Product spectrum

1 UDP + N-acetyl-alpha-D-galactosaminyl-1,3-[alpha-L-fucosyl-1,2]-D-galactose [1–6]
2 ?
3 ?
4 ?
5 ?

Inhibitor(s)

Uridine [6]; CDP [6]; Fluorescein mercuriacetate [6]; N-Ethylmaleimide (weak) [6]; GDP (weak) [6]; UMP [2, 6]; UDP [5, 6]; UDPgalactose [5]; 2'-Fucosyllactose (substrate inhibition above 1 mM) [5]; Lacto-N-fucopentaose I (substrate inhibition above 2 mM) [5]

Cofactor(s)/prosthetic group(s)/activating agents

Metal compounds/salts

Mn^{2+} (maximal activity with 10 mM Mn^{2+} [6], 15–30 mM required for maximal activity [3]) [3, 6]

Turnover number (min^{-1})

Specific activity (U/mg)

0.2 (plasma) [4]; 5.7 (lung) [4]; 7 [6]; 30 [2]; More [3]

K_m-value (mM)

0.013 (UDP-GalNAc) [5]; 0.270 (2'-fucosyllactose) [5]; 0.350 (lacto-N-fucopentaose I) [5]

pH-optimum

6.0 (assay at) [2]; 6.5–7.0 [3]; 6.6 [5]; 7.0–7.4 [6]

pH-range

Temperature optimum (°C)
    37 (assay at) [2, 5]

Temperature range (°C)

---

3 ENZYME STRUCTURE

Molecular weight
    90000–100000 (human, gel filtration) [3]
    100000 (pig, zonal centrifugation, gel filtration) [2]

Subunits
    Dimer (2 × 46000, pig, sedimentation equilibrium of reduced and carboxy-
    methylated enzyme [2], 2 × 52000, human, SDS-PAGE of carboxymaleyl
    enzyme [3]) [2, 3]
    ? (x × 35000, human, SDS-PAGE with and without 2-mercaptoethanol [5],
    x × 40000, human, SDS-PAGE [4, 6], under reducing and nonreducing con-
    ditions [4]) [4–6]

Glycoprotein/Lipoprotein
    Glycoprotein (N-linked carbohydrate chains [4]) [2, 4]

---

4 ISOLATION/PREPARATION

Source organism
    Human (women [1], blood type A or AB [1], blood group A1 [3], blood
    group A [4–6]) [1, 3–6]; Pig [2]

Source tissue
    Lung [4]; Milk [1]; Blood plasma (blood group A1 [3], A [4, 5]) [3–5]; Sub-
    maxillary gland [2]; Kidney [4]; Gut mucosal tissue [6]

Localization in source

Purification
    Pig [2]; Human (partial [4], blood group A [4–6]) [3–6]

Crystallization
    –

Cloned
    –

Renatured
    –

---

5 STABILITY

pH

Temperature (°C)
37 (30 min, 5% loss of activity) [5]

Oxidation

Organic solvent

General stability information
Enzyme is not stable in crude plasma [3]; Irreversible inactivation in absence of Triton X-100 [2]; Loss of activity after freezing in buffer of pH 7.5 [2]

Storage
4°C, pH 7.5, half-life is about 3 months [2]; 4°C, rapid loss of activity of purified enzyme [3]; 4°C, 0.05 M Tris-HCl, pH 7.4, 2 mM $MnCl_2$, 1 mM EDTA, 0.01% w/v Triton X-100, 0.2 M NaCl, 1 mM dithiothreitol, 0.03% w/v NaN_3, 10% loss of activity per month [5]; –60°C, 2 mM $MnCl_2$, 1 mM EDTA, 0.02 mM UDP, stable for up to 30 days [3]; –70°C, 0.025 M MES buffer, pH 6.0, 0.001 M EDTA, 0.03% Triton X-100, after removal of most of the buffer by dialysis, stable for 1 year [2]; –80°C, 20% loss of activity after 1 year [5]; –20°C, rapid loss of activity [5]; 4°C, 25% loss of activity after storage of plasma for 2 weeks [3]

6 CROSSREFERENCES TO STRUCTURE DATABANKS

PIR/MIPS code

Brookhaven code

7 LITERATURE REFERENCES

[1] Kobata, A., Grollman, E.F., Ginsburg, V.: Arch. Biochem. Biophys.,124,609–612 (1968)
[2] Schwyzer, M., Hill, R.L.: J. Biol. Chem.,252,2338–2345 (1977)
[3] Nagai, M., Dave, V., Kaplan, B.E., Yoshida, A.: J. Biol. Chem.,253,377–379 (1987)
[4] Clausen, H., White, T., Takio, K., Titani, K., Stroud, M., Holmes, E., Karkov, J., Thim, L., Hakomori, S.: J. Biol. Chem.,265,1139–1145 (1990)
[5] Takeya, A., Hosomi, O., Ishiura, M.: J. Biochem.,107,360–368 (1990)
[6] Navaratnam, N., Findlay, J.B.C., Keen, J.N., Watkins, W.M.: Biochem. J.,271,93–98 (1990)

1 NOMENCLATURE

EC number
2.4.1.41

Systematic name
UDP-N-acetyl-D-galactosamine:polypeptide N-acetylgalactosaminyltrans-ferase

Recommended name
Polypeptide N-acetylgalactosaminyltransferase

Synonyms
Protein-UDP acetylgalactosaminyltransferase
UDP-GalNAc:polypeptide N-acetylgalactosaminyl transferase [1]
UDP-N-acetylgalactosamine:kappa-casein polypeptide N-acetylgalactos-aminyltransferase [2]
Acetylgalactosaminyltransferase, uridine diphosphoacetylgalactosamine-glycoprotein
Glycoprotein acetylgalactosaminyltransferase
Polypeptide-N-acetylgalactosamine transferase
UDP-acetylgalactosamine-glycoprotein acetylgalactosaminyltransferase
UDP-acetylgalactosamine:peptide-N-galactosaminyltransferase
UDP-GalNAc:polypeptide N-acetylgalactosaminyltransferase
UDP-N-acetyl-alpha-D-galactosamine:polypeptide N-acetylgalactosaminyl-transferase
UDP-N-acetylgalactosamine-glycoprotein N-acetylgalactosaminyltransferase
UDP-N-acetylgalactosamine-protein N-acetylgalactosaminyltransferase
UDP-N-acetylgalactosamine:polypeptide N-acetylgalactosaminyltransferase
UDP-N-acetylgalactosamine:protein N-acetylgalactosaminyl transferase

CAS Reg. No.
9075-15-4

---

2 REACTION AND SPECIFICITY

Catalysed reaction
UDP-N-acetyl-D-galactosamine + polypeptide →
→ UDP + N-acetyl-D-galactosaminyl-polypeptide

Reaction type
Hexosyl group transfer

Natural substrates

UDP-N-acetyl-D-galactosamine + polypeptide (catalyzes the first step in biosynthesis of O-linked oligosaccharides in many glycoproteins) [2]

Substrate spectrum

1 UDP-N-acetyl-D-galactosamine + polypeptide (only UDP-GalNAc serves as sugar donor (colostrum enzyme [6]) [1, 6], not: UDP-Gal [4], UDP-GlcNAc [4], acceptors: apomucin [1, 2], A1 protein [1], kappa-casein [1, 2], apofetuin [1], apoantifreeze glycoproteins [1], asialo mucin [2], mucin-like synthetic peptides [4], bovine submaxillary mucin core protein [8], myelin basic protein [8], synthetic polypeptides with sequence identical or similar to those found in porcine mucin or human erythropoietin [3], synthetic peptides containing a triprolyl carboxyl to a threonine residue, threonine cannot be glycosylated without a carboxyl triprolyl sequence, the alpha amino acid group of the threonine must be blocked, the nature of the group NH_2-terminal to the threonine effects the kinetics of the reaction, one residue can be between the threonyl and the triprolyl sequence [7], acceptor substrate specificity is dependent on the amino acid sequence adjacent to serine and threonine residues [3], specifically transfers N-acetylgalactosamine from UDP-GalNAc to the hydroxyl group of threonine, devoid of transferase activity towards serine-containing peptides [4], L-threonine specifically O-glycosylated in alpha-configuration [8], Thr-Pro-Pro-Pro-sequence requires a peptide length of five or more for significant acceptor activity [8]) [1–8]

2 More (not: lacto-N-fucopentaose I [1], ceramide dihexoside [1], ceramide trihexoside [1], globoside [1], $alpha_{S1}$-casein [2], $alpha_{S2}$-casein [2], beta-casein [2], alpha-lactalbumin [2], beta-lactalbumin [2], bovine serum albumin [2]) [1, 2]

Product spectrum

1 UDP + N-acetyl-D-galactosaminyl-polypeptide
2 ?

Inhibitor(s)

$alpha_{S1}$-Casein (inhibtion of GalNAc transfer to kappa-casein) [2]; CTP [2]; GTP [2]; CDP (not [1]) [2]; GDP (not [1], weak) [2]; AMP [2]; CMP [2]; GMP [2]; Helix pomatia lectin [2]; Soybean lectin [2]; Mn^{2+} (metal-free enzyme is inactive, Mn^{2+} is the best activator, maximal activity at 5 mM, half-maximal activity at 0.4 mM $MnCl_2$, 20 and 100 mM result in 10% and 80% inhibition) [4]; UMP-butanesulfonic anhydride [5]; UMP-1-octanesulfonic anhydride [5]; Uridine 5'-phosphonic (1-hexadecanesulfonic anhydride) [5]; EDTA (Na_2EDTA [8]) [1, 6, 8]; UDP [1, 2]; UTP [1, 2]; UMP [1, 2]; ATP (not [1], weak [2]) [2]; More (not: wheat germ lectin) [2]

Cofactor(s)/prosthetic group(s)/activating agents
DTT (required for maximal activity) [2]; Triton X-100 (required for maximal activity) [2]

Metal compounds/salts
Mn^{2+} (metal-free enzyme is inactive, Mn^{2+} is the best activator, maximal activity at 5 mM, half-maximal activity at 0.4 mM $MnCl_2$, 20 and 100 mM result in 10% and 80% inhibition [4], required, most effective cation [1], required for maximal activity, maximal activity with 4–10 mM [2], required, restores activity after EDTA treatment [6], required, 10 mM [8]) [1, 2, 4, 6, 8]; Co^{2+} (stimulates [1], restores activity after EDTA treatment [6]) [1, 6]; Cu^{2+} (stimulates) [1]; Zn^{2+} (stimulates to a smaller extent than Mn^{2+} [2], no reactivation of metal-free enzyme [4]) [2]; Ca^{2+} (stimulates to a smaller extent than Mn^{2+} [2], no reactivation of metal-free enzyme [4]) [2]; Mg^{2+} (stimulates to a smaller extent than Mn^{2+} [2], no reactivation of metal-free enzyme [4]) [2]; Ba^{2+} (stimulates to a smaller extent than Mn^{2+}) [2]; Ni^{2+} (partially restores activity after EDTA treatment) [6]; Cd^{2+} (partially restores activity after EDTA treatment) [6]

Turnover number (min^{-1})

Specific activity (U/mg)
0.39 [1]; 8.63 [2]; 1.65 [4]

K_m-value (mM)
0.0045 (Pro-Pro-Asp-Ala-Ala-Ser-Ala-Ala-Pro-Leu-Arg) [4]; 0.006 (UDP-Gal-NAc (+ ovine apomucin)) [4]; 0.008 (UDP-GalNAc) [6]; 0.0083 (Val-Leu-Gly-Thr-Ala-Ala-Val) [4]; 0.0129 (Pro-Pro-Asp-Val-Val-Ser-Val-Val-Pro-Leu-Arg) [4]; 0.0140 (Val-Leu-Gly-Ala-Thr-Ala-Val) [4]; 0.0149 (Asp-Ala-Ala-Ser-Ala-Ala-Pro-Leu) [4]; 0.0162 (UDP-GalNAc) [2]; 0.0180 (Pro-Pro-Asp-Ala-Ser-Ser-Ser-Ala-Pro-Leu-Arg) [4]; 0.0186 (Val-Leu-Gly-Thr-Thr-Ala-Val) [4]; 0.0295 (Gln-Ala-Ala-Gly-Thr-Ser-Gly-Ala-Gly-Pro-Gly) [4]; 0.0417 (UDP-GalNAc) [1]; 0.11 (UDP-GalNAc) [8]; 0.40 (D-Arg-Thr-Pro-Pro-Pro, myelin basic protein) [7]; 0.52 (Pro-Thr-Ala-Pro-Pro-Pro) [7]; 1.33 (Ac-Thr-Pro-Pro-Pro) [7]; 2.0 (Ac-Asn-Leu-Thr-Pro-Pro-Pro) [7]; 3.0 (Val-Thr-Pro-Arg-Thr-Pro-Pro-Pro) [4]; 7.2 (Val-Lys-Thr-Glu-Ala-Thr-Thr-Phe-Ile) [4]; 18.6 (Val-Leu-Gly-Thr-Ala-Val) [4]; More (2.5 mg/ml: deglycosylated bovine submaxillary mucin [6], 1.01 mg/ml: ovine apomucin [4], 2.54 mg/ml: porcine apomucin [4], 0.192 mg/ml: apomucin [2], 1.15 mg/ml: kappa-subcomponent 1 [2], 5.1 mg/ml: kappa-subcomponent 7 [2]) [2, 4, 6]

pH-optimum
6.5–7.5 (Tris, imidazole and MES buffer, kappa-casein) [2]; 6.8–8.2 [4]; 7.2–8.6 (colostrum enzyme) [6]; 7.2 (assay at) [1, 2]; 7.5 [1, 8]

pH-range
6.8–9.5 (4.5: no activity, 6.8: gradual decrease of activity below, 9.5: 16% of activity maximum) [4]

Temperature optimum (°C)
37 (assay at) [1, 8]

Temperature range (°C)

3 ENZYME STRUCTURE

Molecular weight
55000 (rat, ascites fluid from hepatoma AH 66 cells, gel filtration) [1]
70000 (bovine colostrum, gel filtration, BW 5147 mouse lymphoma cells, gel filtration) [6]
200000 (bovine, gel filtration) [2]

Subunits
Monomer (1 × 55000, rat, SDS-PAGE in presence of 2-mercaptoethanol [1], 1 × 70000, bovine colostrum, BW 5147 mouse lymphoma cells, SDS-PAGE [6]) [1, 6]

Glycoprotein/Lipoprotein
Glycoprotein (colostrum enzyme, 2 N-linked oligosaccharides on most enzyme molecules, both oligosaccharides are of the complex type, but some molecules contain one complex type and one high mannose type) [6]

4 ISOLATION/PREPARATION

Source organism
Rat [1]; Bovine (lactating cow [2]) [2, 6, 7]; Pig [3–5, 8]; Mouse [6]

Source tissue
Mammary gland [2]; Submaxillary gland [3–5, 7]; BW 5147 mouse lymphoma cells [6]; Tracheal epithelium [7]; Ascites fluid from hepatoma AH 66 cells [1]

Localization in source
Soluble (bovine colostrum) [6]; Golgi apparatus [2]; Membrane (bound [1, 4, 8]) [1, 4, 7, 8]

Purification
Rat (ascites fluid from hepatoma AH 66 cells) [1]; Bovine (lactating cow [2]) [2, 6]; Pig [4]; Mouse (BW 5147 mouse lymphoma cells) [6]

Crystallization
–

4

Cloned

–

Renatured

–

5 STABILITY

pH

Temperature (°C)
37 (8 h, 40% loss of activity) [4]; 4–35 (1 h, less than 20% loss of activity) [2]; 45 (1 h, 50% loss of activity) [2]; 55 (1 h, complete loss of activity) [2]

Oxidation

Organic solvent

General stability information

Storage
–20°C, stable for 32 months [1]; –20°C, 20 mM imidazole-HCl, pH 7.2, 0.1% Triton X-100, 5 mM $MnCl_2$, 60% w/v glycerol, stable for 1 year [4]; 4°C, 20 mM imidazole-HCl, pH 7.2, 0.1% Triton X-100, 5 mM $MnCl_2$, 25% w/v glycerol, 10% loss of activity after 1 month [4]

6 CROSSREFERENCES TO STRUCTURE DATABANKS

PIR/MIPS code
PIR2:A45987 (bovine)

Brookhaven code

7 LITERATURE REFERENCES

[1] Sugiura, M., Kawasaki, T., Yamashina, I.: J. Biol. Chem.,257,9501–9507 (1982)
[2] Takeuchi, M., Yoshikawa, M., Sasaki, R., Chiba, R.: Agric. Biol. Chem.,49,1059–1069 (1985)
[3] Wang, Y., Agrwal, N., Eckhardt, A.E., Stevens, R.D., Hill, R.L.: J. Biol. Chem.,268, 22979–22983 (1993)
[4] Wang, Y., Abernethy, J.L., Eckhardt, A.E., Hill, R.L.: J. Biol. Chem.,267,12709–12716 (1992)
[5] Hatanaka, K., Slama, J.T., Elbein, A.D.: Biochem. Biophys. Res. Commun.,175, 668–672 (1991)
[6] Elhammer, A., Kornfeld, S.: J. Biol. Chem.,261,5249–5255 (1986)
[7] Briand, J.P., Andrews, S.P., Cahill, E., Conway, N.A., Young, J.D.: J. Biol. Chem.,256, 12205–12207 (1981)
[8] Cottrell, J.M., Hall, R.L., Sturton, R.G., Kent, P.W.: Biochem. J.,283,299–305 (1992)

1 NOMENCLATURE

EC number
2.4.1.43

Systematic name
UDP-D-galacturonate:1,4-alpha-poly-D-galacturonate 4-alpha-D-galacturono-syltransferase

Recommended name
Polygalacturonate 4-alpha-galacturonosyltransferase

Synonyms
Galacturonosyltransferase, uridine diphosphogalacturonate-polygalacturo-nate alpha-
UDP galacturonate-polygalacturonate alpha-galacturonosyltransferase
Uridine diphosphogalacturonate-polygalacturonate alpha-galacturonosyl-transferase

CAS Reg. No.
37277-53-5

---

2 REACTION AND SPECIFICITY

Catalysed reaction
UDP-D-galacturonate + (1,4-alpha-D-galacturonosyl)$_n$ →
→ UDP + (1,4-alpha-D-galacturonosyl)$_{n+1}$

Reaction type
Hexosyl group transfer

Natural substrates

Substrate spectrum
1 UDP-D-galacturonic acid + an alpha-1,4-linked D-galacturonic acid chain [1]

Product spectrum
1 UDP + polygalacturonic acid chain [1]

Inhibitor(s)
ZnCl$_2$ (weak) [1]; CaCl$_2$ (weak) [1]; CuSO$_4$ (strong) [1]; HgCl$_2$ (strong) [1]; UMP (strong) [1]; UDP (strong) [1]; UTP (strong) [1]; Diphosphate (weak) [1]

Cofactor(s)/prosthetic group(s)/activating agents
Sucrose (0.4 M, 5fold increase of activity) [1]; Bovine serum albumin (necessary in homogenization media) [1]

Metal compounds/salts
$MnCl_2$ (stimulation, optimal concentration: 1.7 mM) [1]; $MnSO_4$ (stimulation) [1]; KCl (slight stimulation) [1]; $MgCl_2$ (slight stimulation) [1]; NH_4Cl (slight stimulation) [1]; NaCl (slight stimulation) [1]; $CoCl_2$ (slight stimulation) [1]; Na_2SO_4 (slight stimulation) [1]

Turnover number (min^{-1})

Specific activity (U/mg)

K_m-value (mM)
0.0017 (UDP-D-galacturonic acid) [1]

pH-optimum
6.0 [1]

pH-range
5.3–7.0 (33% of maximal activity at pH 5.3, 35% of maximal activity at pH 7.0) [1]

Temperature optimum (°C)
30 [1]

Temperature range (°C)

3 ENZYME STRUCTURE

Molecular weight

Subunits

Glycoprotein/Lipoprotein
–

4 ISOLATION/PREPARATION

Source organism
Phaseolus aureus [1]

Source tissue
Seedling [1]

Localization in source

Purification
Phaseolus aureus (partial) [1]

Crystallization

–

Cloned

–

Renatured

–

5 STABILITY

pH

Temperature (°C)
  37 (5 min, complete inactivation) [1]; 30 (20 min, complete inactivation) [1];
  0 (24 h, complete inactivation) [1]

Oxidation

Organic solvent

General stability information

Storage
  –18°C, about 3 weeks, complete inactivation [1]; –18°C, 1,7 mM $MnCl_2$,
  11 days, 35% loss of activity [1]; –18°C, without $MnCl_2$, 11 days, 60% loss of
  activity [1]

6 CROSSREFERENCES TO STRUCTURE DATABANKS

PIR/MIPS code

Brookhaven code

7 LITERATURE REFERENCES

[1] Villemez, C.L., Swanson, A.L., Hassid, W.Z.: Arch. Biochem. Biophys.,116,446–452
  (1966)

1 NOMENCLATURE

EC number
2.4.1.44

Systematic name
UDPgalactose:lipopolysaccharide 3-alpha-D-galactosyltransferase

Recommended name
Lipopolysaccharide 3-alpha-galactosyltransferase

Synonyms
Uridine diphosphate galactose:lipopolysaccharide alpha-3-galactosyltrans-
ferase [1]
UDP-galactose:lipopolysaccharide alpha,3-galactosyltransferase [2]
Galactosyltransferase, uridine diphosphogalactose-lipopolysaccharide
alpha,3-
Galactosyltransferase, lipopolysaccharide alpha, 3-
UDP-galactose:polysaccharide galactosyltransferase
More (cf. EC 2.4.1.56, EC 2.4.1.58 and EC 2.4.1.73)

CAS Reg. No.
9073-98-7

2 REACTION AND SPECIFICITY

Catalysed reaction
UDPgalactose + lipopolysaccharide →
→ UDP + 1,3-alpha-D-galactosyl-lipopolysaccharide

Reaction type
Hexosyl group transfer

Natural substrates
UDPgalactose + lipopolysaccharide (transfers D-galactosyl residues to
D-glucose in the partially completed core of lipopolysaccharide [1, 2], catalyzes one of the reactions involved in biosynthesis of cell envelope lipopolysaccharide of the organism [1], reaction in vivo is very sensitive to
subtle changes in the fatty acid structure of the membrane phospholipid [3])
[1–3]

Substrate spectrum

1 UDPgalactose + lipopolysaccharide (highly specific for UDPgalactose [1], lipopolysaccharides: lacking the alpha-3-galactosyl residue of the core lipopolysaccharide, from Salmonella typhimurium strains G-30 and SL1060 [1]) [1–3]

Product spectrum

1 UDP + 1,3-alpha-D-galactosyl-lipopolysaccharide [1–3]

Inhibitor(s)

Nonsubstrate lipopolysaccharide (competitive, Salmonella typhimurium SL1032, TV119, LT2 lipopolysaccharides) [1]

Cofactor(s)/prosthetic group(s)/activating agents

Phospholipid (required for activity [1, 2], phosphatidylethanolamine most effective [2]) [1, 2]

Metal compounds/salts

$MgCl_2$ (no activity in absence of divalent cations, highest activity with $MgCl_2$) [1]; $MnCl_2$ (no activity in absence of divalent cations, 20% of the activation with $MgCl_2$) [1]; $CaCl_2$ (no activity in absence of divalent cations, 20% of the activation with $MgCl_2$) [1]

Turnover number (min^{-1})

Specific activity (U/mg)

More [1]

K_m-value (mM)

0.074 (UDPgalactose) [1]

pH-optimum

8.5–9.0 [1]

pH-range

Temperature optimum (°C)

37 (assay at) [1]

Temperature range (°C)

---

3 ENZYME STRUCTURE

Molecular weight

Subunits

Glycoprotein/Lipoprotein
More (unidentified lipid-soluble component which contains no phosphorus is bound to the enzyme, it causes aggregation of the enzyme, but no role of the lipid in enzyme reaction has yet been established) [1]

4 ISOLATION/PREPARATION

Source organism
Salmonella typhimurium (G-30A [1], LT2 [2]) [1, 2]; E. coli (unsaturated fatty acid auxotrophs) [3]

Source tissue
Cell [1]

Localization in source
Membrane (bound) [2]

Purification
Salmonella typhimurium [1]

Crystallization
–

Cloned
–

Renatured
–

5 STABILITY

pH
8.5 (4°C, several days, inactivation) [1]

Temperature (°C)

Oxidation

Organic solvent

General stability information
One cycle of freezing and thawing causes marked loss of activity [1]

Storage
4°C, pH 6.8, 0.02–0.04 mg/ml protein, stable for at least 1 month [1]

6 CROSSREFERENCES TO STRUCTURE DATABANKS

PIR/MIPS code

Brookhaven code

7 LITERATURE REFERENCES

[1] Endo, A., Rothfield, L.: Biochemistry,8,3500–3507 (1969)
[2] Müller, E., Hinckley, A., Rothfield, L.: J. Biol. Chem.,247,2614–2622 (1972)
[3] Beacham, I.R., Silbert, D.F.: J. Biol. Chem.,248,5310–5318 (1973)

1 NOMENCLATURE

EC number
2.4.1.45

Systematic name
UDPgalactose:2-(2-hydroxyacyl)sphingosine 1-beta-D-galactosyltransferase

Recommended name
2-Hydroxyacylsphingosine 1-beta-galactosyltransferase

Synonyms
Galactosyltransferase, uridine diphosphogalactose-2-hydroxyacylsphingosine
UDPgalactose-2-hydroxyacylsphingosine galactosyltransferase [5]
UDPgalactose:ceramide galactosyltransferase [2, 3, 5]
UDPgalactose:2–2-hydroxyacylsphingosine galactosyltransferase [4]

CAS Reg. No.
37277-54-6

2 REACTION AND SPECIFICITY

Catalysed reaction
UDPgalactose + 2-(2-hydroxyacyl)sphingosine →
→ UDP + 1-(beta-D-galactosyl)-2-(2-hydroxyacyl)sphingosine

Reaction type
Hexosyl group transfer

Natural substrates
UDPgalactose + 2-(2-hydroxyacyl)sphingosine (i.e. 2-hydroxy fatty acid
ceramide, synthesis of cerebroside) [1, 2]

Substrate spectrum
1 UDPgalactose + 2-(2-hydroxyacyl)sphingosine (i.e. 2-hydroxy fatty acid
 ceramide [1–3], highly specific [1, 2]) [1–3]

Product spectrum
1 1-(beta-D-Galactosyl)-2-(2-hydroxyacyl)sphingosine + UDP (i.e. 2-hydroxy
 fatty acid galactosylceramide [1]) [1–3]

Inhibitor(s)
Octanoyl-D-threo-p-nitro-1-phenyl-2-amino-1,3-propanediol [3]; Protein
inhibitor (from brain, kidney, spleen and liver of rat and other animals) [3];
Sphingosine (DL- [2]) [2, 3]; Sodium deoxycholate [3]; Sodium taurocholate
[3]; Sodium taurodeoxycholate [3]; Detergents [4]

Cofactor(s)/prosthetic group(s)/activating agents
Phospholipids (addition to incubation mixture containing Triton X-100 increases activity) [3]

Metal compounds/salts
Mg^{2+} (stimulates [2, 4], dialyzed enzyme has requirement for divalent cation [2], Mg^{2+} required rather than Mn^{2+} [4]) [2, 4]; Mn^{2+} (stimulates [2, 4], dialyzed enzyme has requirement for divalent cation [2], Mg^{2+} required rather than Mn^{2+} [4]) [2, 4]; Ca^{2+} (stimulates, dialyzed enzyme has requirement for divalent cation) [2]

Turnover number (min^{-1})

Specific activity (U/mg)
0.002733 [3]; More (assay method [4]) [4, 5]

K_m-value (mM)
0.027–0.034 (UDPgalactose) [3]; 0.04 (UDPgalactose) [2]; 0.11 (ceramide) [2]; More (K_m value for ceramide is influenced by the dispersion state of this lipid, in presence of 2 mM Triton X-100 biphasic kinetic is observed with increasing ceramide concentration, yielding two apparent K_m values of 0.018 and 0.15 mM) [3]

pH-optimum
7.0 (assay at) [2]; 7.7 (Bicine buffer) [2]; 7.8–8.2 [3]; 8.5 [4]

pH-range

Temperature optimum (°C)
37 (assay at) [2, 4]

Temperature range (°C)

3 ENZYME STRUCTURE

Molecular weight

Subunits

Glycoprotein/Lipoprotein
Lipoprotein (intact phospholipids required for full activity [5]) [3, 5]

4 ISOLATION/PREPARATION

Source organism
Mouse [1]; Chicken [2]; Rat [3–5]

Source tissue
Brain [1–5]; Embryo (maximal activity in 19–20 day-old embryos) [2]

2

Localization in source
  Microsomes [1, 3, 5]

Purification
  Rat [3, 5]

Crystallization
  –

Cloned
  –

Renatured
  –

---

5 STABILITY

pH

Temperature (°C)

Oxidation

Organic solvent

General stability information
  Glycerol stabilizes [3, 5]; In the presence of glycerol the enzyme can be frozen and rethawed several times without appreciable loss of stability [3]; Stable to repeated freezing and thawing [4]

Storage
  –20°C, 50% w/v glycerol, about 30% loss of activity after 1 month [3, 5]

---

6 CROSSREFERENCES TO STRUCTURE DATABANKS

PIR/MIPS code

Brookhaven code

---

7 LITERATURE REFERENCES

[1] Morell, P., Radin, N.S.: Biochemistry,8,506–512 (1969)
[2] Basu, S., Schultz, A.M., Basu, M., Roseman, S.: J. Biol. Chem.,246,4272–4279 (1971)
[3] Neskovic, N.M., Mandel, P., Gatt, S.: Methods Enzymol.,71,521–527 (1981) (Review)
[4] Constantino-Ceccarini, E., Cestelli, A.: Methods Enzymol.,72,384–391 (1981) (Review)
[5] Neskovic, N.M., Sarlieve, L.L., Mandel, P.: Biochim. Biophys. Acta,429,342–351 (1976)

---

1 NOMENCLATURE

EC number
2.4.1.46

Systematic name
UDPgalactose:1,2-diacylglycerol 3-beta-D-galactosyltransferase

Recommended name
1,2-Diacylglycerol 3-beta-galactosyltransferase

Synonyms
Galactosyltransferase, uridine diphosphogalactose-1,2-diacylglycerol
UDP-galactose:diacylglycerol galactosyltransferase [3]
MGDG synthase [6]
UDP galactose-1,2-diacylglycerol galactosyltransferase
UDP-galactose-diacylglyceride galactosyltransferase
Uridine diphosphogalactose-1,2-diacylglycerol galactosyltransferase

CAS Reg. No.
37277-55-7

---

2 REACTION AND SPECIFICITY

Catalysed reaction
UDPgalactose + 1,2-diacylglycerol →
→ UDP + 3-beta-D-galactosyl-1,2-diacylglycerol

Reaction type
Hexosyl group transfer

Natural substrates
UDPgalactose + 1,2-diacylglycerol (enzyme is responsible for the biosynthesis of monogalactosyl diglyceride in brain and may function significantly in myelination [1], involed in biosynthesis of monogalactosyldiacyl-
glycerol [6]) [1, 6]

Substrate spectrum
1 UDPgalactose + 1,2-diacylglycerol (requires the 1,2-isomer of the
  diglyceride substrate and prefers diglycerides with long-chain-saturated
  fatty acid constituents [1]) [1–4]
2 UDPgalactose + 1,2-dipalmitoylglycerol [1, 2]
3 UDPgalactose + 1,2-didecanoylglycerol [1]
4 UDPgalactose + 1,2-dioleoylglycerol [1, 3, 4]
5 UDPgalactose + 1,2-dilinoleoylglycerol [1]
6 UDPgalactose + 1,2-distearoylglycerol [4]

Product spectrum

1 UDP + 3-beta-D-galactosyl-1,2-diacylglycerol (i.e. 1,2-di-O-acyl-(beta-D-galactopyranosyl)-sn-glycerol [1]) [1–4]
2 UDP + 3-beta-D-galactosyl-1,2-dipalmitoylglycerol
3 UDP + 3-beta-D-galactosyl-1,2-didecanoylglycerol
4 UDP + 3-beta-D-galactosyl-1,2-dioleoylglycerol
5 UDP + 3-beta-D-galactosyl-1,2-dilinoleoylglycerol
6 UDP + 3-beta-D-galactosyl-1,2-distearoylglycerol

Inhibitor(s)

Zn^{2+} (10 mM: inhibition, 1 mM: stimulation [4]) [3, 4]; Cu^{2+} [2]; UDP [3, 4, 6]; Cd^{2+} [4]; Fe^{2+} (stimulation [2]) [4]

Cofactor(s)/prosthetic group(s)/activating agents

Myristic acid (stimulates) [2]; Palmitic acid (stimulates) [2]; SDS (stimulates) [2]; Lipid (addition of lipids extracted from chloroplast membranes is necessary to reveal activity of highly delipidated enzyme fraction, acidic glycerolipids, especially phosphatidylglycerol, are the best activators [6], no activation [3]) [6]

Metal compounds/salts

Mg^{2+} (addition of cation stimulates, order of efficiency: $Mn^{2+} > Co^{2+} > Mg^{2+} > Fe^{2+} > Ca^{2+} > Ni^{2+}$, $K^+ > Na^+$ [2], stimulates [1, 4], cation not required [4]) [1, 2, 4]; Mn^{2+} (stimulates, cation not required [4], addition of cation stimulates, order of efficiency: $Mn^{2+} > Co^{2+} > Mg^{2+} > Fe^{2+} > Ca^{2+} > Ni^{2+}$, $K^+ > Na^+$ [2]) [2, 4]; Co^{2+} (addition of cation stimulates, order of efficiency: $Mn^{2+} > Co^{2+} > Mg^{2+} > Fe^{2+} > Ca^{2+} > Ni^{2+}$, $K^+ > Na^+$) [2]; Fe^{2+} (addition of cation stimulates, order of efficiency: $Mn^{2+} > Co^{2+} > Mg^{2+} > Fe^{2+} > Ca^{2+} > Ni^{2+}$, $K^+ > Na^+$ [2], inhibition [4]) [2]; Ca^{2+} (addition of cation stimulates, order of efficiency: $Mn^{2+} > Co^{2+} > Mg^{2+} > Fe^{2+} > Ca^{2+} > Ni^{2+}$, $K^+ > Na^+$) [2]; Ni^{2+} (addition of cation stimulates, order of efficiency: $Mn^{2+} > Co^{2+} > Mg^{2+} > Fe^{2+} > Ca^{2+} > Ni^{2+}$, $K^+ > Na^+$) [2]; K^+ (addition of cation stimulates, order of efficiency: $Mn^{2+} > Co^{2+} > Mg^{2+} > Fe^{2+} > Ca^{2+} > Ni^{2+}$, $K^+ > Na^+$) [2]; Na^+ (addition of cation stimulates, order of efficiency: $Mn^{2+} > Co^{2+} > Mg^{2+} > Fe^{2+} > Ca^{2+} > Ni^{2+}$, $K^+ > Na^+$) [2]; Zn^{2+} (10 mM: inhibition, 1 mM: stimulation) [4]

Turnover number (min^{-1})

Specific activity (U/mg)

0.272 [3]; More [5]

K_m-value (mM)

0.040 (UDPgalactose) [4]; 0.087 (UDPgalactose (+ dioleoylglycerol)) [2]; 0.1 (UDPgalactose) [6]; 1 (diacylglycerol) [6]

pH-optimum

5.0–7.0 [2]; 7 (around) [4]

pH-range
   5.5–9 (5.5: 55% of activity maximum, 9.0: 85% of activity maximum) [4]

Temperature optimum (°C)
   30 (assay at, activity maximum in long-term incubation) [4]; 37 (assay at)
   [2]; 50 (activity maximum in short-term incubation) [4]

Temperature range (°C)

---

3 ENZYME STRUCTURE

Molecular weight
   22000 (Spinacia oleracea, gel filtration) [3]

Subunits
   Monomer (1 × 22000, Spinacia oleracea, SDS-PAGE) [3]

Glycoprotein/Lipoprotein
   –

---

4 ISOLATION/PREPARATION

Source organism
   Rat (maximal activity in 14–18 days old rats) [1]; Bifidobacterium bifidum
   (var. pennsylvanicus) [2]; Spinacia oleracea [3–6]

Source tissue
   Brain [1]; Leaf [3]

Localization in source
   Microsomes [1]; Membrane (bound) [2]; Chloroplast envelope (integral
   membrane protein [3]) [3–6]

Purification
   Spinacia oleracea (partial [5, 6]) [3, 5, 6]

Crystallization
   –

Cloned
   –

Renatured
   –

5 STABILITY

pH

Temperature (°C)

Oxidation

Organic solvent

General stability information

Storage

6 CROSSREFERENCES TO STRUCTURE DATABANKS

PIR/MIPS code

Brookhaven code

7 LITERATURE REFERENCES

[1] Wenger, D.A., Petipas, J.W., Pieringer, R. A.: Biochemistry,7,3700–3707 (1968)
[2] Veerkamp, J.H.: Biochim. Biophys. Acta,348,23–34 (1974)
[3] Teucher, T., Heinz, E.: Planta,184,319–326 (1991)
[4] Heemskerk, J.W.M., Jacobs, F.H.H., Scheijen, M.A.M., Helsper, J.P.F.G.,
    Wintermans, J.F.G.M.: Biochim. Biophys. Acta,918,189–203 (1987)
[5] Coves, J., Block, M.A., Joyard, J., Douce, R.: FEBS Lett.,208,401–406 (1986)
[6] Coves, J., Joyard, J., Douce, R.: Proc. Natl. Acad. Sci. USA,85,4966–4970 (1988)

1 NOMENCLATURE

EC number
2.4.1.47

Systematic name
UDPgalactose:N-acylsphingosine D-galactosyltransferase

Recommended name
N-Acylsphingosine galactosyltransferase

Synonyms
Galactosyltransferase, uridine diphosphogalactose-acylsphingosine
UDP galactose-N-acylsphingosine galactosyltransferase
Uridine diphosphogalactose-acylsphingosine galactosyltransferase

CAS Reg. No.
37277-56-8

2 REACTION AND SPECIFICITY

Catalysed reaction
UDPgalactose + N-acylsphingosine →
→ UDP + D-galactosylceramide

Reaction type
Hexosyl group transfer

Natural substrates

Substrate spectrum
1 UDPgalactose + ceramide (erythro-ceramides, unsaturated ceramides
and bound sphingosines are more active than threo-ceramides, saturated
and free sphingosines) [1]

Product spectrum
1 UDP + ceramide galactose [1]

Inhibitor(s)

Cofactor(s)/prosthetic group(s)/activating agents

Metal compounds/salts

Turnover number (min^{-1})

Specific activity (U/mg)

K$_m$-value (mM)

pH-optimum
   7.8 (assay at) [1]

pH-range

Temperature optimum (°C)
   37 (assay at) [1]

Temperature range (°C)

---

3 ENZYME STRUCTURE

Molecular weight

Subunits

Glycoprotein/Lipoprotein
   –

---

4 ISOLATION/PREPARATION

Source organism
   Rat [1]

Source tissue
   Brain [1]

Localization in source
   Microsomes [1]; More (activity can also be found to a minor degree in
   mitochondrial fraction) [1]

Purification
   Rat (partial) [1]

Crystallization
   –

Cloned
   –

Renatured
   –

5 STABILITY

pH

Temperature (°C)

Oxidation

Organic solvent

General stability information

Storage

6 CROSSREFERENCES TO STRUCTURE DATABANKS

PIR/MIPS code

Brookhaven code

7 LITERATURE REFERENCES

[1] Fujino, Y., Nakano, M.: Biochem. J.,113,573–575 (1969)

1 NOMENCLATURE

EC number
2.4.1.48

Systematic name
GDPmannose:heteroglycan 2-(or 3-)-alpha-D-mannosyltransferase

Recommended name
Heteroglycan alpha-mannosyltransferase

Synonyms
Mannosyltransferase, guanosine diphosphomannose-heteroglycan alpha-
GDP mannose alpha-mannosyltransferase
Guanosine diphosphomannose-heteroglycan alpha-mannosyltransferase

CAS Reg. No.
37277-57-9

2 REACTION AND SPECIFICITY

Catalysed reaction
GDPmannose + heteroglycan →
→ GDP + 1,2(or 1,3)-alpha-D-mannosylheteroglycan

Reaction type
Hexosyl group transfer

Natural substrates

Substrate spectrum
1 GDPmannose + heteropolysaccharide (containing mannosyl, galactosyl, xylosyl units) [1]
2 GDPmannose + mannose [2]
3 GDPmannose + methyl-alpha-D-mannoside [2]

Product spectrum
1 GDP + 1,2(or 1,3)-alpha-D-mannosylheteropolysaccharide
2 GDP + O-alpha-D-mannosyl(1, 2)-mannose [2]
3 ?

Enzyme Handbook © Springer-Verlag Berlin Heidelberg 1996
Duplication, reproduction and storage in data banks are only
allowed with the prior permission of the publishers

Inhibitor(s)

Cofactor(s)/prosthetic group(s)/activating agents

Metal compounds/salts
  Mn^{2+} (20 mM, required, 100% activity [1], essential for reaction, cannot be replaced by Mg^{2+} [2]) [1, 2]; Mg^{2+} (20 mM, required, 35% of activity compared to Mn^{2+}) [1]; Co^{2+} (20 mM, required, 16% of activity compared to Mn^{2+}) [1]; Zn^{2+} (20 mM, required, 16% of activity compared to Mn^{2+}) [1]; Ca^{2+} (20 mM, required, 15% of activity compared to Mn^{2+}) [1]; Ni^{2+} (20 mM, required, 4% of activity compared to Mn^{2+}) [1]; Cu^{2+} (20 mM, required, about 1% of activity compared to Mn^{2+}) [1]; More (absolute requirement for a divalent cation) [1]

Turnover number (min^{-1})

Specific activity (U/mg)

K_m-value (mM)
  0.14 (GDPmannose) [1]; 0.25 (GDPmannose) [2]

pH-optimum
  7.5 [1]; 8 [2]

pH-range
  6–10 (20% of maximal activity at pH 6 and pH 10) [1]; 6–8.5 (50% of maximal activity at pH 6, 90% of maximal activity at pH 8.5) [2]

Temperature optimum (°C)
  25 (assay at) [1]

Temperature range (°C)

3 ENZYME STRUCTURE

Molecular weight

Subunits

Glycoprotein/Lipoprotein
  –

4 ISOLATION/PREPARATION

Source organism
  Saccharomyces cerevisiae (strain 66.24) [2]; Cryptococcus laurentii var. flavescens (NRRL Y-1401) [1]

Source tissue
   Cell [1, 2]

Localization in source
   Membranes (cell wall associated) [1]

Purification
   Cryptococcus laurentii var. flavescens (partial) [1]; Saccharomyces cerevisiae (partial) [2]

Crystallization
   –

Cloned
   –

Renatured
   –

---

5 STABILITY

pH

Temperature (°C)

Oxidation

Organic solvent

General stability information
   Repeated freezing and thawing reduces activity [1]

Storage
   –20°C, stable for at least 4 weeks [1]

---

6 CROSSREFERENCES TO STRUCTURE DATABANKS

PIR/MIPS code

Brookhaven code

---

7 LITERATURE REFERENCES

[1] Ankel, H., Ankel, E., Schutzbach, J.S.: J. Biol. Chem.,245,3945–3955 (1970)
[2] Lehle, L., Tanner, W.: Biochim. Biophys. Acta,350,225–235 (1974)

1 NOMENCLATURE

EC number
2.4.1.49

Systematic name
1,4-beta-D-Oligo-D-glucan:orthophosphate alpha-D-glucosyltransferase

Recommended name
Cellodextrin phosphorylase

Synonyms
beta-1,4-Oligoglucan:orthophosphate glucosyltransferase
Phosphorylase, cellodextrin

CAS Reg. No.
37277-58-0

2 REACTION AND SPECIFICITY

Catalysed reaction
$(1,4$-beta-D-Glucosyl$)_n$ + phosphate $\rightarrow$
$\rightarrow (1,4$-beta-D-glucosyl$)_{n-1}$ + alpha-D-glucose 1-phosphate

Reaction type
Hexosyl group transfer

Natural substrates
More (involved in the catabolism of cellulose) [1]

Substrate spectrum
1 Phosphate + cellotriose (r) [1]
2 Phosphate + cellotetraose (r) [1]
3 Phosphate + cellopentaose (r) [1]
4 Phosphate + cellohexaose (r) [1]
5 More (no phosphorolysis of cellobiose, cellulose, laminaritriose, melezitose or raffinose. Overview of cellobiose-analogues as glycosyl acceptors) [1]

Product spectrum
1 Cellobiose + glucose 1-phosphate [1]
2 Cellotriose + glucose 1-phosphate [1]
3 Cellotetraose + glucose 1-phosphate [1]
4 Cellopentaose + glucose 1-phosphate [1]
5 ?

Enzyme Handbook © Springer-Verlag Berlin Heidelberg 1996
Duplication, reproduction and storage in data banks are only
allowed with the prior permission of the publishers

Inhibitor(s)

Cofactor(s)/prosthetic group(s)/activating agents
  Reducing agents (e.g. cysteine, dithiothreitol, reduced glutathione, absolute requirement) [1]

Metal compounds/salts
  More (no activation with 1 mM and 10 mM concentration of Mg^{2+}, Mn^{2+}, Ca^{2+}, Zn^{2+}, Sn^{2+}, Al^{3+}, Cu^{2+}, Ni^{2+}, Co^{2+}, Fe^{2+}) [1]

Turnover number (min^{-1})

Specific activity (U/mg)
  11.33 [1]

K_m-value (mM)
  0.13 (phosphate (+ cellotriose)) [1]; 0.19 (phosphate (+ cellotetraose)) [1]; 0.24 (phosphate (+ cellopentaose)) [1]; 0.26 (phosphate (+ cellohexaose)) [1]; 0.37 (cellohexaose) [1]; 1 (cellotriose, cellotetraose, cellopentaose) [1]; 1.2 (cellobiose (+ glucose 1-phosphate)) [1]; 4.7 (glucose 1-phosphate (+ cellobiose)) [1]

pH-optimum
  7.5 [1]

pH-range
  5.5–9 (10% of maximal activity at pH 5.5, 15% of maximal activity at pH 9, Tris-acetate buffer) [1]

Temperature optimum (°C)
  37 (assay at) [1]

Temperature range (°C)

3 ENZYME STRUCTURE

Molecular weight

Subunits

Glycoprotein/Lipoprotein
  –

4 ISOLATION/PREPARATION

Source organism
  Clostridium thermocellum (strain 651) [1]

Source tissue
Cell [1]

Localization in source

Purification
Clostridium thermocellum (partial) [1]

Crystallization
–

Cloned
–

Renatured
–

5 STABILITY

pH

Temperature (°C)

Oxidation
O_2-sensitive, enzyme is inactive in the absence of cysteine or DTT [1]

Organic solvent

General stability information
Absolute requirement for a reducing agent [1]; Cysteine, 50 mM, stabilizes [1]; DTT, 40 mM, stabilizes [1]

Storage
–5°C, at least 4 months [1]; –5°C, crude enzyme preparation, at least 6 months [1]

6 CROSSREFERENCES TO STRUCTURE DATABANKS

PIR/MIPS code

Brookhaven code

7 LITERATURE REFERENCES

[1] Sheth, K., Alexander, J.K.: J. Biol. Chem.,244,457–464 (1969)

3

1 NOMENCLATURE

EC number
2.4.1.50

Systematic name
UDPgalactose:procollagen-5-hydroxy-L-lysine D-galactosyltransferase

Recommended name
Procollagen galactosyltransferase

Synonyms
Hydroxylysine galactosyltransferase
Galactosyltransferase, uridine diphosphogalactose-collagen
Collagen galactosyltransferase
Collagen hydroxylysyl galactosyltransferase
UDP galactose-collagen galactosyltransferase
Uridine diphosphogalactose-collagen galactosyltransferase
UDPgalactose:5-hydroxylysine-collagen galactosyltransferase [1]
More (cf. EC 2.4.1.66)

CAS Reg. No.
9028-07-3

---

2 REACTION AND SPECIFICITY

Catalysed reaction
UDPgalactose + procollagen 5-hydroxy-L-lysine →
→ UDP + procollagen 5-(D-galactosyloxy)-L-lysine

Reaction type
Hexosyl group transfer

Natural substrates
UDPgalactose + procollagen 5-hydroxy-L-lysine (involved in collagen biosynthesis, post-translational modification [1], probably involved in synthesis
of carbohydrate units in complement) [1]

Substrate spectrum
1 UDPgalactose + procollagen 5-hydroxy-L-lysine (acceptor specificity: free
epsilon-amino group in hydroxylysyl-residues [1], triple helix conformation
of native collagen at 30°C prevents glycosylation [1, 4], native collagen
is a more efficient substrate than glycopeptides derived by collagenase-
digestion [5], UDPgalactose cannot be replaced by UDPglucose, UDP-N-
acetylglucosamine [2, 3], UDP-N-acetylgalactosamine [3], UDPxylose,

GDPmannose, GDPfucose [2], galactosyl-acceptors are acid-soluble calf skin collagen [5], pig skin collagen (sugar residues removed), bovine anterior lens capsule collagen (carbohydrate units removed) [1], bovine renal glomerular basement membranes (carbohydrate units removed, native less effective [5]) [1, 5], gelatinized insoluble calf skin collagen (less effective [4]) [1, 4, 8], bovine Achilles tendon collagen (glucose and galactose residues removed) [3], and dialyzable peptides thereof by collagenase digestion [8], ovalbumin, fetuin (sialic acid and galactose removed), bovine [2, 3] and ovine [3] submaxillary glycoprotein (sialic acid removed, not porcine [2]) [3], alpha$_1$-glycoprotein (sialic acid and galactose removed) [2], peptides derived from citrate soluble collagen by collagenase digestion (longer peptides are better substrates) [4]. No acceptors are free hydroxylysine [1, 2, 5, 7, 8], galactosylhydroxylysine [1], galactose, glucose, glucosamine, N-acetylglucosamine, lysine, hydroxyproline, threonine, serine [2], transferrin, native fetuin [2, 3], native guinea pig collagen, glucose-free guinea pig skin collagen, native alpha$_1$-glycoprotein [2], thyroglobulin unit B [5]) [1–8]

2 UDPgalactose + heat-denatured citrate-soluble skin collagen (best substrate [4], collagen from rat [1, 4, 6] or guinea pig [2]) [1, 2, 4, 6]

3 UDPgalactose + ichthyocol (fish collagen, best substrate, peptides derived from collagenase- and pronase-digest are less effective) [5]

Product spectrum

1 UDP + procollagen 5-(D-galactosyloxy)-L-lysine [1]

2 ?

3 ?

Inhibitor(s)

UDP [1, 3, 7]; EDTA [2]; p-Hydroxymercuribenzoate [3]; p-Substituted mercuribenzoate (strong) [4]; Pb^{2+} [3]; Cu^{2+} [3]; Co^{2+} (6 mM) [7]; Mn^{2+} (above 2.5 mM) [7]; Hydroxylysine (weak) [7]; Galactosylhydroxylysine [7]; Concanavalin A (methyl-alpha-D-glucopyranoside, not galactose, protects) [7]; UMP [7]; UDPgalacturonate [7]; UDPglucuronate [7]; ATP [7]; ADP (not [3]) [7]; AMP [7]; CDP [7]; GDP [7]; Gelatinized collagen (above 60 mg/ml) [7]; Renal basement membrane collagen (glucose-free) [7]; More (no inhibition by Hg^{2+}, acetylsalicylic acid, D-glucosamine, GSH [3], lysine, free monosaccharides, lactose, sucrose [7]) [3, 7]

Cofactor(s)/prosthetic group(s)/activating agents

Dithiothreitol (activation) [4]; Triton X-100 (stimulation) [3, 4]

Metal compounds/salts

Mn^{2+} (requirement, 2 mM [1], 2–2.5 mM [7], 5–10 mM [5, 8], 15 mM [3]) [1–8]; Co^{2+} (activation, 50% [3], 5.3% [8] as effective as Mn^{2+}, not [5], inhibits at 6 mM [7]) [2, 3, 8]; Mg^{2+} (activation, 50% [3], 20% [5], 12.2% [8] as effective as Mn^{2+}, not [2, 7]) [3, 5, 8]; Ca^{2+} (activation, 60% [5], 20% [3]

as effective as Mn^{2+}, not [2, 8]) [3, 5]; Cd^{2+} (activation, 20% as effective as Mn^{2+} [8], not [2, 7]) [3]; Fe^{2+} (activation, 2.7% as effective as Mn^{2+} [8], not [7]) [8]; More (no activation by folate [8], Fe^{3+}, Ba^{2+} [3], Zn^{2+} [3, 5, 8], Na^+, K^+, Pb^{2+} [2], Cu^{2+} [2, 5, 8], Ni^{2+} [5, 8]) [2, 3, 5, 8]

Turnover number (min^{-1})

Specific activity (U/mg)
  0.00093 [4]; 0.0147 (chicken embryo) [1, 6]

K_m-value (mM)
  More (kinetic study [8], K_m-values: 2–4 mg/ml (denatured citrate-soluble rat skin collagen) [1, 4], 15–35 mg/ml (gelatinized calf skin collagen) [1], 16 mg/ml (gelatinized collagen) [7], 150 (gelatinized collagen) [8]) [1, 4, 7, 8]; 0.0057 (UDPgalactose) [2]; 0.02–0.023 (UDPgalactose) [7]; 0.028 (calf skin collagen, value based on actual amount of galactosylhydroxylysyl-residues in calf skin collagen) [5]; 0.03 (UDPgalactose) [8]; 0.049 (UDPgalactose) [5]; 0.1–0.2 (hydroxylysyl acceptor sites) [1]; 0.28 (ichthyocol, value based on actual amount of galactosylhydroxylysyl-residues in ichthyocol) [5]

pH-optimum
  6 [2]; 6–6.5 (2 pH-optima: 6–6.5 and 7.5–8) [5]; 7 [8]; 7–7.4 [1]; 7.5 [3]; 7.5–8 (2 pH-optima: 6–6.5 and 7.5–8) [5]

pH-range
  5.5–9.5 (about half-maximal activity at pH 5.5 and 9.5) [3]

Temperature optimum (°C)
  37 (assay at) [1–8]

Temperature range (°C)

---

3 ENZYME STRUCTURE

Molecular weight

Subunits

Glycoprotein/Lipoprotein
  Glycoprotein [7]

---

4 ISOLATION/PREPARATION

Source organism
  Human (fetus [5]) [1–3, 5]; Chicken [1, 4, 6–8]; Rat [1]; Guinea pig [1, 2]

Source tissue
Embryo (chicken, rat [1]) [1, 2, 4–8]; Cartilage (cartilaginous ends of limb bone rudiments [2]) [2, 4, 8]; Bone (rat) [1]; HeLa cells [2]; Lung fibroblasts (cell suspension culture, diploid, WI-38 or IMR90) [5]; Kidney (rat) [1]; Liver (rat) [1]; Skin (guinea pig, human, cultured fibroblasts [1]) [1, 2]; Blood platelets (human) [1, 3]; Serum [1]

Localization in source
Membrane-associated (plasma-membrane [3]) [3, 4]; Endoplasmic reticulum [1]; Soluble [1]

Purification
Human (partial, solubilized with Triton X-100) [5]; Chicken (partial [1, 4, 6], solubilized with Triton X-100 [4, 7] or Nonidet P-40 [7], affinity chromatography [1]) [1, 4, 6, 7]; Guinea pig (skin, partial) [2]

Crystallization
–

Cloned
–

Renatured
–

---

5 STABILITY

pH

Temperature (°C)

Oxidation

Organic solvent
Ethylene glycol, 50% v/v, stable to [7]

General stability information
Unstable in crude extracts [1]; Unstable during purification [1, 4]; DTT stabilizes during purification [7]; Gel filtration inactivates [6]; Glycerol, 25% v/v, does not stabilize [7]

Storage
–20°C, several weeks in 50% v/v ethylene glycol [7]; Unstable to storage at 4°C, urea, galactose, Triton X-100, glycerol or ethylene glycol does not stabilize [7]

6 CROSSREFERENCES TO STRUCTURE DATABANKS

PIR/MIPS code

Brookhaven code

7 LITERATURE REFERENCES

[1] Kivirikko, K.I., Myllylä, R.: Methods Enzymol.,82,245–304 (1982) (Review)
[2] Bosmann, H.B., Eylar, E.H.: Biochem. Biophys. Res. Commun.,33,340–346 (1968)
[3] Barber, A.J., Jamieson, G.A.: Biochim. Biophys. Acta,252,546–552 (1971)
[4] Risteli, L., Myllylä, R., Kivirikko, K.I.: Biochem. J.,155,145–153 (1976)
[5] Carnicero, H.H., Adamany, A.M., England, S.: Arch. Biochem. Biophys.,210,678–690 (1981)
[6] Risteli, L., Myllylä, R., Kivirikko, K.I.: Eur. J. Biochem.,67,197–202 (1976)
[7] Risteli, L.: Biochem. J.,169,189–196 (1978)
[8] Myllylä, R., Risteli, L., Kivirikko, K.I.: Eur. J. Biochem.,52,401–410 (1975)

1 NOMENCLATURE

EC number
2.4.1.52

Systematic name
UDPglucose:poly(glycerol-phosphate) alpha-D-glucosyltransferase

Recommended name
Poly(glycerol-phosphate) alpha-glucosyltransferase

Synonyms
Glucosyltransferase, uridine diphosphoglucose-poly(glycerol-phosphate) alpha-
UDP glucose-poly(glycerol-phosphate) alpha-glucosyltransferase
Uridine diphosphoglucose-poly(glycerol-phosphate) alpha-glucosyltransferase

CAS Reg. No.
37277-60-4

---

2 REACTION AND SPECIFICITY

Catalysed reaction
UDPglucose + poly(glycerol phosphate) $\rightarrow$
$\rightarrow$ UDP + alpha-D-glucosyl-poly(glycerol phosphate)

Reaction type
Hexosyl group transfer

Natural substrates

Substrate spectrum
1 UDPglucose + poly(glycerophosphate) [1]
2 dTDPglucose + poly(glycerophosphate) [1]

Product spectrum
1 UDP + alpha-D-glucosyl-poly(glycerol phosphate)
2 ?

Inhibitor(s)
More (vancomycin, novobionin, crystal violet are no inhibitors) [1]; Mn^{2+}
(inhibitory in presence of optimal Mg^{2+}-concentration) [1]

Cofactor(s)/prosthetic group(s)/activating agents

Metal compounds/salts
Mg^{2+} (requirement) [1]; Ca^{2+} (requirement, as active as Mg^{2+}) [1]; Mn^{2+} (requirement, much less effective than Mg^{2+} and inhibitory in presence of optimal Mg^{2+} concentration) [1]

Turnover number (min^{-1})

Specific activity (U/mg)

K_m-value (mM)
0.04 (UDPglucose) [1]; 0.1–2 (poly(glycerophosphate), depending on chain length) [1]

pH-optimum
8.0 [1]

pH-range
5.5–9.5 (15% of maximal activity at pH 5.5, 60% of maximal activity at pH 9.5) [1]

Temperature optimum (°C)
37 (assay at) [1]

Temperature range (°C)

3 ENZYME STRUCTURE

Molecular weight

Subunits

Glycoprotein/Lipoprotein
–

4 ISOLATION/PREPARATION

Source organism
Bacillus subtilis (NCTC 3610) [1]

Source tissue
Cell [1]

Localization in source
Membrane [1]

Purification
Bacillus subtilis (partial) [1]

Crystallization

–

Cloned

–

Renatured

–

5 STABILITY

pH

Temperature (°C)

Oxidation

Organic solvent

General stability information

Storage

6 CROSSREFERENCES TO STRUCTURE DATABANKS

PIR/MIPS code

Brookhaven code

7 LITERATURE REFERENCES

[1] Glaser, L., Burger, M.M.: J. Biol. Chem.,239,3187–3191 (1964)

1 NOMENCLATURE

EC number
2.4.1.53

Systematic name
UDPglucose:poly(ribitol-phosphate) beta-D-glucosyltransferase

Recommended name
Poly(ribitol-phosphate) beta-glucosyltransferase

Synonyms
Glucosyltransferase, uridine diphosphoglucose-poly(ribitol-phosphate) beta-
UDP glucose-poly(ribitol-phosphate) beta-glucosyltransferase
Uridine diphosphoglucose-poly(ribitol-phosphate) beta-glucosyltransferase
UDP-D-glucose polyribitol phosphate glucosyl transferase [1]
UDP-D-glucose:polyribitol phosphate glucosyl transferase [1]

CAS Reg. No.
37277-61-5

2 REACTION AND SPECIFICITY

Catalysed reaction
UDPglucose + poly(ribitol phosphate) →
→ UDP + beta-D-glucosyl-poly(ribitol phosphate)

Reaction type
Hexosyl group transfer

Natural substrates

Substrate spectrum
1 UDP-D-glucose + poly(ribitol phosphate) [1]

Product spectrum
1 UDP + glucosyl-poly(ribitol phosphate) [1]

Inhibitor(s)

Cofactor(s)/prosthetic group(s)/activating agents

Metal compounds/salts
$MgCl_2$ (absolute requirement for the addition of a divalent cation, optimal
concentration: 0.05 M) [1]; $CaCl_2$ (absolute requirement for the addition of a
divalent cation, optimal concentration: 0.01 M) [1]; $MnCl_2$ (absolute requirement for the addition of a divalent cation, optimal concentration 0.01 M) [1]

Turnover number (min^{-1})

Specific activity (U/mg)

K_m-value (mM)
0.001 (poly(ribitol phosphate), molar concentration of the polymer) [1];
0.017–0.08 (poly(ribitol phosphate), concentration as ribitol units) [1]; 0.12
(UDP-D-glucose) [1]

pH-optimum
7.0 [1]

pH-range
6–8 (62% of maximal activity at pH 6, 70% of maximal activity at pH 8) [1]

Temperature optimum (°C)
37 (assay at) [1]

Temperature range (°C)

3 ENZYME STRUCTURE

Molecular weight

Subunits

Glycoprotein/Lipoprotein
–

4 ISOLATION/PREPARATION

Source organism
Bacillus subtilis (W-23) [1]

Source tissue
Cell [1]

Localization in source
Cell wall bound [1]

Purification
Bacillus subtilis (W-23, partial) [1]

Crystallization
–

Cloned
–

Renatured
–

5 STABILITY

pH

Temperature (°C)

Oxidation

Organic solvent

General stability information

Storage
   −20°C, stable for several months [1]

---

6 CROSSREFERENCES TO STRUCTURE DATABANKS

PIR/MIPS code

Brookhaven code

---

7 LITERATURE REFERENCES

[1] Chin, T., Burger, M.M., Glaser, L.: Arch. Biochem. Biophys.,116,358–367 (1966)

1 NOMENCLATURE

EC number
2.4.1.54

Systematic name
GDPmannose:undecaprenyl-phosphate D-mannosyltransferase

Recommended name
Undecaprenyl-phosphate mannosyltransferase

Synonyms
Mannosyltransferase, guanosine diphosphomannose-undecaprenyl phosphate
GDP mannose-undecaprenyl phosphate mannosyltransferase
GDP-D-mannose:lipid phosphate transmannosylase [2]

CAS Reg. No.
37277-62-6

---

2 REACTION AND SPECIFICITY

Catalysed reaction
GDPmannose + undecaprenyl phosphate →
→ GDP + D-mannosyl-1-phosphoundecaprenol

Reaction type
Hexosyl group transfer

Natural substrates
GDPmannose + phospholipid (polyprenol type, involved in glycophospholipid biosynthesis) [3]

Substrate spectrum
1 GDPmannose + undecaprenyl phosphate (r, in the reverse reaction GMP is no substrate [3]) [1, 3]
2 GDPmannose + phytanol phosphate (poor substrate is phytol phosphate, no substrates are lauryl or myristyl sulfate, phytol or phytanol, GDPglucose or UDPxylose) [2]
3 GDPmannose + ficaprenyl phosphate (mannosylated at about 50% the rate of undecaprenyl phosphate [1], no substrates are phosphatidylglycerol, phosphatidylethanolamine, phosphatidylserine or Triton X-100 [3]) [1, 3]
4 UDPglucose + ficaprenyl phosphate (r, in the reverse reaction UMP is no substrate) [3]

Product spectrum
1 GDP + D-mannosyl-1-phosphoundecaprenol [1, 3]
2 ?
3 ?
4 ?

Inhibitor(s)
Triton X-100 [3]

Cofactor(s)/prosthetic group(s)/activating agents
Phytanol phosphate (stimulation) [2]; Phosphatidylglycerol (requirement) [1];
Detergents (activation, e.g. phosphatidylethanolamine, SDS, Triton X-100,
Cutscum, Nonidet P-40) [1];

Metal compounds/salts
Mg^{2+} (requirement, 15 mM [3]) [1]; Ca^{2+} (activation, can replace Mg^{2+} to
some extent) [1]; Mn^{2+} (activation, can replace Mg^{2+} to some extent) [1];
More (no activation by Co^{2+}) [1]

Turnover number (min^{-1})

Specific activity (U/mg)

K_m-value (mM)
0.00067 (GDPmannose, cotton) [3]; 0.04 (ficaprenyl phosphate, cotton) [3]

pH-optimum
6.9–7.5 (0.06 M Tris-maleate buffer) [1]; 7.5 (Tris buffer, cotton) [3]

pH-range
6.2–8.6 (about half-maximal activity at pH 6.2 and 8.6) [1]; 6.2–9.0 (cotton,
Tris buffer, about half-maximal activity at pH 6.2 and 63% of maximal activity
at pH 9.0) [3]

Temperature optimum (°C)
40 (assay at) [2]

Temperature range (°C)

3 ENZYME STRUCTURE

Molecular weight

Subunits

Glycoprotein/Lipoprotein
—

4 ISOLATION/PREPARATION

Source organism
  Micrococcus lysodeikticus [1]; Cotton [3]; Phaseolus aureus [2]; Mycobacterium smegmatis [3]

Source tissue
  Cell [1, 3]; Hypocotyls [2]; Fibers [3]

Localization in source
  Membrane-bound [1–3]

Purification

Crystallization
  –

Cloned
  –

Renatured
  –

5 STABILITY

pH

Temperature (°C)
  50 (above, rapid inactivation) [1]

Oxidation

Organic solvent

General stability information

Storage

6 CROSSREFERENCES TO STRUCTURE DATABANKS

PIR/MIPS code

Brookhaven code

7 LITERATURE REFERENCES

[1] Lahav, M., Chiu, T.H., Lennarz, W.J.: J. Biol. Chem.,244,5890–5898 (1969)
[2] Clark, A.F., Villemez, C.L.: FEBS Lett.,32,84–86 (1973)
[3] Forsee, W.T., Elbein, A.D.: J. Biol. Chem.,248,2858–2867 (1973)

1 NOMENCLATURE

EC number
2.4.1.56

Systematic name
UDP-N-acetyl-D-glucosamine:lipopolysaccharide N-acetyl-D-glucosaminyl-transferase

Recommended name
Lipopolysaccharide N-acetylglucosaminyltransferase

Synonyms
Acetylglucosaminyltransferase, uridine diphosphoacetylglucosamine-lipopolysaccharide
UDP-N-acetylglucosamine-lipopolysaccharide N-acetylglucosaminyltrans-ferase
Uridine diphosphoacetylglucosamine-lipopolysaccharide acetylglucos-aminyltransferase
More (cf. EC 2.4.1.44, EC 2.4.1.58 and EC 2.4.1.73)

CAS Reg. No.
37277-64-8

---

2 REACTION AND SPECIFICITY

Catalysed reaction
UDP-N-acetyl-D-glucosamine + lipopolysaccharide →
→ UDP + N-acetyl-D-glucosaminyllipopolysaccharide

Reaction type
Hexosyl group transfer

Natural substrates
UDP-N-acetyl-D-glucosamine + lipopolysaccharide (transfers N-ace-tylglucosaminyl residues to a D-galactose residue in the partially completed lipopolysaccharide core, the results are consistent with the postulated attachment of N-acetylglucosamine to glucose, but do not exclude linkage to another residue) [1]

Substrate spectrum
1 UDP-N-acetyl-D-glucosamine + lipopolysacccharide [1]

Product spectrum
1 UDP + N-acetyl-D-glucosaminyllipopolysaccharide [1]

Inhibitor(s)

Cofactor(s)/prosthetic group(s)/activating agents

Metal compounds/salts

Turnover number (min^{-1})

Specific activity (U/mg)

K_m-value (mM)

pH-optimum

pH-range

Temperature optimum (°C)

Temperature range (°C)

3 ENZYME STRUCTURE

Molecular weight

Subunits

Glycoprotein/Lipoprotein
–

4 ISOLATION/PREPARATION

Source organism
Salmonella typhimurium [1]

Source tissue

Localization in source
More (cell-wall membrane fraction) [1]

Purification

Crystallization
–

Cloned
–

Renatured
–

5 STABILITY

pH

Temperature (°C)

Oxidation

Organic solvent

General stability information

Storage

---

6 CROSSREFERENCES TO STRUCTURE DATABANKS

PIR/MIPS code

Brookhaven code

---

7 LITERATURE REFERENCES

[1] Osborn, M.J., D'Ari, L.: Biochem. Biophys. Res. Commun.,16,568–575 (1964)

Enzyme Handbook © Springer-Verlag Berlin Heidelberg 1996
Duplication, reproduction and storage in data banks are only
allowed with the prior permission of the publishers

1 NOMENCLATURE

EC number
2.4.1.57

Systematic name
GDPmannose:1-phosphatidyl-myo-inositol alpha-D-mannosyltransferase

Recommended name
Phosphatidyl-myo-inositol alpha-mannosyltransferase

Synonyms
Mannosyltransferase, guanosine diphosphomannose-phosphatidyl-inositol alpha-
GDP mannose-phosphatidyl-myo-inositol alpha-mannosyltransferase
Guanosine diphosphomannose-phosphatidyl-inositol alpha-mannosyltrans-
ferase

CAS Reg. No.
37277-65-9

2 REACTION AND SPECIFICITY

Catalysed reaction
Transfers one or more alpha-D-mannose units from GDPmannose to posi-
tions 2,6 and others in 1-phosphatidyl-myo-inositol

Reaction type
Hexosyl group transfer

Natural substrates

Substrate spectrum
1 GDPmannose + phosphatidyl-myo-inositol (from yeast) [1]

Product spectrum
1 GDP + phosphatidyl-myo-inositol monomannoside [1]

Inhibitor(s)

Cofactor(s)/prosthetic group(s)/activating agents

Metal compounds/salts

Enzyme Handbook © Springer-Verlag Berlin Heidelberg 1996
Duplication, reproduction and storage in data banks are only
allowed with the prior permission of the publishers

Turnover number (min⁻¹)

Specific activity (U/mg)

K_m-value (mM)

pH-optimum

pH-range

Temperature optimum (°C)
  37 (assay at) [1]

Temperature range (°C)

---

3 ENZYME STRUCTURE

Molecular weight

Subunits

Glycoprotein/Lipoprotein
  –

---

4 ISOLATION/PREPARATION

Source organism
  Propionibacterium shermanii (ATCC 9614) [1]

Source tissue
  Cell [1]

Localization in source

Purification
  Propionibacterium shermanii (partial) [1]

Crystallization
  –

Cloned
  –

Renatured
  –

5 STABILITY

pH

Temperature (°C)

Oxidation

Organic solvent

General stability information

Storage

6 CROSSREFERENCES TO STRUCTURE DATABANKS

PIR/MIPS code

Brookhaven code

7 LITERATURE REFERENCES

[1] Brennan, P., Ballou, C.E.: Biochem. Biophys. Res. Commun.,30,69–75 (1968)

1 NOMENCLATURE

EC number
2.4.1.58

Systematic name
UDPglucose:lipopolysaccharide glucosyltransferase

Recommended name
Lipopolysaccharide glucosyltransferase I

Synonyms
UDPglucose:lipopolysaccharide glucosyltransferase I [2]
Glucosyltransferase, uridine diphosphoglucose-lipopolysaccharide
Glucosyltransferase, lipopolysaccharide
More (cf. EC 2.4.1.44, EC 2.4.1.56 and EC 2.4.1.73)

CAS Reg. No.
9074-00-4

---

2 REACTION AND SPECIFICITY

Catalysed reaction
UDPglucose + lipopolysaccharide →
→ UDP + D-glucosyllipopolysaccharide

Reaction type
Hexosyl group transfer

Natural substrates

Substrate spectrum
1 UDPglucose + lipopolysaccharide (UDPglucose is the only effective glucosyl donor [2], glucose is transferred only into the glucose-deficient lipopolysaccharide obtained from Salmonella typhimurium SL1032, a strain lacking the glucosyl transferase system [2], transfers glucosyl residues to the backbone portion of lipopolysaccharide) [1, 2]

Product spectrum
1 UDP + D-glucosyllipopolysaccharide [1, 2]

Inhibitor(s)
Iodoacetamide (5 mM, weak) [2]; More (no detectable inhibition by non-substrate lipopolysaccharides obtained from other mutant strains of Salmonella typhimurium, G30G, G30A, TV119, LT2) [2]

Cofactor(s)/prosthetic group(s)/activating agents

Phospholipid (required, phosphatidylethanolamine most effective, phosphatidylethanolamine containing unsaturated or cyclopropane acyl groups is more effective than phosphatidylethanolamine containing saturated acyl groups) [2]

Metal compounds/salts

Mg^{2+} (divalent cation required, order of effectiveness: Mg^{2+} > Ba^{2+} > Ca^{2+} > Co^{2+}, relative activities: 1.0, 0.77, 0.65, 0.31) [2]; Ba^{2+} (divalent cation required, order of effectiveness: Mg^{2+} > Ba^{2+} > Ca^{2+} > Co^{2+}, relative activities: 1.0, 0.77, 0.65, 0.31) [2]; Ca^{2+} (divalent cation required, order of effectiveness: Mg^{2+} > Ba^{2+} > Ca^{2+} > Co^{2+}, relative activities: 1.0, 0.77, 0.65, 0.31) [2]; Co^{2+} (divalent cation required, order of effectiveness: Mg^{2+} > Ba^{2+} > Ca^{2+} > Co^{2+}, relative activities: 1.0, 0.77, 0.65, 0.31) [2]

Turnover number (min^{-1})

Specific activity (U/mg)

More [2]

K_m-value (mM)

0.03 (lipopolysaccharide, in presence of optimal concentrations of phosphatidylethanolamine) [2]; 0.1 (lipopolysaccharide, in absence of phosphatidylethanolamine) [2]; 0.33 (UDPglucose) [2]

pH-optimum

7.5–8.5 [2]

pH-range

Temperature optimum (°C)

37 (assay at) [2]

Temperature range (°C)

3 ENZYME STRUCTURE

Molecular weight

Subunits

Glycoprotein/Lipoprotein

More (no detectable phospholipid or lipopolysaccharide in the purified enzyme) [2]

4 ISOLATION/PREPARATION

Source organism
   Salmonella typhimurium [1, 2]

Source tissue

Localization in source
   More (most of the activity is located in the cell-wall fraction) [1]; Membrane
   [2]

Purification
   Salmonella typhimurium [2]

Crystallization
   –

Cloned
   –

Renatured
   –

5 STABILITY

pH

Temperature (°C)

Oxidation

Organic solvent

General stability information
   Freezing and thawing: rapid loss of activity [1]

Storage
   4°C, 20% glycerol, retains full activity for about 5 days, 10% loss of activity
   per week thereafter [2]

6 CROSSREFERENCES TO STRUCTURE DATABANKS

PIR/MIPS code

Brookhaven code

7 LITERATURE REFERENCES

[1] Rothfield, L., Osborn, M.J., Horecker, B.L.: J. Biol. Chem.,239,2788–2795 (1964)
[2] Müller, E., Hinckley, A., Rothfield, L.: J. Biol. Chem.,247,2614–2622 (1972)

1 NOMENCLATURE

EC number
2.4.1.60

Systematic name
CDPabequose:D-mannosyl-L-rhamnosyl-D-galactose-1-diphospholipid
D-abequosyltransferase

Recommended name
Abequosyltransferase

Synonyms
Abequosyltransferase, trihexose diphospholipid
Trihexose diphospholipid abequosyltransferase

CAS Reg. No.
37277-67-1

---

2 REACTION AND SPECIFICITY

Catalysed reaction
CDPabequose + D-mannosyl-L-rhamnosyl-D-galactose-1-diphospholipid $\rightarrow$
$\rightarrow$ CDP + D-abequosyl-D-mannosyl-rhamnosyl-D-galactose-1-diphospholipid

Reaction type
Hexosyl group transfer

Natural substrates

Substrate spectrum
1 CDPabequose + mannosyl-rhamnosyl-galactose-1-diphospholipid
 (CDPabequose is CDP-3,6-dideoxy-D-galactose) [1]
2 CDPtyvelose + mannosyl-rhamnosyl-galactose-1-diphospholipid
 (CDPtyvelose is CDP-3,6-dideoxy-D-mannose, 20% of activity with
 CDPabequose) [1]

Product spectrum
1 CDP + abequosyl-mannosyl-rhamnosyl-galactose-1-diphospholipid [1]
2 ?

Inhibitor(s)

Cofactor(s)/prosthetic group(s)/activating agents

Metal compounds/salts

Turnover number (min^{-1})

Specific activity (U/mg)

K_m-value (mM)

pH-optimum
    8.5 (assay at) [1]

pH-range

Temperature optimum (°C)
    22 (assay at) [1]

Temperature range (°C)

---

3 ENZYME STRUCTURE

Molecular weight

Subunits

Glycoprotein/Lipoprotein
    –

---

4 ISOLATION/PREPARATION

Source organism
    Salmonella typhimurium (strain G30) [1]

Source tissue
    Cell [1]

Localization in source

Purification
    Salmonella typhimurium (partial) [1]

Crystallization
    –

Cloned
    –

Renatured
    –

5 STABILITY

pH

Temperature (°C)
    37 (10 min, 87% loss of activity) [1]; 29 (10 min, 40% loss of activity) [1]

Oxidation

Organic solvent

General stability information

Storage

6 CROSSREFERENCES TO STRUCTURE DATABANKS

PIR/MIPS code

Brookhaven code

7 LITERATURE REFERENCES

[1] Osborn, M.J., Weiner, I.M.: J. Biol. Chem.,243,2631–2639 (1968)

1 NOMENCLATURE

EC number
2.4.1.62

Systematic name
UDPgalactose:N-acetyl-D-galactosaminyl-(N-acetylneuraminyl)-D-galactosyl-D-glucosyl-N-acylsphingosine beta-1,3-D-galactosyltransferase

Recommended name
Ganglioside galactosyltransferase

Synonyms
Galactosyltransferase, uridine diphosphogalactose-ceramide
UDPgalactose-ceramide galactosyltransferase
UDP galactose-LAC Tet-ceramide alpha-galactosyltransferase
UDP-galactose-GM2 galactosyltransferase
Uridine diphosphogalactose-GM2 galactosyltransferase
Uridine diphosphate D-galactose:glycolipid galactosyltransferase [3]
UDP-galactose:N-acetylgalactosaminyl-(N-acetylneuraminyl) galactosyl-glucosyl-ceramide galactosyltransferase [5]
UDP-galactose-GM2 ganglioside galactosyltransferase [5]
GM1-synthase [6]

CAS Reg. No.
37217-28-0

2 REACTION AND SPECIFICITY

Catalysed reaction
UDPgalactose + N-acetyl-D-galactosaminyl-(N-acetylneuraminyl)-D-galactosyl-1,4-beta-D-glucosyl-N-acylsphingosine →
→ UDP + D-galactosyl-1,3-beta-N-acetyl-D-galactosaminyl-(N-acetylneuraminyl)-D-galactosyl-D-glucosyl-N-acylsphingosine

Reaction type
Hexosyl group transfer

Natural substrates
UDPgalactose + N-acetyl-D-galactosaminyl-(N-acetylneuraminyl)-D-galactosyl-1,4-beta-D-glucosyl-N-acylsphingosine (involved in biosynthesis of gangliosides) [6]

Substrate spectrum

1 UDPgalactose + N-acetyl-D-galactosaminyl-(N-acetylneuraminyl)-D-galac-
tosyl-1,4-beta-D-glucosyl-N-acylsphingosine (i.e. GM2 or Tay-Sachs ganglioside [3], donor and acceptor specificity [5], best acceptor [3, 4], no
substrates are GM3, GM1, GD1a and GT1 gangliosides, glucocerebro-
side, lactosylceramide, trihexosylceramide [5]) [3–7]

Product spectrum

1 D-Galactosyl-1,3-beta-N-acetyl-D-galactosaminyl-(N-acetylneuraminyl)-D-galactosyl-D-glucosyl-N-acylsphingosine + UDP (i.e.
GM1 ganglioside or monosialoganglioside) [3, 5]

Inhibitor(s)

Triton X-100 (in excess, phospholipids protect) [2]; EDTA [3, 5, 6]; GM2
(0.35 mM and above [5], above 1.83 mM [4]) [4, 5]; Gangliosides and
ganglioside components (most active: disialoganglioside) [4]; UDPgalactose (above 0.2 mM) [6]; Endogenous membrane protein (heat-stable, protease sensitive, non-dialyzable) [7]; N-Acetylneuraminic acid (with or without
tetrahexosylceramide) [4]; N-Acetylneuraminyllactose (weak) [4]; Brij 56 [6];
Tween 80 [6]; Cardiolipin [6]; SDS [6]; More (no inhibition by fetuin,
glucosylceramide, lactosylceramide or trihexosylceramide [4], bovine serum albumin [7]) [4, 7]

Cofactor(s)/prosthetic group(s)/activating agents

Cytosolic peptide (activation, MW 25000, heat-labile, protease-sensitive and
non-diffusible, not bovine serum albumin, rat serum lipoproteins, liver or rat
cytosol, cytosolic or microsomal phospholipids) [1]; Bile salts (activation)
[1]; Detergents (requirement, solubilized enzyme, e.g. Cutscum, Lubrol PX,
Triton CF-54, octylglucoside [6], Triton X-100, 1 mM, inhibition at higher concentration [2]) [2, 6]

Metal compounds/salts

Mn^{2+} (requirement, K_m-value: 2.2 mM [4], Co^{2+} or Mg^{2+} can substitute to
some extent [6]) [3–6]; More (no activation by Mg^{2+} [3–5], Ca^{2+} [3, 4], Zn^{2+}
[3], Ni^{2+}, Cu^{2+} [3–5], Co^{2+}, Na^+, Li^+ [4], K^+, Al^{3+} [4, 5]) [3–5]

Turnover number (min^{-1})

Specific activity (U/mg)

K_m-value (mM)

More (kinetic properties) [2]; 0.01 (N-acetyl-D-galactosaminyl-(N-acetylneuraminyl)-D-galactosyl-1,4-beta-D-glucosyl-N-acylsphingosine i.e.
GM2 ganglioside) [6]; 0.012 (UDPgalactose) [5]; 0.094 (N-acetyl-D-galac-
tosaminyl-(N-acetylneuraminyl)-D-galactosyl-1,4-beta-D-glucosyl-N-acyl-
sphingosine i.e. GM2 ganglioside) [5]; 0.14 (UDPgalactose) [6]; 0.18 (N-ace-
tyl-D-galactosaminyl-(N-acetylneuraminyl)-D-galactosyl-1,4-beta-D-glucosyl-
N-acylsphingosine i.e. GM2 ganglioside) [4]

pH-optimum
6.5–7.2 (broad) [5]; 6.6–7.0 [6]; 7.2–7.3 (adult frog brain) [4]

pH-range

Temperature optimum (°C)
37 (assay at) [6]

Temperature range (°C)

---

3 ENZYME STRUCTURE

Molecular weight

Subunits

Glycoprotein/Lipoprotein
–

---

4 ISOLATION/PREPARATION

Source organism
Rat (Sprague-Dawley [5]) [1–3, 5–7]; Chicken (embryonic) [3]; Pig (fetus)
[3]; Rana pipiens (frog) [4]

Source tissue
Brain [1–5, 7]; Liver [6]

Localization in source
Particulate [3]; Microsomes [4, 5]; Mitochondria [5]; Nucleus (adult rats) [5];
Golgi apparatus [6]

Purification
Rat (partial) [2]

Crystallization
–

Cloned
–

Renatured
–

5 STABILITY

pH

Temperature (°C)

Oxidation

Organic solvent

General stability information

Storage
    −20°C, several months [3]

---

6 CROSSREFERENCES TO STRUCTURE DATABANKS

PIR/MIPS code

Brookhaven code

---

7 LITERATURE REFERENCES

[1] Sanyal, S.N.: Indian J. Exp. Biol.,25,606–612 (1987)
[2] Neskovic, N.M., Mandel, P., Gatt, S.: Adv. Exp. Med. Biol.,101 (Enzymes Lipid Metab.) ,613–630 (1978)
[3] Basu, S., Kaufman, B., Roseman, S.: J. Biol. Chem.,240, PC4115-PC4117 (1965)
[4] Yip, M.C.M., Dain, J.A.: Biochem. J.,118,247–252 (1970)
[5] Bellman Yip, G., Dain, J.A.: Biochim. Biophys. Acta,206,252–260 (1970)
[6] Senn, H.-J., Wagner, M., Decker, K.: Eur. J. Biochem.,135,231–236 (1983)
[7] Costantino-Ceccarini, E., Suzuki, K.: J. Biol. Chem.,253,340–342 (1978)

1 NOMENCLATURE

EC number
2.4.1.63

Systematic name
UDPglucose:2-hydroxy-2-methylpropanenitrile beta-D-glucosyltransferase

Recommended name
Linamarin synthase

Synonyms
Glucosyltransferase, uridine diphosphoglucose-ketone
UDP glucose ketone cyanohydrin glucosyltransferase
UDPglucose:ketone cyanohydrin beta-glucosyltransferase [1]
Uridine diphosphoglucose-ketone cyanohydrin glucosyltransferase

CAS Reg. No.
37277-68-2

---

2 REACTION AND SPECIFICITY

Catalysed reaction
UDPglucose + 2-hydroxy-2-methylpropanenitrile →
→ UDP + linamarin

Reaction type
Hexosyl group transfer

Natural substrates
UDPglucose + acetone cyanohydrin (pathway in cyanogenic glucoside biosynthesis) [1]

Substrate spectrum
1 UDPglucose + 2-hydroxy-2-methylpropanenitrile (i.e. acetone cyano-hydrin, the cyanohydrins of butanone and pentan-3-one may also be substrates. Strict donor specificity: no substrates are ADPglucose, CDPglucose, GDPglucose, IDPglucose or TDPglucose) [1]
2 UDPglucose + 2-hydroxy-2-methylbutanenitrile [1]
3 UDPglucose + 2-hydroxy-2-ethylbutanenitrile [1]

Product spectrum
1 UDP + linamarin [1]
2 UDP + 2-(beta-D-glucopyranosyloxy)-2-methylbutanenitrile
3 UDP + 2-(beta-D-glucopyranosyloxy)-2-ethylbutanenitrile

Inhibitor(s)

Cofactor(s)/prosthetic group(s)/activating agents
More (no requirement for 2-mercaptoethanol or DTT, 0.01–100 mM) [1]

Metal compounds/salts
More (no requirement for Mg^{2+}) [1]

Turnover number (min^{-1})

Specific activity (U/mg)
0.0238 [1]

K_m-value (mM)

pH-optimum
8.6 [1]

pH-range
7.2–9.3 (about half-maximal activity at pH 7.2 and 9.3, due to substrate stability Tris buffer preferred over glycine-NaOH buffer) [1]

Temperature optimum (°C)
30 (assay at) [1]

Temperature range (°C)

3 ENZYME STRUCTURE

Molecular weight

Subunits

Glycoprotein/Lipoprotein
–

4 ISOLATION/PREPARATION

Source organism
Linum usitatissimum (flax) [1]

Source tissue
Seedlings [1]

Localization in source

Purification
Linum usitatissimum (partial) [1]

Crystallization
–

Cloned

–

Renatured

–

5 STABILITY

pH

Temperature (°C)

Oxidation

Organic solvent

General stability information
2-Mercaptoethanol stabilizes during purification [1]

Storage
Glycerol, ethanol, ammonium sulfate, DTT or albumin does not stabilize the enzyme upon storage at –18°C or room temperature [1]

6 CROSSREFERENCES TO STRUCTURE DATABANKS

PIR/MIPS code

Brookhaven code

7 LITERATURE REFERENCES

[1] Hahlbrock, K., Conn, E.E.: J. Biol. Chem.,245,917–922 (1970)

1 NOMENCLATURE

EC number
2.4.1.64

Systematic name
alpha,alpha-Trehalose:orthophosphate beta-D-glucosyltransferase

Recommended name
alpha,alpha-Trehalose phosphorylase

Synonyms
Phosphorylase, trehalose
Trehalose phosphorylase

CAS Reg. No.
37205-59-7

---

2 REACTION AND SPECIFICITY

Catalysed reaction
alpha,alpha-Trehalose + phosphate →
→ D-glucose + beta-D-glucose 1-phosphate

Reaction type
Hexosyl group transfer

Natural substrates
alpha,alpha-Trehalose + phosphate (involved in trehalose catabolism) [4]

Substrate spectrum
1 alpha,alpha-Trehalose + phosphate (r, trehalose cannot be replaced by maltose, sucrose, laminaribiose, cellobiose [4, 6], lactose, melibiose [4], glycogen, starch, laminarin [6], in the reverse direction glucose cannot be replaced by D-mannose, D-galactose, D-fructose, D-2-deoxyglucose, laminaribiosyl beta-1,4-glucose, maltose, laminaribiose, D-glucose 6-phosphate, beta-glucose 1-phosphate cannot be replaced by alpha-glucose 1-phosphate (with the exception of Flammulina velupites enzyme [6]) [4, 5], alpha-xylose 1-phosphate, alpha-mannose 1-phosphate, D-fructose 1-phosphate, alpha-galactose 1-phosphate [4], no substrate: fructose 2,6-diphosphate [2]) [1–6]
2 D-6-Deoxyglucose + beta-D-glucose 1-phosphate (glucosylated at 93% the rate of glucose glucosylation) [4]
3 Xylose + beta-D-glucose 1-phosphate (glucosylated at 23% the rate of glucose glucosylation) [4]
4 Trehalose + arsenate [4]

Product spectrum

1 D-Glucose + D-glucose 1-phosphate (beta-D-glucose 1-phosphate [1–4], alpha-D-glucose 1-phosphate [6]) [1–4, 6]
2 Glucosyl-1,1-(6-deoxyglucose) + phosphate [4]
3 Glucosyl-1,1-xylose + phosphate [4]
4 Glucose + arsenate [4]

Inhibitor(s)

Fructose 2,6-diphosphate (0.001–0.005 mM, inhibition kinetics) [2]; Na^+ [3]; More (no inhibition by adenine nucleotides (1 mM), glucose 1-phosphate or glucose 6-phosphate (5 mM), alpha-glucose 1,6-diphosphate (0.1 mM), fructose 6-phosphate (0.5 mM) or fructose 1,6-diphosphate (1 mM)) [2]

Cofactor(s)/prosthetic group(s)/activating agents

Metal compounds/salts

Turnover number (min^{-1})

Specific activity (U/mg)

0.214 [4]; 1.63 [3]

K_m-value (mM)

5 (phosphate) [6]; 6 (beta-glucose 1-phosphate) [4]; 9.4 (phosphate) [4]; 32 (glucose) [4]; 33 (trehalose) [4]; 47 (alpha-glucose 1-phosphate) [6]; 75 (trehalose) [6]; 630 (glucose) [6]

pH-optimum

6 (trehalose synthesis) [3]; 6.3 (trehalose synthesis) [4, 6]; 7 (phosphorolysis) [3, 4, 6]

pH-range

5.7–7.8 (about half-maximal activity at pH 5.7 and 7.8, trehalose synthesis) [4]; 6–8 (about half-maximal activity at pH 6 and 8, phosphorolysis) [4]

Temperature optimum (°C)

40 [3]

Temperature range (°C)

3 ENZYME STRUCTURE

Molecular weight

344000 (Euglena gracilis, sucrose gradient centrifugation) [4]

Subunits

Glycoprotein/Lipoprotein

—

4 ISOLATION/PREPARATION

Source organism
Euglena gracilis (var. bacillaris [1, 3, 4], strain SM-ZK [2]) [1–4]; Glycine max (soy bean, Merril cv. Beeson 80) [5]; Flammulina velutipes (basidiomycete, stock 721-B1) [6]

Source tissue
Cell [1–4]; Nodules (nodulaid inoculum: Bradyrhizobium japonicum) [5]; Fruit body [6]; Mycelium [6]

Localization in source
Soluble [1]

Purification
Euglena gracilis (partial) [2–4]

Crystallization
–

Cloned
–

Renatured
–

5 STABILITY

pH
6–8 (native enzyme, stable) [3]; 6.5–7.5 (90% of maximal activity retained after 66 h, immobilized enzyme) [3]

Temperature (°C)
40 (below, at least 30 min stable, pH 7) [3]

Oxidation

Organic solvent

General stability information
Substrates stabilize [3]; alpha- or beta-Glucose 1-phosphate, 2 mM, stabilizes [4]; Dialysis against water, 5 mM NaCl, glucose 6-phosphate, EDTA or imidazole-HCl buffer, pH 7, inactivates, not reversible by phosphate buffer, alpha- or beta-glucose 1-phosphate, ATP-Mg^{2+} or SH-reagents [4]; Gel filtration on Sephadex G-25 with or without SH-reagents inactivates [4]; Glycerol, 25%, stabilizes to some extent [4]; Phosphate buffer, 2 mM, pH 7, stabilizes [4]; Freeze-thawing, stable to in the presence of glucose 1-phosphate or phosphate buffer [4]

Storage

−14°C, at least 5 months in the presence of 2 mM alpha- or beta-glucose 1-phosphate, in 2 mM phosphate buffer, pH 7, with several freeze-thawing cycles [4]

6 CROSSREFERENCES TO STRUCTURE DATABANKS

PIR/MIPS code

Brookhaven code

7 LITERATURE REFERENCES

[1] Belocopitow, E., Maréchal, L.R.: Biochim. Biophys. Acta,198,151–154 (1970)
[2] Miyatake, K., Kuramoto, Y., Kitaoka, S.: Biochem. Biophys. Res. Commun.,122, 906–911 (1984)
[3] Murao, S., Nagano, H., Ogura, S., Nishino, T.: Agric. Biol. Chem.,49,2113–2118 (1985)
[4] Maréchal, L.R., Belocopitow, E.: J. Biol. Chem.,247,3223–3228 (1972)
[5] Salminen, S.O., Streeter, J.G.: Plant Physiol.,81,538–541 (1986)
[6] Kitamoto, Y., Akashi, H., Tanaka, H., Mori, N.: FEMS Microbiol. Lett.,55,147–150 (1988)

1 NOMENCLATURE

EC number
2.4.1.65

Systematic name
GDP-L-fucose:beta-D-galactosyl-N-acetyl-D-glucosaminyl-R-4-L-fucosyltrans-
ferase

Recommended name
Galactoside 3(4)-L-fucosyltransferase

Synonyms
Fucosyltransferase, guanosine diphosphofucose-beta-acetyl-
glucosaminylsaccharide 4-alpha-L-
alpha(1,3/1,4) Fucosyltransferase III
alpha-(1→4)-L-Fucosyltransferase
alpha-4-L-Fucosyltransferase
beta-Acetylglucosaminylsaccharide fucosyltransferase
Blood-group substance Lea-dependent fucosyltransferase
Fucosyltransferase, guanosine diphosphofucose-glycoprotein 4-alpha-
Guanosine diphosphofucose-glycoprotein 4-alpha-L-fucosyltransferase
Lewis blood group alpha1→3/4 fucosyltransferase
Lewis alpha1→3/4 fucosyltransferase
Blood group Lewis alpha-4-fucosyltransferase
FucT-II [2]
(Lea)-dependent alpha-3/4-fucosyltransferase [3]
Lewis alpha-3/4-fucosyltransferase [5]
Lewis(Le) blood group gene-dependent alpha-3/4-L-fucosyltransferase [6]
More (cf. EC 2.4.1.69 and EC 2.4.1.152)

CAS Reg. No.
37277-69-3

---

2 REACTION AND SPECIFICITY

Catalysed reaction
GDP-L-fucose + 1,3-beta-D-galactosyl-N-acetyl-D-glucosaminyl-R →
→ GDP + 1,3-beta-D-galactosyl-(alpha-1,4-L-fucosyl)-N-ace-
tyl-D-glucosaminyl-R

Reaction type
Hexosyl group transfer

Natural substrates

GDP-L-fucose + 1,3-beta-D-galactosyl-N-acetyl-D-glucosaminyl-R (catalyzes the addition of L-fucose to both the O-3 and O-4 position of subterminal N-acetyl-D-glucosamine residues, also catalyzes the formation of NeuAc-alpha(2–3)Galbeta(1–3)[Fucalpha(1–4)]GlcNAc, sialylated Lewis$_a$ determinants from NeuAcalpha(2–3)Galbeta(1–3)GlcNAc) [6]

Substrate spectrum

1 GDP-L-fucose + 1,3-beta-D-galactosyl-N-acetyl-D-glucosaminyl-R (acceptor substrate: broad specificity [2], e.g. oligosaccharides containing the nonreducing terminal sequence Galbeta(1–3)GlcNAc or Galbeta(1–4)GlcNAc [1], lacto- and neolacto-series acceptors [2], Fucalpha(1–2)Galbeta(1–3)GlcNAcbeta(1–3)Galbeta(1–4)Glc [1], Galbeta(1–4)GlcNAc [1], lacto-N-neotetraose (Galbeta(1–4)GlcNAcbeta(1–3)Galbeta(1–4)Glc) [1], Galbeta(1–3)GlcNAc [1], asialotransferrin (Galbeta(1–4)GlcNAcbeta(1–2)Man-) [1], Galbeta(1–3)GlcNAcbeta(1–4)Glc [1], Fucalpha(1–2)Galbeta(1–4)GlcNAcbeta-OR [5], NeuAcalpha(2–3)Galbeta(1–4)GlcNAcbeta-OR [5], Fucalpha(1–2)Galbeta(1–3)GlcNAcbeta-OR [5], NeuAcalpha(2–3)Galbeta(1–3)GlcNAcbeta-OR [5], precursor alpha-4 activity with lacto-N-biose 1 and alpha-3 activity with 2'-fucosyllactose, but relatively little alpha-3 activity with N-acetyllactosamine [3], precursor Lc3 structure (Lc3, lactosylceramide, is GlcNAcbeta(1–3)Galbeta(1–4)Glcbeta(1–1)Cer) [2]) [1–6]

Product spectrum

1 GDP + 1,3-beta-D-galactosyl-(alpha-1,4-L-fucosyl)-N-acetyl-D-glucosaminyl-R (precursor Lc3 structure (Lc3, lactosylceramide, is GlcNAcbeta(1–3)Galbeta(1–4)Glcbeta(1–1)Cer), the product is composed exclusively of a structure containing a fucose linked to the 3-position of the internal Glc residue [2]) [1–6]

Inhibitor(s)

N-Ethylmaleimide [1, 4, 5]

Cofactor(s)/prosthetic group(s)/activating agents

Metal compounds/salts

Mn^{2+} (nonessential activation by several divalent cations: Mn^{2+}, Ca^{2+}, Co^{2+}, Zn^{2+}, Mg^{2+}, Ni^{2+}, Cd^{2+}, Ba^{2+}, activation by Mn^{2+} is maximal at 5 mM [1], activates [5]) [1, 5]; Mg^{2+} (nonessential activation by several divalent cations: Mn^{2+}, Ca^{2+}, Co^{2+}, Zn^{2+}, Mg^{2+}, Ni^{2+}, Cd^{2+}, Ba^{2+}, activation by Mg^{2+} is maximal at 20 mM) [1]; Ca^{2+} (nonessential activation by several divalent cations: Mn^{2+}, Ca^{2+}, Co^{2+}, Zn^{2+}, Mg^{2+}, Ni^{2+}, Cd^{2+}, Ba^{2+}) [1]; Co^{2+} (nonessential activation by several divalent cations: Mn^{2+}, Ca^{2+}, Co^{2+}, Zn^{2+}, Mg^{2+}, Ni^{2+}, Cd^{2+}, Ba^{2+}) [1]; Zn^{2+} (nonessential activation by several divalent cations: Mn^{2+},

Ca^{2+}, Co^{2+}, Zn^{2+}, Mg^{2+}, Ni^{2+}, Cd^{2+}, Ba^{2+}) [1]; Ni^{2+} (nonessential activation by several divalent cations: Mn^{2+}, Ca^{2+}, Co^{2+}, Zn^{2+}, Mg^{2+}, Ni^{2+}, Cd^{2+}, Ba^{2+}) [1]; Ba^{2+} (nonessential activation by several divalent cations: Mn^{2+}, Ca^{2+}, Co^{2+}, Zn^{2+}, Mg^{2+}, Ni^{2+}, Cd^{2+}, Ba^{2+}) [1]; Cd^{2+} (nonessential activation by several divalent cations: Mn^{2+}, Ca^{2+}, Co^{2+}, Zn^{2+}, Mg^{2+}, Ni^{2+}, Cd^{2+}, Ba^{2+}) [1]

Turnover number (min^{-1})

Specific activity (U/mg)
    More [1]

K$_m$-value (mM)

pH-optimum
    6.0–7.2 (gall-bladder) [5]; 7–7.8 [1]

pH-range

Temperature optimum (°C)

Temperature range (°C)

3 ENZYME STRUCTURE

Molecular weight

Subunits
    ? (x × 65000, rat, SDS-PAGE) [4]

Glycoprotein/Lipoprotein
    –

4 ISOLATION/PREPARATION

Source organism
    Human [1–3, 5, 6]; Rat [4]

Source tissue
    Milk [1, 3, 5, 6]; Small intestine [4]; Gall-bladder [5]; Kidney [5]

Localization in source

Purification
    Human [1, 3]

Crystallization
    –

Cloned
    –

Renatured

–

---

5 STABILITY

pH

Temperature (°C)
   50 (half-life: 5–9 min, milk enzyme, 28 min, gall-bladder enzyme, 46 min, kidney enzyme) [5]; 59 (10 min, about 70% loss of activity) [1]

Oxidation

Organic solvent

General stability information

Storage
   –20°C, 50% glycerol, protein concentration 0.03 mg/ml [1]

---

6 CROSSREFERENCES TO STRUCTURE DATABANKS

PIR/MIPS code
   PIR2:A36669 (human)

Brookhaven code

---

7 LITERATURE REFERENCES

[1] Prieels, J.-P., Monnom, D., Dolmans, M., Beyer, T.A., Hill, R.L.: J. Biol. Chem.,256, 10456–10463 (1981)
[2] Holmes, E.H.: Glycobiology,3,77–81 (1993)
[3] Johnson, P.H., Watkins, W.M.: Glycoconjugate J.,9,241–249 (1992)
[4] Martin, A., Biol, M.C., Richard, M., Louisot, P.: Comp. Biochem. Physiol., Comp. Biochem.,87B,725–731 (1987)
[5] Mollicone, R., Gibaud, A., Francois, A., Ratcliffe, M., Oriol, R.: Eur. J. Biochem., 191,169–176 (1990)
[6] Johnson, P.H., Watkins, W.M.: Biochem. Soc. Trans.,13,1119–1120 (1985)

1 NOMENCLATURE

EC number
2.4.1.66

Systematic name
UDPglucose:5-(D-galactosyloxy)-L-lysine-procollagen D-glucosyltransferase

Recommended name
Procollagen glucosyltransferase

Synonyms
Glucosyltransferase, uridine diphosphoglucose-collagen
Collagen glucosyltransferase
Collagen hydroxylysyl glucosyltransferase
Galactosylhydroxylysyl glucosyltransferase
UDP-glucose-collagen glucosyltransferase
Uridine diphosphoglucose-collagen glucosyltransferase
More (cf. EC 2.4.1.50)

CAS Reg. No.
9028-08-4

2 REACTION AND SPECIFICITY

Catalysed reaction
UDPglucose + 5-(D-galactosyloxy)-L-lysine-procollagen →
→ UDP + 1,2-D-glucosyl-5-D-(galactosyloxy)-L-lysine-procollagen

Reaction type
Hexosyl group transfer

Natural substrates
UDPglucose + 5-(D-galactosyloxy)-L-lysine-procollagen (glycosylation in the course of (pro)-collagen biosynthesis [9], together with EC 2.4.1.50 involved in biosynthesis of glucosylgalactosyl-hydroxylysine units of collagens, basement membranes and serum glycoproteins [14]) [9, 14]

Substrate spectrum
1 UDPglucose + 5-(D-galactosyloxy)-L-lysine-procollagen (highly specific, absolute requirement: free epsilon-amino group in hydroxylysyl-residues [18], requires receptor protein with terminal galactose residues [9], acceptor specificity [10, 18], glucosyl-acceptors: ichthyocol, ichthyocol glycopeptides (obtained from collagenase digestion), bovine glomerular basement membranes [14], guinea pig skin collagen [2], calf skin gelatin

[5, 7, 8, 10, 14], rat [7, 10, 11] or calf [9] skin citrate-soluble collagen (heat-denatured [10, 11]), acetic acid soluble collagen (from arterial calf tissue [9], bovine tendon [17]) [9, 17], fetuin (minus sialic acid, galactose and N-acetylglucosamine), transferrin (minus Fe^{3+} and sialic acid) [15], peptides prepared from collagen (from bovine Achilles tenton [8]) by collagenase-digestion [6, 8], incomplete hexasaccharide chains of bovine Achilles tenton collagen (glucosyl residues removed by mild acid hydrolysis) [15], purified alpha-chain of chicken skin collagen [17], non-functional substrates, such as alpha$_1$-glycoprotein and fetuin can replace collagen to some extent [2]. No substrates are hydroxylysyl- or glucosylgalactosyl hydroxylysyl-residues [18], ADPglucose [13, 17], GDPglucose [17], UDP-N-acetylglucosamine and UDP-N-acetylgalactosamine [13], porcine [2] or bovine [15] submaxillary glycoprotein, galactose, N-acetylgalactosamine [2], ovalbumin, transferrin, ceruloplasmin, human albumin, fibrinogen, haptoglobin [15], glucose free bovine glomerular basement membranes, calf thyroglobulin [14], deaminated collagen, fetuin [15]) [1–3, 5–18]

2 UDPglucose + denatured form of citrate-soluble calf skin collagen (best substrate, and its alpha$_1$-chain and beta12-component) [9]

3 UDPglucose + galactosyl sphingosine (in vitro) [10, 18]

4 UDPglucose + galactosyl hydroxylysine (in vitro) [12, 14, 16–18]

5 TDPglucose + galactosyl hydroxylysine (glucosylation at 60% the rate of UDPglucose) [17]

Product spectrum

1 UDP + 1,2-D-glucosyl-5-D-(galactosyloxy)-L-lysine-procollagen [7, 8]

2 ?

3 UDP + glucosylgalactosyl sphingosine [10]

4 UDP + glucosylgalactosyl hydroxylysine [12, 14, 16, 17]

5 TDP + glucosylgalactosyl hydroxylysine [17]

Inhibitor(s)

EDTA [2]; Carminic acid (mechanism) [5]; UDP (product inhibition) [11, 15, 17]; Acetylsalicylic acid [15]; Glucosamine [15]; p-Substituted mercuribenzoate (substrates plus Mn^{2+} partially protect) [7]; Zn^{2+} (kinetics) [11]; p-Hydroxymercuribenzoate [15]; p-Chloromercuribenzoate [17]; Sucrose [17]; Cu^{2+} (5–10 mM, Mn^{2+}-activated enzyme) [11]; Cu^{2+} (5–10 mM, Mn^{2+}-activated enzyme) [11]; Cd^{2+} (5–10 mM, Mn^{2+}-activated enzyme) [11]; Ni^{2+} (5–10 mM, Mn^{2+}-activated enzyme) [11]; Ca^{2+} (5–10 mM, Mn^{2+}-activated enzyme) [11]; Fe^{2+} (at high concentration) [11]; Co^{2+} (at high concentration) [11]; ADP [15]; AMP [15]; cAMP [15]; UTP [17]; More (no inhibition by Mg^{2+} [11], glucose, N-acetylglucosamine, Triton X-100, delta-hydroxylysine [15], glucosamine [15, 17], NEM, glucosylgalactosylhydroxylysine [17]) [11, 15, 17]

Cofactor(s)/prosthetic group(s)/activating agents

Dithioerythritol (activation, 1 mM) [7]; Bovine serum albumin (activation) [11]; More (no activation by folate) [8]

Metal compounds/salts

Mn^{2+} (requirement, 0.01 mM [11], 0.2–2.0 mM [18], 5 mM [14], 10 mM [7–9], 10–15 mM [16], 15 mM [15], 25–80 mM (soluble enzyme is inhibited above 50 mM) [17], 2 Mn^{2+}-ions per enzyme molecule, mechanism [11, 18]) [2–18]; Mg^{2+} (activation, 30 mM [7], can replace Mn^{2+} with about 20% [8, 15], 38% [14], 50–60% [9] efficiency, respectively, not [11]) [2, 7, 8, 9, 14, 15, 17]; Co^{2+} (activation, also in the presence of Mn^{2+} [11], 2.5 mM [7], can replace Mn^{2+} with 20% [15], about 30% [7, 8], 80% [11] efficiency, respectively, at low concentration, inhibits at higher concentration [11], not [14]) [2, 7, 8, 11, 15, 17, 18]; Ca^{2+} (activation, 10 mM [7, 14], can replace Mn^{2+} with less than 2% [8], 20% [15], 50–60% [9] efficiency, respectively, not [17]) [2, 7–9, 14, 15]; Ni^{2+} (activation, soluble enzyme [17], not [8, 14]) [17]; Fe^{2+} (activation [11, 18], also in the presence of Mn^{2+}, at low concentration [18], not [8, 14], inhibits at high concentration [11]) [11, 18]; More (no activation by Na^+, K^+, Fe^{3+}, Ba^{2+} [15], Zn^{2+}, Cu^{2+} [15, 17], Cd^{2+} [14, 15, 17]) [14, 15, 17]

Turnover number (min^{-1})

Specific activity (U/mg)

More [9]; 0.031 (enzyme from whole embryo) [7]; 0.046 [12]; 0.065 [6]; 0.09–0.1 [16]; 0.43 [13]; 0.57 [10, 18]

K_m-value (mM)

More (K_m-values: 0.5–1.0 mg/ml (denatured citrate soluble rat skin collagen) [18], 4.5 mg/ml (gelatin) [5], 7–14 mg/ml (gelatinized calf skin collagen) [8], 1.5 mg/ml (alpha-chain of chicken skin collagen) [17], kinetic study [8, 17]) [5, 8, 17]; 0.002 (galactosyl hydroxylysine in purified alpha-chain of chicken skin collagen, value based on actual amount of galactosyl hydroxylysine-residues of alpha-chain) [17]; 0.0021 (calf skin collagen) [14]; 0.0026 (galactosyl hydroxylysine) [14]; 0.003–0.006 (galactosyl hydroxylysyl acceptor sites) [18]; 0.005 (UDPglucose, membrane-bound enzyme) [17]; 0.006–0.007 (UDPglucose, at high Mn^{2+}-concentration) [11]; 0.0074 (UDP-glucose) [16]; 0.0167 (UDPglucose) [5]; 0.025 (UDPglucose, at low Mn^{2+}-concentration) [11]; 0.029 (TDPglucose) [17]; 0.03–0.048 (UDPglucose) [8, 14]; 0.167 (calf skin collagen) [9]; 0.28 (acetic acid soluble collagen from calf arterial tissue) [9]; 0.33 (beta12-component) [9]; 0.63 (alpha-chain) [9]; 2.75 (galactosyl hydroxylysine) [16]; 3.8 (galactosyl hydroxylysine, membrane-bound enzyme) [17]

pH-optimum

5.7 [15]; 5.8 (skin) [2]; 6–6.5 (2 pH-optima: 6–6.5 and 7.5–8) [14]; 6.5–7.5
(soluble enzyme) [17]; 6.5–8.0 [17]; 7.0 [8, 16]; 7–8 (membrane-bound enzyme) [17]; 7.5–8.0 (2 pH-optima: 6–6.5 and 7.5–8) [14]; 8.3 (2 pH-optima:
8.3 and 9.9) [9]; 9.9 (2 pH-optima: 8.3 and 9.9) [9]

pH-range

4.0–11.5 (about half-maximal activity at pH 4.0 and 11.5) [9]; 5.2–6.5 (about
half-maximal activity at pH 5.2 and 6.5) [15]

Temperature optimum (°C)

37 (skin) [2]

Temperature range (°C)

---

3 ENZYME STRUCTURE

Molecular weight

50000–55000 (chicken, gel filtration) [6]
52000–54000 (chicken, gel filtration) [7, 12]
72000 (bovine, gel filtration) [9]
More (amino acid analysis) [10]

Subunits

? (x × 72000–78000, chicken, SDS-PAGE) [13, 18]

Glycoprotein/Lipoprotein

–

---

4 ISOLATION/PREPARATION

Source organism

Guinea pig (male embryonic [2]) [1, 2, 18]; Rat [5, 18]; Chicken (white leg-
horn [6–8, 10–13]) [3, 4, 6–8, 10–13, 18]; Bovine (calf) [9, 18]; Human
[14–18]

Source tissue

Skin (human, guinea pig [18], cultured fibroblasts [18]) [1, 2, 18]; Cartilage
(embryonic, cartilaginous ends of limb bone rudiments [2]) [2, 7, 8]; Embryo
(chicken, rat [18]) [2, 6, 7, 10–13, 18]; Liver (rat [18]) [3, 4, 18]; Kidney (rat)
[5, 18]; Arterial tissue (aorta thoracica) [9]; Lung fibroblasts (fetal, WI–38
and IMR-90 diploid fibroblasts, in cell suspension culture) [14]; Blood platelets (human [18]) [15–18]; Peritoneal macrophages (rat) [18]; Blood (distribution, not: erythrocytes) [16]; Bone (rat) [18]

Localization in source
Membrane-bound (plasma-membrane [15, 17], 5–10% of total activity [17])
[2–4, 15–17]; Soluble (90% of total activity [17]) [16–18]; Endoplasmic reticulum [3]; Golgi apparatus [3]

Purification
Rat (partial [18]) [5, 18]; Chicken (partial [3, 4, 6–8], solubilized with 0.5 M
NaCl [13], Nonidet P-40 [3, 4, 13], Triton X-100 [4, 7, 13], CHAPS, Brij-35 [4],
affinity chromatography [6, 10, 12, 13]) [3, 4, 6–8, 10–13]; Guinea pig (skin,
partial, solubilized with Triton X-100) [2, 18]; Bovine (partial) [9, 18]; Human
(partial, solubilized with Triton X-100 [14]) [14, 15, 18]

Crystallization
–

Cloned
–

Renatured
–

5 STABILITY

pH

Temperature (°C)
20 (10% loss of activity within 18 h, soluble enzyme, after DEAE-ion
exchange chromatography) [17]; 58 (denaturation) [9]

Oxidation

Organic solvent

General stability information
Freezing, concentrated enzyme solution, stable to [7]; Freeze-thawing cycles, soluble enzyme, extremely stable to [17]; Repeated thawing inactivates [13]; Repeated thawing, stable to, human serum enzyme [18]; Urea,
2 M, increases solubilization rate [13]; Concentrated enzyme solutions,
above 0.2 mg/ml, lose most of the activity [13, 18]; Chromatography on
Biogel A-1.5M decreases stability [17]; Stable in crude tissue extracts [18]

Storage
–40°C, soluble enzyme after Biogel A-1.5M chromatography, 2–4 weeks
[17]; –40°C, membrane-preparation, 1–2 months [17]; –20°C, enzyme concentration 0.1 mg/ml, 30% loss of activity within several months [13, 18];
4°C, enzyme concentration 0.1 mg/ml, 30% loss of activity within 1 week
[13, 18]; 4°C, in 0.1 M Tris-buffer, pH 7.2, 0.2 M NaCl, 10% loss of activity
within 1 week [9]; 4°C, dilute enzyme solution, $t_{1/2}$: less than 1 week [7]

6 CROSSREFERENCES TO STRUCTURE DATABANKS

PIR/MIPS code

Brookhaven code

7 LITERATURE REFERENCES

[1] Butler, W.T., Cunningham, L.W.: J. Biol. Chem.,241,3882–3888 (1966)
[2] Bosmann, H.B., Eylar, E.H.: Biochem. Biophys. Res. Commun.,30,89–94 (1968)
[3] Bortolato, M., Radisson, J., Azzar, G., Got, R.: Int. J. Biochem.,24,243–248 (1992)
[4] Bortolato, M., Azzar, G., Farjanel, J., Got, R.: Int. J. Biochem.,23,897–900 (1991)
[5] Chang, A.Y., Noble, R.E.: Int. J. Biochem.,14,691–694 (1982)
[6] Risteli, L., Myllylä, R., Kivirikko, K.I.: Eur. J. Biochem.,67,197–202 (1976)
[7] Myllylä, R., Risteli, L., Kivirikko, K.I.: Eur. J. Biochem.,61,59–67 (1976)
[8] Myllylä, R., Risteli, L., Kivirikko, K.I.: Eur. J. Biochem.,52,401–410 (1975)
[9] Henkel, W., Buddecke, E.: Hoppe-Seyler's Z. Physiol. Chem.,356,921–928 (1975)
[10] Anttinen, H., Myllylä, R., Kivirikko, K.I.: Biochem. J.,175,737–742 (1978)
[11] Myllylä, R., Anttinen, H., Kivirikko, K.I.: Eur. J. Biochem.,101,261–269 (1979)
[12] Anttinen, H., Kivirikko, K.I.: Biochim. Biophys. Acta,429,750–758 (1976)
[13] Myllylä, R., Anttinen, H., Risteli, L., Kivirikko, K.I.: Biochim. Biophys. Acta,480, 113–121 (1977)
[14] Carnicero, H.H., Adamany, A.M., Englard, S.: Arch. Biochem. Biophys.,210, 678–690 (1981)
[15] Barber, A.J., Jamieson, G.A.: Biochim. Biophys. Acta,252,533–545 (1971)
[16] Leunis, J.C., Smith, D.F., Nwokoro, N., Fishback, B.L., Wu, C., Jamieson, G.A.: Biochim. Biophys. Acta,611,79–86 (1980)
[17] Smith, D.F., Wu, C., Jamieson, G.A.: Biochim. Biophys. Acta,483,263–278 (1977)
[18] Kivirikko, K.I., Myllylä, R.: Methods Enzymol.,82,245–304 (1982) (Review)

1 NOMENCLATURE

EC number
2.4.1.67

Systematic name
1-alpha-D-Galactosyl-myo-inositol:raffinose galactosyltransferase

Recommended name
Galactinol-raffinose galactosyltransferase

Synonyms
Stachyose synthetase
Galactosyltransferase, galactinol-raffinose
More (cf. EC 2.4.1.123)

CAS Reg. No.
37277-70-6

2 REACTION AND SPECIFICITY

Catalysed reaction
1-alpha-D-Galactosyl-myo-inositol + raffinose →
→ myo-inositol + stachyose

Reaction type
Hexosyl group transfer

Natural substrates
1-alpha-D-Galactosyl-myo-inositol + raffinose (involved in biosynthesis of
sucrosyl-oligosaccharides) [3]

Substrate spectrum
1 1-alpha-D-Galactosyl-myo-inositol + raffinose (i.e. galactinol, high donor
specificity: UDPgalactose or melibiose are no galactosyl-donors, p-nitro-
phenyl-alpha-D-galactopyranoside can replace galactinol in vitro (50%
efficiency) [2], poor substrates are glucose, galactose, lactose, no substrates are glycerol [1], sucrose, maltose, fructose, cellobiose, gentiobiose, melizitose, trehalose [1, 2], maltotriose, manninotriose [2]) [1–3]
2 1-alpha-D-Galactosyl-myo-inositol + melibiose (galactosylation at 25% [1],
10% [2] the rate of raffinose) [1, 2]

Product spectrum
1 myo-Inositol + stachyose [1–3]
2 myo-Inositol + manninotriose (i.e. alpha-D-galactosyl-6-O-D-galac-
tosyl-alpha-D-glucose) [2]

Inhibitor(s)

Co^{2+} [2]; Hg^{2+} [2]; Mn^{2+} [2]; Ni^{2+} [2]; Zn^{2+} [2]; myo-Inositol (strong) [2]; Melibiose (i.e. 6-O-alpha-D-galactosyl-alpha-D-glucose, non-competitive to raffinose, weak) [2]; Tris buffer [2]; Ag^+ (weak) [2]; Ca^{2+} (weak) [2]; Cu^{2+} (weak) [2]; Mg^{2+} (weak) [2]; More (no inhibitors: 5 mM EDTA, 3 mM MoO_4^{2-}, K^+, Na^+, Fe^{3+}, stachyose, galactose) [2]

Cofactor(s)/prosthetic group(s)/activating agents

Metal compounds/salts

Turnover number (min^{-1})

Specific activity (U/mg)
0.000833 [3]; 0.0185 [2]

K_m-value (mM)
0.33 (galactinol, hydrolysis) [2]; 4.6 (raffinose (+ galactinol)) [2]; 5.2 (melibiose (+ galactinol)) [2]; 7.7 (galactinol (+ raffinose)) [2]

pH-optimum
6.5–6.9 (plateau) [2]

pH-range
5.4–7.6 (about half-maximal activity at pH 5.4 and 7.6) [2]

Temperature optimum (°C)
30 (assay at) [2, 3]

Temperature range (°C)

3 ENZYME STRUCTURE

Molecular weight

Subunits

Glycoprotein/Lipoprotein

–

4 ISOLATION/PREPARATION

Source organism
Phaseolus vulgaris [1]; Cucurbita pepo (var. melopepo f. torticolis Bailey) [2]; Cucumis melo (cv. inodorus) [3]

Source tissue
Seeds (ripening) [1]; Leaf (mature [2], distribution [3]) [2, 3]

Localization in source
  Cytoplasm [2]

Purification
  Phaseolus vulgaris (partial) [1]; Cucurbita pepo (partial) [2]

Crystallization
  –

Cloned
  –

Renatured
  –

---

5 STABILITY

pH

Temperature (°C)

Oxidation

Organic solvent

General stability information

Storage
  4°C, 0.1 M sodium phosphate buffer, 20 mM 2-mercaptoethanol, up to 45
  days [2]

---

6 CROSSREFERENCES TO STRUCTURE DATABANKS

PIR/MIPS code

Brookhaven code

---

7 LITERATURE REFERENCES

[1] Tanner, W.: Ber. Dtsch. Bot. Ges.,80,111 (1967)
[2] Gaudreault, P.-R., Webb, J.A.: Phytochemistry,20,2629–2633 (1981)
[3] Schmitz, K., Holthaus, U.: Planta,169,529–535 (1986)

1 NOMENCLATURE

EC number
2.4.1.68

Systematic name
GDP-L-fucose:glycoprotein (L-fucose to asparagine-linked N-acetylglucosamine of N-acetyl-beta-D-glucosaminyl-1,2-alpha-D-mannosyl-1,3-(R$_1$-alpha-1,6)-beta-D-mannosyl-1,4-beta-N-acetyl-D-glucosaminyl-1,4-N-acetyl-D-glucosaminyl)asparagine 6-alpha-L-fucosyltransferase

Recommended name
Glycoprotein 6-alpha-L-fucosyltransferase

Synonyms
GDP-fucose-glycoprotein fucosyltransferase
GDP-L-Fuc:N-acetyl-beta-D-glucosaminide alpha1→6fucosyltransferase [1]
Fucosyltransferase, guanosine diphosphofucose-glycoprotein
GDP-fucose glycoprotein fucosyltransferase
GDP-L-fucose-glycoprotein fucosyltransferase
Glycoprotein fucosyltransferase
Guanosine diphosphofucose-glycoprotein fucosyltransferase
GDPfucose glycoprotein fucosyltransferase

CAS Reg. No.
9033-08-3

---

2 REACTION AND SPECIFICITY

Catalysed reaction
GDP-L-fucose + N^4-(N-acetyl-beta-D-glucosaminyl-1,2-alpha-D-mannosyl-1,3-(R$_1$-alpha-1,6)-beta-D-mannosyl-beta-N-acetyl-1,4-D-glucosaminyl-1,4-N-acetyl-D-glucosaminyl)asparagine →
→ GDP + N^4-(N-acetyl-beta-D-glucosaminyl-1,2-alpha-D-mannosyl-1,3-(R$_1$-alpha-1,6)-beta-D-mannosyl-1,4-beta-N-acetyl-D-glucosaminyl-1,4-(alpha-L-fucosyl-1,6)-N-acetyl-D-glucosaminyl)asparagine

Reaction type
Hexosyl group transfer

Natural substrates

Substrate spectrum
1 GDPfucose + asialo-agalactotransferrin glycopeptide [1]

Product spectrum
1 Fucose alpha-1,6 bound to the asparagine-linked N-acetylglucosamine of
the asialo-agalactotransferrin glycopeptide + GDP [1]

Inhibitor(s)

Cofactor(s)/prosthetic group(s)/activating agents

Metal compounds/salts

Turnover number (min^{-1})

Specific activity (U/mg)

K_m-value (mM)

pH-optimum

pH-range

Temperature optimum (°C)
37 (assay at) [1]

Temperature range (°C)

3 ENZYME STRUCTURE

Molecular weight

Subunits

Glycoprotein/Lipoprotein
–

4 ISOLATION/PREPARATION

Source organism
Human [1]

Source tissue
Skin fibroblasts [1]

Localization in source

Purification

Crystallization
–

Cloned
–

Renatured

–

5 STABILITY

pH

Temperature (°C)

Oxidation

Organic solvent

General stability information

Storage

6 CROSSREFERENCES TO STRUCTURE DATABANKS

PIR/MIPS code

Brookhaven code

7 LITERATURE REFERENCES

[1] Voynow, J.A., Scanlin, T.F., Glick, M.C.: Anal. Biochem.,168,367–373 (1988)

1 NOMENCLATURE

EC number
    2.4.1.69

Systematic name
    GDP-L-fucose:beta-D-galactosyl-R 2-alpha-L-fucosyltransferase

Recommended name
    Galactoside 2-alpha-L-fucosyltransferase

Synonyms
    Fucosyltransferase, guanosine diphosphofucose-galactoside 2-L-
    alpha-(1→2)-L-Fucosyltransferase
    alpha-2-Fucosyltransferase
    alpha-2-L-Fucosyltransferase
    Blood-group substance H-dependent fucosyltransferase
    Fucosyltransferase, guanosine diphosphofucose-glycoprotein 2-alpha-
    Fucosyltransferase, guanosine diphosphofucose-lactose
    GDP fucose-lactose fucosyltransferase
    Guanosine diphospho-L-fucose-lactose fucosyltransferase
    Guanosine diphosphofucose-beta-D-galactosyl-alpha-2-L-fucosyltransferase
    Guanosine diphosphofucose-galactosylacetylglucosaminylgalactosyl-
    glucosylceramide alpha-L-fucosyltransferase
    Guanosine diphosphofucose-glycoprotein 2-alpha-L-fucosyltransferase
    H-Gene-encoded beta-galactoside alpha1→2fucosyltransferase [8]
    Secretor-type beta-galactoside alpha1→2fucosyltransferase [9]
    Blood group H alpha-2-fucosyltransferase
    beta-Galactoside alpha1→2fucosyltransferase [1]
    GDP-L-fucose:lactose fucosyltransferase [6]
    EC 2.4.1.89 (the action on glycolipid was previously listed as EC 2.4.1.89)
    More (cf. EC 2.4.1.65 and EC 2.4.1.152)

CAS Reg. No.
    56093-23-3

---

2 REACTION AND SPECIFICITY

Catalysed reaction
    GDP-L-fucose + beta-D-galactosyl-R →
    → GDP + alpha-L-fucosyl-1,2-beta-D-galactosyl-R (either a rapid equilibrium
    random kinetic mechanism or a steady state ordered mechanism with
    GDPfucose binding first [2])

Reaction type
Hexosyl group transfer

Natural substrates
GDP-L-fucose + tetraglycosylceramide (i.e. galactosyl-beta-1,4-N-ace-
tylglucosaminyl-beta-1,3-galactosyl-beta-1,4-glucosylceramide, to form the
blood group H-related glycosphingolipid) [5]

Substrate spectrum
1 GDP-L-fucose + beta-D-galactosyl-R (specificity: overview [2, 7, 13], overview of glycoprotein acceptors [12], e.g.: Galbeta(1–3)GalNAc [2], Gal-
  beta(1–3)GlcNAc [2], Galbeta(1–3)GlcNAcbeta(1–3)Galbeta(1–4)Glc [2],
  Galbeta(1–4)GlcNAcbeta(1–3)Galbeta(1–4)Glc [2], lactose [6, 10, 12,
  13], phenyl-beta-D-galactopyranoside [7], lacto-N-tetraose [12],
  lacto-N-biose I [12, 13], N-acetyllactosamine [12, 13], lacto-N-fucopentaose II [12], beta-methyl-D-galactopyranoside [12], phenyl-beta-D-gal-
  actoside [13], D-galactose [13], di-N-acetylchitobiose [13], fetuin [13],
  asialofetuin [10, 13], enzyme forms Fuc-alpha(1–2)Gal linkage with
  oligosaccharides, glycoproteins and glycolipids which contain nonreducing terminal galactzose residues and shows no absolute specificity for a
  particular penultimate residue or for the linkage between the galactose
  and the penultimate residue [2], enzyme transfers fucose equally well to
  Galbeta(1–3)GlcNAc and Galbeta(1–4)GlcNAc substrates, Gal-
  beta(1–3)GalNAc structures are less efficient acceptors [4], secretor-type
  alpha(1–2)fucosyltransferase shows significant lower affinity than H-en-
  zyme for phenyl-beta-D-galactosylpyranoside and GDP-fucose as well as
  for Galbeta(1–4)GlcNAc oligosaccharides, Galbeta(1–3)GlcNAc and Gal-
  beta(1–3)GalNAc oligosaccharide acceptors are preferred by the secretor-type enzyme [9]) [1–13]
2 GDP-L-fucose + tetraglycosylceramide (i.e. galactosyl-beta-1,4-N-ace-
  tylglucosaminyl-beta-1,3-galactosyl-beta-1,4-glucosylceramide) [5]
3 More (enzyme also has GDPfucose hydrolase activity) [1]

Product spectrum
1 GDP + alpha-L-fucosyl-1,2-beta-D-galactosyl-R
2 ?
3 ?

Inhibitor(s)
Lactose (competitive to antifreeze glycoprotein) [2]; N-Ethylmaleimide (3
mM [11]) [11, 13, 15]; GDP [11–14]; GTP [12, 14]; GDPmannose [12, 14];
GMP (competitive to GDPfucose) [2, 12–14]; p-Chloromercuribenzoate [13];
dGMP [14]; IMP (weak) [14]; Guanosine (weak) [14]; 2′,3′-cyclic GMP
(weak) [14]; IDP [13]; XMP [14]; XDP [14]; Ni^{2+} [2]; Cd^{2+} [2]; Zn^{2+} [2]; Cu^{2+}
[2]; Hg^{2+} [2]; Ba^{2+} [2]

Cofactor(s)/prosthetic group(s)/activating agents
GTP (stimulates [6], inhibits [12, 14]) [6]; GDPmannose (stimulates [6], inhibits [12, 14]) [6]; UTP (stimulates) [6, 12]; UDP (stimulates) [6]; UDP-N-acetyl-D-glucosamine (stimulates) [6]; UDP-D-glucose (stimulates) [6]; CTP (stimulates) [12]; ATP (stimulates) [12]

Metal compounds/salts
Mn^{2+} (stimulates) [2]; Ca^{2+} (stimulates) [2]; Co^{2+} (stimulates) [2]; Mg^{2+} (stimulates [2], required [6, 11]) [2, 6, 11]

Turnover number (min^{-1})

Specific activity (U/mg)
More [1, 8, 9]

K_m-value (mM)

pH-optimum
6 (in presence of 40 mM Mg^{2+} or 20 mM Mn^{2+}) [12]; 6.4 (cacodylate-HCl buffer) [5]; 6.5–7.0 [4]; 7.2 (lactose, bone marrow enzyme) [10]; 7.5 (lactose, serum enzyme) [10]

pH-range
6.4–8.3 (lactose, bone marrow enzyme, about 50% of activity maximum at pH 6.4 and 8.3) [10]

Temperature optimum (°C)
37 (assay at) [1, 6]

Temperature range (°C)

---

3 ENZYME STRUCTURE

Molecular weight
75000–80000 (pig, gel filtration) [1]
148000–150000 (human, gel filtration) [8, 9]
158000 (human, gel filtration) [4]

Subunits
Monomer (1 × 60000, pig, SDS-PAGE) [1]
? (x × 70000, rat, SDS-PAGE [15], x × 50000, human, SDS-PAGE in presence of beta-mercaptoethanol [9]) [9, 15]
More (polymeric form of covalently bound subunits) [8]

Glycoprotein/Lipoprotein
Glycoprotein [8, 9]

4 ISOLATION/PREPARATION

Source organism
Pig (H blood group [1]) [1, 2]; Rat [12, 14, 15]; Human (2 enzymes from cervical epithelium, pH optimum: 6.0 and 7.2 [13], H-gene-encoded beta-galactoside alpha1→2fucosyltransferase [8], secretor-type beta-galactoside alpha1→2fucosyltransferase [9]) [3, 4, 7–11, 13]; Bovine [5]; Mongrel dog [6]

Source tissue
Serum (of donors of all ABO blood-groups examined, except those of the rare Ob (Bombay) and Bh phenotypes [7]) [7–10]; Bone marrow [10]; Cervical epithelium [13]; Submaxillary gland (only from individuals who are secretors of ABH blood-group [7]) [1, 2, 7]; Stomach mucosa [7]; Small intestine (mucosa [12, 14]) [12, 14, 15]; Plasma [4, 11]; Spleen [5]; Mammary tissue [6]; Milk [3]

Localization in source
Membrane [5]; Microsomes [12]; More (bone marrow enzyme is in part particle bound) [10]

Purification
Pig [1]; Rat [15]; Human [3, 4, 8, 9]; Mongrel dog [6]

Crystallization
–

Cloned
–

Renatured
–

---

5 STABILITY

pH

Temperature (°C)

Oxidation

Organic solvent

General stability information
Inactivation by repeated freezing and thawing [1]

Storage
4°C, 0.025 M sodium cacodylate, pH 6.0, 0.5–2.0 M NaCl, protein concentration greater than 0.1 mg/ml, 0.02% NaN_3, 50% loss of activity after 8 months [1]; –20°C, stable for at least 1 year [1]; 4°C, 36 h, 50% loss of activity [6]

6 CROSSREFERENCES TO STRUCTURE DATABANKS

PIR/MIPS code

PIR3:S46494 (rat); PIR2:A36047 (human); PIR2:S46493 (rat); PIR3:S51582 (rat)

Brookhaven code

7 LITERATURE REFERENCES

[1] Beyer, T.A., Sadler, J.E., Hill, R.L.: J. Biol. Chem.,255,5364–5372 (1980)
[2] Beyer, T.A., Hill, R.L.: J. Biol. Chem.,255,5373–5379 (1980)
[3] Johnson, P.H., Watkins, W.M.: Glycoconjugate J.,9,241–249 (1992)
[4] Kyprianou, P., Betteridge, A., Donald, A.S.R., Watkins, W.M.: Glycoconjugate J., 7,573–588 (1990)
[5] Basu, S., Basu, M., Chien, J.-L.: J. Biol. Chem.,250,2956–2962 (1975)
[6] Grollman, A.P.: Methods Enzymol.,8,351–353 (1966) (Review)
[7] Chester, M.A., Yates, A.D., Watkins, W.M.: Eur. J. Biochem.,69,583–592 (1976)
[8] Sarnesto, A., Köhlin, T., Thurin, J., Blaszczyk-Thurin, M.: J. Biol. Chem.,265, 15067–15075 (1990)
[9] Sarnesto, A., Köhlin, T., Hindsgaul, O., Thurin, J., Blaszczyk-Thurin, M.: J. Biol. Chem.,267,2737–2744 (1992)
[10] Pacuszka, T., Koscielak, J.: FEBS Lett.,41,348–351 (1974)
[11] Chou, T.H., Murphy, C., Kessel, D.: Biochem. Biophys. Res. Commun.,74, 1001–1006 (1977)
[12] Bella, A., Kim, Y.S.: Arch. Biochem. Biophys.,147,753–761 (1971)
[13] Scudder, P.R., Chantler, E.N.: Biochim. Biophys. Acta,660,128–135 (1981)
[14] Bella, A., Kim, Y.S.: Biochem. J.,125,1157–1158 (1971)
[15] Martin, A., Biol, M.C., Richard, M., Louisot, P.: Comp. Biochem. Physiol. B Comp. Biochem.,87B,725–731 (1987)

1 NOMENCLATURE

EC number
2.4.1.70

Systematic name
UDP-N-acetyl-D-glucosamine:poly(ribitol-phosphate) N-acetyl-
D-glucosaminyltransferase

Recommended name
Poly(ribitol-phosphate) N-acetylglucosaminyltransferase

Synonyms
Acetylglucosaminyltransferase, uridine diphosphoacetylglucos-
amine-poly(ribitol phosphate)
UDP acetylglucosamine-poly(ribitol phosphate) acetylglucosaminyltransferase
Uridine diphosphoacetylglucosamine-poly(ribitol phosphate) acetylglucosaminyltransferase

CAS Reg. No.
37277-71-7

---

2 REACTION AND SPECIFICITY

Catalysed reaction
UDP-N-acetyl-D-glucosamine + poly(ribitol phosphate) →
→ UDP + N-acetyl-D-glucosaminyl-poly(ribitol phosphate)

Reaction type
Hexosyl group transfer

Natural substrates

Substrate spectrum
1 UDP-N-acetyl-D-glucosamine + poly(ribitol phosphate) (D-alanyl-N-ace-
tyl-alpha-galactosamine polyribitol phosphate is a less efficient acceptor,
poor donor substrate is UDPglucose, no substrates are teichoic acid,
D-ribitol 5-phosphate or partial alkali hydrolysates of teichoic acid or
N-acetyl-galactosamine polyribitol phosphate, UDP-N-acetyl-galactosamine, N-acetyl-galactosamine and N-acetyl-galactosamine phosphate) [1]

Product spectrum
1 UDP + N-acetyl-D-glucosaminyl-poly(ribitol phosphate) [1]

Inhibitor(s)
  Cu^{2+} [1]; Fe^{2+} [1]

Cofactor(s)/prosthetic group(s)/activating agents
  Mg^{2+} (requirement, 8 mM) [1]; Ca^{2+} (slight activation) [1]; Co^{2+} (slight activation) [1]

Metal compounds/salts

Turnover number (min^{-1})

Specific activity (U/mg)

K_m-value (mM)
  0.06 (poly(ribitol phosphate)) [1]; 0.7 (UDP-N-acetylglucosamine) [1]

pH-optimum
  7.5–8.5 [1]

pH-range

Temperature optimum (°C)
  37 (assay at) [1]

Temperature range (°C)

---

3 ENZYME STRUCTURE

Molecular weight

Subunits

Glycoprotein/Lipoprotein
  –

---

4 ISOLATION/PREPARATION

Source organism
  Staphylococcus aureus (strains H or Duncan: primarily N-acetyl-beta-glucosamine transfer, strain 3528: N-acetyl-alpha-glucosamine transfer, strain Copenhagen: both transferases) [1]

Source tissue
  Cell [1]

Localization in source

Purification

Crystallization
  –

Cloned

–

Renatured

–

5 STABILITY

pH

Temperature (°C)

Oxidation

Organic solvent

General stability information

Storage
    Frozen, N-acetyl-alpha-glucosamine transferase, not N-acetyl-beta-glucosa-
    mine transferase, a few months [1]

6 CROSSREFERENCES TO STRUCTURE DATABANKS

PIR/MIPS code

Brookhaven code

7 LITERATURE REFERENCES

[1] Nathenson, S.G., Ishimoto, N., Strominger, J.L.: Methods Enzymol.,8,426–429 (1966)
    (Review)

1 NOMENCLATURE

EC number
2.4.1.71

Systematic name
UDPglucose:arylamine N-D-glucosyltransferase

Recommended name
Arylamine glucosyltransferase

Synonyms
Glucosyltransferase, uridine diphosphoglucose-arylamine
UDP glucose-arylamine glucosyltransferase
Uridine diphosphoglucose-arylamine glucosyltransferase

CAS Reg. No.
37277-72-8

---

2 REACTION AND SPECIFICITY

Catalysed reaction
UDPglucose + an arylamine →
→ UDP + an N-D-glucosylarylamine

Reaction type
Hexosyl group transfer

Natural substrates
UDPglucose + an arylamine (involved in N-glucosylarylamines biosynthesis, may be participate in arylamine herbicide metabolism) [1]

Substrate spectrum
1 UDPglucose + 3,4-dichloroaniline (best glucosyl acceptor, no glucosyl donors are GDPglucose, CDPglucose, ADPglucose, UDPmannose) [1]
2 UDPglucose + 4-chloroaniline (71% efficiency compared to 3,4-dichloro-aniline) [1]
3 UDPglucose + 3-chloroaniline (31% efficiency compared to 3,4-dichloro-aniline) [1]
4 UDPglucose + 4-bromoaniline (28% efficiency compared to 3,4-dichloro-aniline) [1]
5 UDPglucose + 3-nitroaniline [1]
6 UDPglucose + 2,4-dichloroaniline [1]
7 UDPglucose + 2,5-dichloroaniline [1]
8 UDPglucose + 2,3-dichloroaniline [1]
9 UDPglucose + 3,5-dichloroaniline [1]

10 UDPglucose + 2,4-dibromoaniline [1]
11 UDPglucose + 2,5-dibromoaniline [1]
12 UDPglucose + 3-aminobenzoic acid [1]
13 UDPglucose + 5-amino-2,3-dichlorobenzoic acid [1]
14 UDPglucose + 3-amino-2,5-dichlorobenzoic acid [1]
15 TDPglucose + an arylamine [1]

Product spectrum

1 UDP + N-(3,4-dichlorophenyl)-glucosylamine [1]
2 UDP + N-(4-chlorophenyl)-glucosylamine
3 UDP + N-(3-chlorophenyl)-glucosylamine
4 UDP + N-(4-bromophenyl)-glucosylamine
5 UDP + N-(3-nitrophenyl)-glucosylamine
6 UDP + N-(2,4-dichlorophenyl)-glucosylamine
7 UDP + N-(2,5-dichlorophenyl)-glucosylamine
8 UDP + N-(2,3-dichlorophenyl)-glucosylamine
9 UDP + N-(3,5-dichlorophenyl)-glucosylamine
10 UDP + N-(2,4-dibromophenyl)-glucosylamine
11 UDP + N-(2,5-dibromophenyl)-glucosylamine
12 ?
13 ?
14 ?
15 TDP + an N-D-glucosylarylamine [1]

Inhibitor(s)

UDP (linear competitive inhibitor to UDPglucose) [1]; $HgCl_2$ [1]; $CuSO_4$ [1]; PCMB [1]; NEM [1]; o-Iodosobenzoate [1]; Tris buffer [1]; p-Benzoquinone [1]; Sodium arsenate (1 mM, weak) [1]; Catechol (weak) [1]; GSSG (oxidized glutathione, weak) [1]; IAA (weak) [1]; More (no inhibition by EDTA or iodoacetate, 1 mM each) [1]

Cofactor(s)/prosthetic group(s)/activating agents

Metal compounds/salts

Turnover number (min^{-1})

Specific activity (U/mg)

0.000912 [1]

K_m-value (mM)

0.563 (3,4-dichloroaniline) [1]; 1.88 (UDPglucose) [1]

pH-optimum

7.4–7.8 (3,4-dichloroaniline) [1]

pH-range

Temperature optimum (°C)
    25 (assay at) [1]

Temperature range (°C)

---

3 ENZYME STRUCTURE

Molecular weight

Subunits

Glycoprotein/Lipoprotein
    –

---

4 ISOLATION/PREPARATION

Source organism
    Glycine max (soy bean, Merril var. Hawkeye) [1]

Source tissue
    Seeds [1]; Hypocotyls [1]; Cotyledons [1]; Roots [1]; Leaf [1]

Localization in source
    Cytoplasm (soluble) [1]

Purification
    Glycine max (partial) [1]

Crystallization
    –

Cloned
    –

Renatured
    –

---

5 STABILITY

pH

Temperature (°C)
    50 (80% loss of activity within 10 min) [1]

Oxidation
    Acetone and ethanol inactivate [1]

Organic solvent

General stability information
   Ammonium sulfate at high concentrations inactivates [1]; Freeze-thawing,
   stable to [1]

Storage
   $-15°C$, lyophilized, several months [1]; $4°C$, $t_{1/2}$: 48 h [1]

6 CROSSREFERENCES TO STRUCTURE DATABANKS

PIR/MIPS code

Brookhaven code

7 LITERATURE REFERENCES

[1] Frear, D.S.: Phytochemistry,7,381–390 (1968)

1 NOMENCLATURE

EC number
2.4.1.73

Systematic name
UDPglucose:galactosyl-lipopolysaccharide alpha-D-glucosyltransferase

Recommended name
Lipopolysaccharide glucosyltransferase II

Synonyms
Glucosyltransferase, uridine diphosphoglucose-galactosylpolysaccharide
More (cf. EC 2.4.1.44, EC 2.4.1.56 and EC 2.4.1.58)

CAS Reg. No.
51004-27-4

---

2 REACTION AND SPECIFICITY

Catalysed reaction
UDPglucose + lipopolysaccharide →
→ UDP + alpha-D-glucosyllipopolysaccharide

Reaction type
Hexosyl group transfer

Natural substrates
UDPglucose + lipopolysaccharide (transfers glucosyl residues to the
D-galactosyl-D-glucosyl side-chains in the partially completed core of the
lipopolysaccharide) [1]

Substrate spectrum
1 UDPglucose + lipopolysaccharide [1]

Product spectrum
1 UDP + alpha-D-glucosyllipopolysaccharide [1]

Inhibitor(s)

Cofactor(s)/prosthetic group(s)/activating agents

Metal compounds/salts

Turnover number (min^{-1})

Specific activity (U/mg)

K$_m$-value (mM)
    0.009 (UDP-D-glucose) [1]

pH-optimum
    7.5–9.0 [1]

pH-range

Temperature optimum (°C)
    37 (assay at) [1]

Temperature range (°C)

---

3 ENZYME STRUCTURE

Molecular weight

Subunits

Glycoprotein/Lipoprotein
    –

---

4 ISOLATION/PREPARATION

Source organism
    E. coli (J5) [1]

Source tissue
    Cell [1]

Localization in source

Purification

Crystallization
    –

Cloned
    –

Renatured
    –

5 STABILITY

pH

Temperature (°C)

Oxidation

Organic solvent

General stability information

Storage

6 CROSSREFERENCES TO STRUCTURE DATABANKS

PIR/MIPS code

Brookhaven code

7 LITERATURE REFERENCES

[1] Edstrom, R.D., Heath, E.C.: J. Biol. Chem.,242,3581–3588 (1967)

1 NOMENCLATURE

EC number
2.4.1.74

Systematic name
UDPgalactose:glycosaminoglycan D-galactosyltransferase

Recommended name
Glycosaminoglycan galactosyltransferase

Synonyms
Galactosyltransferase, uridine diphosphogalactose-mucopolysaccharide
UDPgalactose:polysaccharide transferase [1]

CAS Reg. No.
51004-28-5

2 REACTION AND SPECIFICITY

Catalysed reaction
UDPgalactose + glycosaminoglycan →
→ UDP + D-galactosylglycosaminoglycan

Reaction type
Hexosyl group transfer

Natural substrates
UDPgalactose + glycosaminoglycan (involved in the biosynthesis of galac-
tose-containing glycosaminoglycan of Dictostelium discoideum) [1]

Substrate spectrum
1 UDPgalactose + glycosaminoglycan (preferred substrate: desialylated
  mucin [2], other galactosyl acceptors are acidic mucopolysaccharide [1],
  desialylated bovine submaxillary mucin type I, ovomucoid, ovalbumin [2])
  [1, 2]

Product spectrum
1 UDP + D-galactosylglycosaminoglycan [1, 2]

Inhibitor(s)
Triton X-100 [2]; ATP (inhibits at high UDPgalactose concentration, stimulates at low UDPgalactose concentration) [2]

Cofactor(s)/prosthetic group(s)/activating agents
2-Mercaptoethanol (stimulation) [2]; ATP (stimulates at low UDPgalactose concentration, inhibits at high UDPgalactose concentration) [2]

Metal compounds/salts
Mn^{2+} (requirement [2], not [1]) [2]; K^+ (stimulation) [1]; More (no activation by Mg^{2+}) [1]

Turnover number (min^{-1})

Specific activity (U/mg)
0.0076 [2]

K_m-value (mM)

pH-optimum
7.5 [2]

pH-range
6.2–8.0 (about half-maximal activity at pH 6.2 and about 90% of maximal activity at pH 8.0) [2]

Temperature optimum (°C)
30 (assay at) [1]; 37 (assay at) [2]

Temperature range (°C)

3 ENZYME STRUCTURE

Molecular weight

Subunits

Glycoprotein/Lipoprotein
–

4 ISOLATION/PREPARATION

Source organism
Dictyostelium discoideum (NC-4, haploid wild-type and its mutant Fr-17, slime mould, not mutants Fr-2 and AGG-204) [1]; Rat (male albino Sprague-Dawley) [2]

Source tissue
Migrating pseudoplasmodia (no activity in vegetative cells) [1]; Mature fruiting bodies [1]; Liver [2]

Localization in source
Golgi membranes [2]

Purification

Crystallization

–

Cloned

–

Renatured

–

5 STABILITY

pH

Temperature (°C)

Oxidation

Organic solvent

General stability information

Storage

6 CROSSREFERENCES TO STRUCTURE DATABANKS

PIR/MIPS code

Brookhaven code

7 LITERATURE REFERENCES

[1] Sussman, M., Osborn, M.J.: Proc. Natl. Acad. Sci. USA,52,81–87 (1964)
[2] Andersson, G.N., Eriksson, L.C.: Biochim. Biophys. Acta,570,239–247 (1979)

1 NOMENCLATURE

EC number
2.4.1.75

Systematic name
UDPgalacturonate beta-D-galacturonosyltransferase (acceptor-unspecific)

Recommended name
UDPgalacturonosyltransferase

Synonyms
p-Nitrophenol conjugating enzyme
Galacturonosyltransferase, uridine diphospho-

CAS Reg. No.
60063-85-6

2 REACTION AND SPECIFICITY

Catalysed reaction
UDPgalacturonate + acceptor →
→ UDP + acceptor beta-D-galacturonide

Reaction type
Hexosyl group transfer

Natural substrates

Substrate spectrum
1 UDPgalacturonic acid + p-nitrophenol [1]

Product spectrum
1 UDP + p-nitrophenyl-beta-D-galacturonide [1]

Inhibitor(s)
Mersalyl (5 mM: complete inactivation, 1 mM: activation) [1]; EDTA (5 mM: slight) [1]; More (p-nitrophenylglucuronide is no inhibitor) [1]

Cofactor(s)/prosthetic group(s)/activating agents
Mersalyl (1 mM: maximal activation, 5 mM: complete inactivation) [1]; UDP-N-acetylglucosamine (5 mM: stimulation) [1]; Triton X-100 (0.06%, stimulation) [1]; Phenylglucuronide (stimulation) [1]; 2-Naphthylglucuronide (stimulation) [1]; More (phospholipase A-treated microsomes in combination with EDTA: slight stimulation) [1]

Metal compounds/salts

Turnover number (min^{-1})

Specific activity (U/mg)

K_m-value (mM)

pH-optimum

pH-range

Temperature optimum (°C)
  37 (assay at) [1]

Temperature range (°C)

---

3 ENZYME STRUCTURE

Molecular weight

Subunits

Glycoprotein/Lipoprotein
  –

---

4 ISOLATION/PREPARATION

Source organism
  Guinea pig [1]

Source tissue
  Liver [1]

Localization in source
  Microsomes [1]

Purification
  Guinea pig (partial) [1]

Crystallization
  –

Cloned
  –

Renatured
  –

5 STABILITY

pH

Temperature (°C)

Oxidation
   Sensitive to SH-reagents [1]

Organic solvent

General stability information

Storage

6 CROSSREFERENCES TO STRUCTURE DATABANKS

PIR/MIPS code

Brookhaven code

7 LITERATURE REFERENCES

[1] Vessey, D.A., Zakim, D.: Biochim. Biophys. Acta,315,43–48 (1973)

1 NOMENCLATURE

EC number
2.4.1.78

Systematic name
UDPglucose:phosphopolyprenol D-glucosyltransferase

Recommended name
Phosphopolyprenol glucosyltransferase

Synonyms
Glucosyltransferase, uridine diphosphoglucose-polyprenol monophosphate
UDPglucose:polyprenol monophosphate glucosyltransferase

CAS Reg. No.
55576-46-0

---

2 REACTION AND SPECIFICITY

Catalysed reaction
UDPglucose + polyprenol phosphate →
→ UDP + polyprenolphosphate-glucose

Reaction type
Hexosyl group transfer

Natural substrates

Substrate spectrum
1 UDPglucose + ficaprenol phosphate (i.e. mixture of C_{50}, C_{55} (mainly) and C_{66} cis- and trans-polyprenols, most effective lipid glucosylacceptor. Phosphate esters of phytol, betulaprenol (mixture of C_{30}-C_{45} cis- and trans-polyprenols), solanesol (C_{45} all-trans polyprenol) and dolichol (mainly C_{95} cis- and trans-polyprenols, containing saturated alpha-iso-prene residue) are poor substrates. UDPgalactose replaces UDPglucose with crude membrane-preparations, but the product is ficaprenolphos-phate-glucose. No substrates are GDPmannose, UDPglucuronic acid and UDP-N-acetylglucosamine) [1]

Product spectrum
1 UDP + ficaprenolphosphate-glucose [1]

Inhibitor(s)
UDP (not UMP) [1]

Cofactor(s)/prosthetic group(s)/activating agents
  Triton X-100 (stimulates solubilized activity) [1]; Taurocholate (stimulates
  solubilized activity) [1]; Deoxycholate (stimulates solubilized activity) [1]

Metal compounds/salts

Turnover number (min^{-1})

Specific activity (U/mg)

K_m-value (mM)

pH-optimum
  7.5 (assay at) [1]

pH-range

Temperature optimum (°C)
  37 (assay at) [1]

Temperature range (°C)

3 ENZYME STRUCTURE

Molecular weight

Subunits

Glycoprotein/Lipoprotein
  –

4 ISOLATION/PREPARATION

Source organism
  Shigella flexneri (strain 2a, not strain Y) [1]

Source tissue
  Cell [1]

Localization in source
  Membrane-bound [1]

Purification

Crystallization
  –

Cloned
  –

Renatured
  –

5 STABILITY

pH

Temperature (°C)

Oxidation

Organic solvent

General stability information
  Triton X-100, taurocholate or deoxycholate stabilizes [1]

Storage

6 CROSSREFERENCES TO STRUCTURE DATABANKS

PIR/MIPS code

Brookhaven code

7 LITERATURE REFERENCES

[1] Jankowski, W., Mánkowski, T., Chojnacki, T.: Biochim. Biophys. Acta,337,153–162 (1974)

Enzyme Handbook © Springer-Verlag Berlin Heidelberg 1996
Duplication, reproduction and storage in data banks are only
allowed with the prior permission of the publishers

1 NOMENCLATURE

EC number

2.4.1.79

Systematic name

UDP-N-acetyl-D-galactosamine:D-galactosyl-1,4-D-galactosyl-
1,4-D-glucosylceramide beta-N-acetyl-D-galactosaminyltransferase

Recommended name

Galactosylgalactosylglucosylceramide beta-D-acetylgalactosaminyltransferase

Synonyms

Acetylgalactosaminyltransferase, uridine diphosphoacetylgalactosamine-galactosylgalactosylglucosylceramide
Globoside synthetase
UDP-N-acetylgalactosamine:globotriaosylceramide beta-3-N-acetylgalactosaminyltransferase [4]

CAS Reg. No.

62213-46-1

---

2 REACTION AND SPECIFICITY

Catalysed reaction

UDP-N-acetyl-D-galactosamine + D-galactosyl-1,4-D-galactosyl-1,4-D-glucosylceramide →
→ UDP + N-acetyl-D-galactosaminyl-1,3-D-galactosyl-1,4-D-galactosyl-1,4-D-glucosylceramide

Reaction type

Hexosyl group transfer

Natural substrates

UDP-N-acetyl-D-galactosamine + D-galactosyl-1,4-D-galactosyl-1,4-D-glucosylceramide (involved in globoside biosynthesis) [1]

Substrate spectrum

1 UDP-N-acetyl-D-galactosamine + D-galactosyl-1,4-D-galactosyl-1,4-D-glucosylceramide (i.e. trihexosylceramide, globotriaosylceramide or Pk antigen, best substrate (from pig heart, human serum, erythrocytes and rabbit erythrocytes [2]) [1, 2], can be replaced to some extent by human kidney globoside or equine erythrocyte hematoside (i.e. N-glycolyl-GM3) [1], globotriaose [4], general structure of preferred sub-

strates: galactosyl-alpha-1,4-galactosyl-O-R, R moiety has little effect on activity [4], acceptor specificity [4], poor substrates: N-acetyl-GM3 [1], UDPgalactose, UDP-N-acetylglucosamine [4], no substrates are UDPglucose, UDPgalactose and UDP-N-acetylglucosamine [2], galactosylceramide, glucosylceramide, lactosylceramide [1], GM3, 2'-fucosyllactose, blood group H substance [4]) [1–4]

Product spectrum

1 UDP + N-acetyl-D-galactosaminyl-1,3-D-galactosyl-1,4-D-galactosyl-1,4-D-glucosylceramide (i.e. globoside) [1–4]

Inhibitor(s)

UDP (competitive to UDP-N-acetylgalactosamine, non-competitive to globotriaosylceramide) [4]; EDTA (reversible [4]) [1, 2, 4]; Sodium deoxycholate [1]; NaN_3 [1]; Iodoacetic acid [4]; IAA [4]; NEM [4]; 4-Hydroxymercuribenzoate (strong) [4]; More (no inhibitors: PCMB, NaF, N-acetylglucosamine, N-acetylgalactosamine) [1]

Cofactor(s)/prosthetic group(s)/activating agents

Dithiothreitol (slight stimulation, 10 mM) [4]; Nonionic detergents (activation, e.g. Triton X-100 or CF-54) [1, 2]; Sodium taurocholate (activation) [2]; Tween 20 (activation, half as effective as taurocholate [2], not [1]) [2]; Myrj 59 (activation, half as effective as taurocholate) [2]; More (no stimulation by GSH [4], Tween 40 [1], SPAN 80 or cationic detergent G3634A [2]) [1, 2, 4]

Metal compounds/salts

Mn^{2+} (requirement, 10 mM [1], other divalent cations can replace Mn^{2+} with 1–10% efficiency [2]) [1, 2, 4]; Co^{2+} (activation, about half as effective as Mn^{2+}) [2, 4]; Fe^{2+} (activation, 19% as effective as Mn^{2+}) [4]; Li^+ (activation, 11% as effective as Mn^{2+}) [4]; More (no activation by Ca^{2+} [1, 4], Zn^{2+}, Cu^{2+}, Mg^{2+}, Ni^{2+} [4]) [1, 4]

Turnover number (min^{-1})

Specific activity (U/mg)

0.000565 [4]

K_m-value (mM)

More (kinetics) [1]; 0.0025 (globotriaosylceramide) [4]; 0.014 (UDP-N-acetylglucosamine) [4]; 0.14 (trihexosylceramide) [1]; 0.2–0.23 (UDP-N-acetylgalactosamine) [1, 2]; 1.7 (trihexosylceramide) [2]

pH-optimum

6.5 [2]; 6.9 (preferred buffer: sodium cacodylate-HCl, less effective are piperazine-N,N'-bis-2-ethanesulfonic acid, HEPES, or Tris-maleate buffer [1]) [1, 4]; 7.2 [3]

pH-range
5.5–8.2 (about half-maximal activity at pH 5.5 and 8.2) [1]; 6.7–7.2 (maximum activity in this range) [4]

Temperature optimum (°C)
37 (assay at) [1, 2]

Temperature range (°C)

---

3 ENZYME STRUCTURE

Molecular weight
120000 (Mongrel dog, PAGE without prior reduction) [4]

Subunits
Dimer ($1 \times 57000 + 1 \times 64000$, Mongrel dog, SDS-PAGE) [4]

Glycoprotein/Lipoprotein
–

---

4 ISOLATION/PREPARATION

Source organism
Guinea pig [1]; Chicken (embryo) [2, 3]; Mongrel dog [4]

Source tissue
Kidney [1]; Spleen [1, 4]; Brain [1–3]; More (tissue distribution) [1]

Localization in source
Microsomes [1, 4]; Mitochondria [1]; Membrane-bound [2, 3]

Purification
Chicken (partial, affinity chromatography) [3]; Mongrel dog (affinity chromatography) [4]

Crystallization
–

Cloned
–

Renatured
–

5 STABILITY

pH

Temperature (°C)

Oxidation

Organic solvent

General stability information
Low ionic strength destabilizes [4]; Diluted enzyme solutions, below 0.1 mg protein/ml, unstable [4]; Repeated freeze-thawing, stable to [4]; Glycerol, 20%, stabilizes [4]

Storage
−80°C, above 0.1 mg protein/ml in 25 mM sodium cacodylate buffer, pH 6.9, 1% Triton X-100 and 20% glycerol, at least 3 months [4]

6 CROSSREFERENCES TO STRUCTURE DATABANKS

PIR/MIPS code

Brookhaven code

7 LITERATURE REFERENCES

[1] Ishibashi, T., Kijimoto, S., Makita, A.: Biochim. Biophys. Acta,337,92–106 (1974)
[2] Chien, J.-L., Williams, T., Basu, S.: J. Biol. Chem.,248,1778–1785 (1973)
[3] Schaeper, R.J., Das, K.K., Zhixiong, L., Basu, S.: Carbohydr. Res.,236,227–244 (1992)
[4] Taniguchi, N., Makita, A.: J. Biol. Chem.,259,5637–5642 (1984)

4

1 NOMENCLATURE

EC number
2.4.1.80

Systematic name
UDPglucose:N-acylsphingosine D-glucosyltransferase

Recommended name
Ceramide glucosyltransferase

Synonyms
UDP-glucose:ceramide glucosyltransferase [4]
Ceramide:UDPGlc glucosyltransferase [5]
Glucosyltransferase, uridine diphosphoglucose-ceramide
Ceramide:UDP-glucose glucosyltransferase
Glucosylceramide synthase
UDP glucose-ceramide glucosyltransferase

CAS Reg. No.
37237-44-8

---

2 REACTION AND SPECIFICITY

Catalysed reaction
UDPglucose + N-acylsphingosine →
→ UDP + D-glucosyl-N-acylsphingosine

Reaction type
Hexosyl group transfer

Natural substrates
UDPglucose + N-acylsphingosine (first step in synthesis of gangliosides [1],
catalyzes the first step during the sequential addition of carbohydrate
moieties for ganglioside biosynthesis [2, 7]) [1, 2, 7]

Substrate spectrum
1 UDPglucose + N-acylsphingosine (i.e. ceramide, other acceptors:
sphingosine [1], dihydrosphingosine (not [2]) [1], N-octanoyl sphingosine
[5], decasphingosine [5], lauroyl amide [5], overview: short-chain ceramide substrates [5]) [1–5]

Product spectrum
1 UDP + D-glucosyl-N-acylsphingosine (i.e. glucosylceramide) [1–5]

Inhibitor(s)
   EDTA [1]; Cu^{2+} [3]; Zn^{2+} [3, 4]; Fe^{2+} [3]; Cd^{2+} [4]

Cofactor(s)/prosthetic group(s)/activating agents
   Detergent (absolute requirement for [1], optimum activity obtained with a
   mixture of Cutsum and Triton X-100 (2:1) [1], activity is reduced by most of
   the detergents tested, only CHAPS and CHAPSO at concentration of 1% w/v
   stimulate [4], CHAPS stimulates [7]) [1, 4, 7]

Metal compounds/salts
   More (no stimulation by metal ions) [1]; Ca^{2+} (stimulates) [3]; Mg^{2+}
   (stimulates) [3, 4, 7]; Mn^{2+} (stimulates) [3, 4, 7]

Turnover number (min^{-1})

Specific activity (U/mg)
   More (assay method) [3, 4]

K_m-value (mM)
   0.0087 (UDPglucose) [4]; 0.022 (UDPglucose) [7]; 0.054 (ceramide) [7];
   0.08 (ceramide) [1]; 0.12 (UDPglucose) [1]; 0.292 (ceramide) [2]

pH-optimum
   6.4–6.5 (MES or Tris-maleate buffer) [4]; 6.5 [7]; 7.0 (assay at) [1]; 7.4 [3]

pH-range

Temperature optimum (°C)
   30 (assay at) [1]; 37 (assay at) [3]

Temperature range (°C)

3 ENZYME STRUCTURE

Molecular weight

Subunits

Glycoprotein/Lipoprotein
   –

4 ISOLATION/PREPARATION

Source organism
   Chicken (13–14 days old) [1]; Rat [1, 2, 4]; Pig [1, 6, 7]; Sheep [1]; Human
   [1]; Mouse [3, 5]; Guinea pig [1]

Source tissue
   Brain [1, 2, 4, 5]; Kidney [3]; Liver [5]; Submaxillary ganglia [6, 7]

Localization in source
Microsomes (associated with membranes, structural integrity of the micro-somal membranes is essential for activity) [2]; Golgi apparatus (mem-branes) [6, 7]; Membrane [2, 4, 6, 7]

Purification
More (solubilization method) [6]

Crystallization
–

Cloned
–

Renatured
–

---

5 STABILITY

pH

Temperature (°C)

Oxidation

Organic solvent

General stability information
NAD^+ stabilizes [3]

Storage
–70°C, enzyme in intact kidneys is stable for at least 90 days [3]; 4°C, enzyme in intact kidneys, 53% loss of activity after overnight storage [3]

---

6 CROSSREFERENCES TO STRUCTURE DATABANKS

PIR/MIPS code

Brookhaven code

---

7 LITERATURE REFERENCES

[1] Basu, S., Kaufman, B., Roseman, S.: J. Biol. Chem.,248,1388–1394 (1973)
[2] Shah, S.N.: Arch. Biochem. Biophys.,159,143–150 (1973)
[3] Shukla, G.S., Radin, N.S.: Arch. Biochem. Biophys.,283,372–378 (1990)
[4] Matsuo, N., Nomura, T., Imokawa, G.: Biochim. Biophys. Acta,1116,97–103 (1992)
[5] Vunnam, R.R., Radin, N.S.: Biochim. Biophys. Acta,573,73–82 (1979)
[6] Durieux, I., Martel, M.B., Got, R.: Biochim. Biophys. Acta,1024,263–266 (1990)
[7] Coste, H., Martel, M.-B., Azzar, G., Got, R.: Biochim. Biophys. Acta,814,1–7 (1985)

1 NOMENCLATURE

EC number
2.4.1.81

Systematic name
UDPglucose:5,7,3',4'-tetrahydroxyflavone 7-O-beta-D-glucosyltransferase

Recommended name
Flavone 7-O-beta-glucosyltransferase

Synonyms
UDPglucose-apigenin beta-glucosyltransferase
UDPglucose-luteolin beta-D-glucosyltransferase
Glucosyltransferase, uridine diphosphoglucose-luteolin
Glucosyltransferase, uridine diphosphoglucose-apigenin 7-O-
Uridine diphosphoglucose-apigenin 7-O-glucosyltransferase
UDP-glucosyltransferase [1]
More (different from EC 2.4.1.91)

CAS Reg. No.
37332-50-6

---

2 REACTION AND SPECIFICITY

Catalysed reaction
UDPglucose + 5,7,3',4'-tetrahydroxyflavone →
→ UDP + 7-O-beta-D-glucosyl-5,7,3',4'-tetrahydroxyflavone

Reaction type
Hexosyl group transfer

Natural substrates
UDPglucose + 5,7,3',4'-tetrahydroxyflavone (involved in biosynthesis of flavone glucosides) [1]

Substrate spectrum
1 UDPglucose + luteolin (best acceptor, no acceptors are isoflavones, cyanidin, p-coumaric acid) [1]
2 UDPglucose + apigenin (i.e. 5,7,4'-trihydroxyflavone, glucosylation at 48% the rate of luteolin, TDPglucose can replace UDPglucose, no glucosyl-donors are ADPglucose, GDPglucose, UDPglucuronic acid, UDPxylose, alpha-D-glucose 1-phosphate) [1]
3 UDPglucose + chrysoeriol (glucosylation at 30% the rate of luteolin) [1]
4 UDPglucose + naringenin (glucosylation at 21% the rate of luteolin) [1]

Product spectrum
  1 UDP + 7-O-beta-D-glucosylluteolin [1]
  2 UDP + 7-O-beta-glucosylapigenin [1]
  3 ?
  4 ?

Inhibitor(s)
  Ethylene monomethylether [1]; Ammonium sulfate (weak) [1]; UDP (weak)
  [1]; More (no inhibition by Mg^{2+}, Ca^{2+}, EDTA, 2-mercaptoethanol, UTP, apigenin (up to 0.5 mM), chrysoeriol (up to 0.5 mM), naringenin (up to 0.5 mM))
  [1]

Cofactor(s)/prosthetic group(s)/activating agents
  Bovine serum albumin (increase of activity) [1]

Metal compounds/salts

Turnover number (min^{-1})

Specific activity (U/mg)

K_m-value (mM)
  0.12 (UDPglucose (+ apigenin)) [1]; 0.26 (TDPglucose (+ apigenin)) [1]

pH-optimum
  7.5 [1]

pH-range
  6.0–9.0 (about half-maximal activity at pH 6.0 and 9.0) [1]

Temperature optimum (°C)
  30 (assay at) [1]

Temperature range (°C)

3 ENZYME STRUCTURE

Molecular weight
  50000 (Petroselinum hortense, gel filtration) [1]

Subunits

Glycoprotein/Lipoprotein
  —

4 ISOLATION/PREPARATION

Source organism
  Petroselinum hortense (parsley) [1]

Source tissue
Cell suspension culture (illuminated with white light for 24 h) [1]

Localization in source
Soluble [1]

Purification
Petroselinum hortense (partial) [1]

Crystallization
–

Cloned
–

Renatured
–

---

5 STABILITY

pH
6.5 (below, rapid loss of activity) [1]; 7.5–8.5 (most stable) [1]; 9.5 (above, rapid loss of activity) [1]

Temperature (°C)

Oxidation

Organic solvent

General stability information

Storage
3°C, $t_{1/2}$: 1 month in 0.05 M Tris-HCl buffer, pH 7.5, 7 mM 2-mercaptoethanol [1]

---

6 CROSSREFERENCES TO STRUCTURE DATABANKS

PIR/MIPS code

Brookhaven code

---

7 LITERATURE REFERENCES

[1] Sutter, A., Ortmann, R., Grisebach, H.: Biochim. Biophys. Acta,258,71–87 (1972)

1 NOMENCLATURE

EC number
2.4.1.82

Systematic name
1-alpha-D-Galactosyl-myo-inositol:sucrose 6-alpha-D-galactosyltransferase

Recommended name
Galactinol-sucrose galactosyltransferase

Synonyms
Galactosyltransferase, galactinol-sucrose
More (cf. EC 2.4.1.123)

CAS Reg. No.
62213-45-0

2 REACTION AND SPECIFICITY

Catalysed reaction
1-alpha-D-Galactosyl-myo-inositol + sucrose →
→ myo-inositol + raffinose

Reaction type
Hexosyl group transfer

Natural substrates
1-alpha-D-Galactosyl-myo-inositol + sucrose (first step in biosynthesis of raffinose sugars) [1]

Substrate spectrum
1 1-alpha-D-Galactosyl-myo-inositol + sucrose (i.e. galactinol, can be replaced in vitro by p-nitrophenylgalactoside, in the absence of acceptor substrate galactinol is hydrolyzed to galactose, no substrate is UDPgalactose) [1]
2 Raffinose + sucrose (r, exchange reaction) [1]

Product spectrum
1 myo-Inositol + raffinose [1]
2 Sucrose + raffinose [1]

Inhibitor(s)

Ag^+ (1 mM, strong) [1]; Al^{3+} (1 mM, strong) [1]; Hg^{2+} (1 mM, strong) [1];
Mn^{2+} [1]; Zn^{2+} (1 mM, strong) [1]; Iodoacetamide [1]; N-Ethylmaleimide [1]

Cofactor(s)/prosthetic group(s)/activating agents

Metal compounds/salts

Turnover number (min^{-1})

Specific activity (U/mg)

0.497 [1]

K_m-value (mM)

1 (sucrose (+ galactinol)) [1]; 2.9 (sucrose (+ raffinose), exchange reaction)
[1]; 7 (galactinol (+ sucrose)) [1]; 10 (raffinose (+ sucrose), exchange reaction) [1]

pH-optimum

7 [1]

pH-range

5.7–7.8 (about half-maximal activity at pH 5.7 and 7.8) [1]

Temperature optimum (°C)

42 [1]

Temperature range (°C)

28–50 (about half-maximal activity at 28°C and 50°C) [1]

---

3 ENZYME STRUCTURE

Molecular weight

80000 (Vicia faber, gel filtration) [1]
100000 (Vicia faber, glycerol density gradient centrifugation) [1]

Subunits

Glycoprotein/Lipoprotein

–

---

4 ISOLATION/PREPARATION

Source organism

Vicia faber [1]

Source tissue

Seeds [1]

Localization in source
  Cytoplasm [1]

Purification
  Vicia faber (partial) [1]

Crystallization
  –

Cloned
  –

Renatured
  –

---

5 STABILITY

pH

Temperature (°C)
  50 (10 min, 80% loss of activity, galactinol and raffinose protect, not
  sucrose) [1]

Oxidation

Organic solvent

General stability information
  Dithioerythritol stabilizes during purification [1]

Storage
  Frozen, purified enzyme, at least 1 month [1]; 4°C, in crude extract, $t_{1/2}$: 3
  days [1]

---

6 CROSSREFERENCES TO STRUCTURE DATABANKS

PIR/MIPS code

Brookhaven code

---

7 LITERATURE REFERENCES

[1] Lehle, L., Tanner, W.: Eur. J. Biochem.,38,103–110 (1973)

1 NOMENCLATURE

EC number
2.4.1.83

Systematic name
GDPmannose:dolichyl-phosphate beta-D-mannosyltransferase

Recommended name
Dolichyl-phosphate beta-D-mannosyltransferase

Synonyms
GDPMan:DolP mannosyltransferase
Dolichyl mannosyl phosphate synthase
Dolichyl-phospho-mannose synthase [7]
GDPmannose:dolichyl-phosphate mannosyltransferase [7]
Mannosyltransferase, guanosine diphosphomannose-dolichol phosphate
Dolichol phosphate mannose synthase
Dolichyl phosphate mannosyltransferase
Dolichyl-phosphate mannose synthase
GDP-mannose-dolichol phosphate mannosyltransferase
GDPmannose-dolichylmonophosphate mannosyltransferase
Mannosylphosphodolichol synthase
Mannosylphosphoryldolichol synthase

CAS Reg. No.
62213-44-9

2 REACTION AND SPECIFICITY

Catalysed reaction
GDPmannose + dolichyl phosphate →
→ GDP + dolichyl D-mannosyl phosphate (sequential mechanism [14])

Reaction type
Hexosyl group transfer

Natural substrates
GDPmannose + dolichyl phosphate (biosynthesis of the asparagine-linked carbohydrate side chains of eukaryotic glycoproteins requires the formation of oligosaccharides linked to a carrier lipid, dolichyl phosphate [13], involved in glycoconjugate biosynthesis [16], synthesis of N-linked glycoproteins [18]) [13, 16, 18]

Substrate spectrum

1 GDPmannose + dolichyl phosphate (acts only on long-chain polyprenyl phosphates and alpha-dihydropolyprenyl phosphates, larger than C_{35}, dependence on alpha-saturation [2], branching of lipid phosphates is essential, transfer of mannose occurs even if only branching at C-3 is present [16], highest activity with C_{55} alpha-dihydropolyprenyl phosphate [2], C_{35} alpha-dihydropolyprenyl phosphate (weak) [2], C_{100} alpha-dihydropolyprenyl phosphate [2], C_{55} polyprenyl phosphate [2], specific for GDPmannose [15], preferred chain length: $C_{55} > C_{100} > C_{80} >> C_{35}$ [2], phytanyl(3,7,11,15-tetramethylhexadecanyl)phosphate (60–70% as effective as natural dolichyl lipid in transfer of mannose from GDPmannose to organic soluble material) [16], S-3-methyloctadecanyl phosphate (25% as effective as natural dolichyl lipid in transfer of mannose from GDPmannose to organic soluble material) [16], one synthase catalyzes the synthesis of both mannosylphosphoryldolichol and mannosylphosphorylretinol in rat liver tissues and Chinese hamster ovary cells [18]) [1–18]
2 More (not: GDP-D-glucose, UDP-D-xylose, UDP-D-glucuronic acid) [15]

Product spectrum

1 GDP + dolichyl D-mannosyl phosphate [1–18]
2 ?

Inhibitor(s)

Phosphatidylcholine (inactive in presence of phosphatidylcholine alone, detergent-free enzyme active in presence of phosphatidylethanolamine and in presence of phospholipid mixtures of phosphatidylethanolamine and phosphatidylcholine when the molar proportion of phosphatidylcholine is 70% or less) [7]; GDP-D-glucose [15]; Nonidet P-40 [7]; Triton X-100 [7]; Octyl beta-glucoside [7]; Deoxycholate [7]; GMP [7, 10, 13–15]; GDP [7, 10, 12–15]; GTP [7]; GDPglucose [7, 15]; ADP (inhibits less than 30% at concentration of 0.06–0.12 mM [7], no effect [14]) [7]; UDP (inhibits less than 30% at concentration of 0.06–0.12 mM) [7]; CDP (inhibits less than 30% at concentration of 0.06–0.12 mM) [7]; TDP (inhibits less than 30% at concentration of 0.06–0.12 mM) [7]; dGDP (inhibits less than 30% at concentration of 0.06–0.12 mM) [7]; UDPglucose (inhibits less than 30% at concentration of 0.06–0.12 mM) [7]; UDP-N-acetylglucosamine (inhibits less than 30% at concentration of 0.06–0.12 mM) [7]; ADPmannose (inhibits less than 30% at concentration of 0.06–0.12 mM) [7]; Tunicamycin [10]; Mn^{2+} (above 5 mM) [1]; N-Ethylmaleimide [3]; Amphomycin [3]; Nonionic detergents [6]; GDP esters (of 2-deoxy-D-glucose, 2-deoxy-2-fluoro-D-mannose, 3-deoxy-D-mannose, 4-deoxy-D-mannose, 6-deoxy-D-mannose) [12]

Cofactor(s)/prosthetic group(s)/activating agents

Phospholipid (no absolute requirement, required for interaction of the enzyme with long-chain polyisoprenol substrate dolichyl phosphate) [6]; Triton X-100 (detergent required for maximal activity, concentration optimum, 0.015%) [19]; Choline-containing lipids (e.g. phosphatidylcholine, lysophosphatidylcholine, sphingomyelin, activate) [13]; Emulgen 909 (reaction optimal at 0.015% v/v, at lower concentrations the reaction is almost abolished, at higher concentration the activity is greatly reduced) [14]

Metal compounds/salts

Mn^{2+} (divalent cation required [1, 7, 10, 13, 15], Mg^{2+} and Mn^{2+} up to 5 mM result in the same activation [1], Mn^{2+} is most effective [15], less effective stimulation than with Mg^{2+} [13], maximal stimulation at 2.5 mM [7], Mg^{2+} > Mn^{2+} >> Ca^{2+} > Co^{2+} > Zn^{2+} [10]) [1, 7, 10, 13, 15]; Mg^{2+} (divalent cation required [1, 7, 10, 13, 15], Mg^{2+} and Mn^{2+} up to 5 mM result in the same activation [1], K_m 1.13 mM [10], Mg^{2+} > Mn^{2+} >> Ca^{2+} > Co^{2+} > Zn^{2+} [10], stimulates, at 2.5 mM 70% of stimulation with Mn^{2+} [7], stimulation maximal at 5 mM [13], $MgCl_2$ stimulates, optimum 3 mM [14], Mg^{2+}, Mn^{2+} and Co^{2+} serve as cofactors [15]) [1, 7, 10, 13–15]; Co^{2+} (stimulates, at 2.5 mM 90% of stimulation with Mn^{2+}) [7]; Ni^{2+} (stimulates, at 2.5 mM 27% of stimulation with Mn^{2+}) [7]; Ca^{2+} (stimulates, at 2.5 mM 5% of stimulation with Mn^{2+}) [7]; More (no stimulation by Zn^{2+}) [7]

Turnover number (min^{-1})

Specific activity (U/mg)

0.0253 [12]; 0.096 [8]; 0.475 [3]; More [7, 14]

K_m-value (mM)

0.00025 (GDP-D-mannose) [15]; 0.0003 (dolichyl phosphate) [7]; 0.00043 (GDPmannose) [14]; 0.00052 (GDPmannose) [12]; 0.00069 (GDPmannose) [7]; 0.00133 (GDPmannose) [13]; 0.00173 (GDPmannose) [10]; 0.0033 (dolichyl phosphate) [15]; 0.007 (GDPmannose) [1]; 0.0143 (dolichyl phosphate) [14]; More (dolichyl phosphate: 0.0017 mg/ml [10], effects of site-directed mutagenesis [17]) [10, 15, 17]

pH-optimum

6.1 (MES buffer or sodium/potassium phosphate buffer) [13]; 7.0 [10, 15]; 7.2 [14]; 7.3 [1]; 7.5 [7]

pH-range

5.8–8.0 (5.8: about 35% of activity maximum, 8.0: about 60% of activity maximum) [15]

Temperature optimum (°C)

20 (assay at) [14]; 27 [15]; 37 (assay at) [7, 13]

Temperature range (°C)
    15–35 (15°C: about 50% of activity maximum, 35°C: about 30% of activity maximum) [15]

3 ENZYME STRUCTURE

Molecular weight

Subunits
    ? (x × 30000, Saccharomyces cerevisiae, SDS-PAGE [3], x × 30000, mouse, SDS-PAGE [8]) [3, 8]

Glycoprotein/Lipoprotein
    −

4 ISOLATION/PREPARATION

Source organism
    Pig [5]; Rat [6, 7, 18, 19]; Mouse [8, 9, 13]; Zea mays (L. inbred A636) [10]; Trichoderma reesei [11]; Chicken [12, 14]; Acanthamoeba castellanii [15]; Saccharomyces cerevisiae [1–3, 16]; Chinese hamster [18]; Hansenula holstii [4]; Yeast (enzyme cloned and expressed in E. coli, wild type and site-directed mutant enzymes) [17]

Source tissue
    Liver [5–9, 13, 14]; Cell [3]; Endosperm culture [10]; Embryo cells [12]; Encysting cultures [15]; Parotid acinar cells [19]; Ovary cells [18]

Localization in source
    Mitochondria (cytosolic face of outer membrane [8, 9], outer membrane [13]) [8, 9, 13, 14]; Membrane (bound) [1]; Microsomes (membrane [10, 12, 19]) [6, 7, 10–12, 19]; Endoplasmic reticulum (enzyme is anchored with both it's N- and C-termini in the membrane, while the main part of the protein is oriented towards the lumes of the ER) [3]

Purification
    Saccharomyces cerevisiae [1, 3]; Rat [7]; Mouse [8, 9, 13]; Chicken [14]; Yeast (enzyme cloned and expressed in E. coli, wild type and site-directed mutant enzymes) [17]

Crystallization
    −

Cloned
    [17]

Renatured
    −

4

5 STABILITY

pH

Temperature (°C)
    30 (inactivation above) [14]; 55 (inactivation) [6]

Oxidation

Organic solvent

General stability information
    Removal of detergents is not prerequisite for stability [3]; Unstable in pres-
    ence of detergents [7]; Heat inactivation can be prevented by addition of
    heat-denatured mitochondrial extracts [14]; Phospholipids and dolichyl
    phosphate stabilize and prevent denaturation on storage at 0°C, combina-
    tion of these lipids enhance the stabilizing effect greatly, sphingomyelin is
    the most effective phospholipid, dolichol can partially substitute dolichyl
    phosphate but worked at higher concentrations [14]

Storage
    4°C, stable for more than 24 h [8]; 4°C, phosphate buffer, overnight, relative-
    ly stable, no loss of activity when stored within liposomes [13]; 0°C, over-
    night, complete loss of activity [14]; −40°C, 1 month, 20% loss of activity
    [14]; −70°C, 2 weeks, stable [15]; 0°C, 50% loss of activity after 3 days [15];
    −20°C, about 30% loss within 2 weeks [15]; −20°C, 1 mM dithiothreitol, 15%
    glycerol, stable for at least 3 months [17]

---

6 CROSSREFERENCES TO STRUCTURE DATABANKS

PIR/MIPS code
    PIR2:A32122 (precursor yeast (Saccharomyces cerevisiae))

Brookhaven code

---

7 LITERATURE REFERENCES

[1] Babczinski, P., Haselbeck, A., Tanner, W.: Eur. J. Biochem.,105,509–515 (1980)
[2] Palamarczyk, G., Lehle, L., Mankowski, T., Chojnacki, T., Tanner, W.: Eur. J.
    Biochem.,105,517–523 (1980)
[3] Haselbeck, A.: Eur. J. Biochem.,181,663–668 (1989)
[4] Bretthauer, R.K., Wu, S., Irwin, W.E.: Biochim. Biophys. Acta,304,736–747 (1973)
[5] Richards, J.B., Hemming, F.W.: Biochem. J.,130,77–93 (1972)
[6] Jensen, J.W., Schutzbach, J.S.: Carbohydr. Res.,149,199–208 (1986)
[7] Jensen, J.W., Schutzbach, J.S.: Eur. J. Biochem.,153,41–48 (1985)
[8] Gasnier, F., Rousson, R., Lerme, F., Vaganay, E., Louisot, P., Gateau-Roesch,
    O.: Eur. J. Biochem.,206,853–858 (1992)

[9] Gasnier, F., Louisot, P., Gateau-Roesch, O.: Biochim. Biophys. Acta,980,339–347 (1989)
[10] Riedell, W.E., Miernyk, J.A.: Plant Physiol.,87,420–426 (1988)
[11] Kruszewska, J., Palamarczyk, G., Kubicek, C.P.: FEMS Microbiol. Lett.,80,81–86 (1991)
[12] McDowell, W., Schwarz, R.T.: FEBS Lett.,243,413–416 (1989)
[13] Gasnier, F., Morelis, R., Louisot, P., Gateau, O.: Biochim. Biophys. Acta,925, 297–304 (1987)
[14] Tomita, Y., Motokawa, Y.: Biochim. Biophys. Acta,842,176–183 (1985)
[15] Carlo, P.L., Villemez, C.L.: Arch. Biochem. Biophys.,198,117–123 (1979)
[16] Wilson, I.B.H., Taylor, J.P., Webberley, M.C., Turner, N.J., Flitsch, S.L.: Biochem. J., 295,195–201 (1993)
[17] Schutzbach, J.S., Zimmerman, J.W., Forsee, W.T.: J. Biol. Chem.,268,24190–24196 (1993)
[18] Stoll, J., Rosenberg, L., Carson, D.D., Lennarz, W.J., Karg, S.S.: J. Biol. Chem.,260, 232–236 (1985)
[19] Banerjee, D.K., Kousvelari, E.E., Baum, B.J.: Proc. Natl. Acad. Sci. USA,84,6389–6393 (1987)

1 NOMENCLATURE

EC number
2.4.1.85

Systematic name
UDPglucose:(S)-4-hydroxymandelonitrile beta-D-glucosyltransferase

Recommended name
Cyanohydrin beta-glucosyltransferase

Synonyms
Glucosyltransferase, uridine diphosphoglucose-p-hydroxymandelonitrile
UDP-glucose-p-hydroxymandelonitrile glucosyltransferase
Uridine diphosphoglucose-cyanohydrin glucosyltransferase
Uridine diphosphoglucose:aldehyde cyanohydrin beta-glucosyltransferase
[1–3]

CAS Reg. No.
55354-52-4

2 REACTION AND SPECIFICITY

Catalysed reaction
UDPglucose + (S)-4-hydroxymandelonitrile →
→ UDP + (S)-4-hydroxymandelonitrile beta-D-glucoside

Reaction type
Hexosyl group transfer

Natural substrates
UDPglucose + (S)-4-hydroxymandelonitrile (involved in dhurrin biosynthesis)
[1]

Substrate spectrum
1 UDPglucose + (S)-4-hydroxymandelonitrile [1–3]
2 UDPglucose + (S)-mandelonitrile (as good as 4-hydroxymandelonitrile) [1]
3 UDPglucose + hydroquinone (at 41% the rate of 4-hydroxymandelonitrile)
  [1]
4 UDPglucose + 4-hydroxybenzyl alcohol (at 36% the rate of 4-hydroxy-
  mandelonitrile) [1]
5 UDPglucose + 4-hydroxybenzaldehyde (poor substrate) [1]
6 UDPglucose + 4-hydroxybenzoic acid (poor substrate) [1]
7 UDPglucose + phenol (poor substrate) [1]
8 More (no substrates are catechol, resorcinol, phloroglucinol, fructose,
  glucose, dhurrin, acetone cyanohydrin, acetaldehyde cyanohydrin) [1]

Product spectrum

1 UDP + (S)-4-hydroxymandelonitrile beta-D-glucopyranoside (i.e. dhurrin) [1–3]
2 ?
3 ?
4 ?
5 ?
6 ?
7 ?
8 ?

Inhibitor(s)

Cofactor(s)/prosthetic group(s)/activating agents

Dithiothreitol (activation, 5 mM) [1]

Metal compounds/salts

More (no activation by Ca^{2+} or Mn^{2+}) [1]

Turnover number (min^{-1})

Specific activity (U/mg)

More [2]; 0.76 [1]

K_m-value (mM)

pH-optimum

8.2–8.5 [1]

pH-range

Temperature optimum (°C)

30 (assay at) [1–3]

Temperature range (°C)

3 ENZYME STRUCTURE

Molecular weight

Subunits

Glycoprotein/Lipoprotein

–

4 ISOLATION/PREPARATION

Source organism

Sorghum bicolor (sudan gras hybrid, Sordan 70, cv. Moench [1], cv. Moench var. Redland x Greenleaf [2, 3]) [1–3]

Source tissue
Seedlings (leaves [1–3], roots [1]) [1–3]; Leaf blade (epidermal and me-
sophyllic tissue) [2, 3]

Localization in source
Plastids (subcellular distribution) [3]

Purification
Sorghum bicolor (partial) [1]

Crystallization
–

Cloned
–

Renatured
–

---

5 STABILITY

pH

Temperature (°C)

Oxidation

Organic solvent

General stability information
Freezing, stable to [1]; DEAE-chromatography decreases stability to
freezing [1]

Storage
0°C, unstable at [1]

---

6 CROSSREFERENCES TO STRUCTURE DATABANKS

PIR/MIPS code

Brookhaven code

---

7 LITERATURE REFERENCES

[1] Reay, P.F., Conn, E.E.: J. Biol. Chem.,249,5826–5830 (1974)
[2] Kojima, M., Poulton, J.E., Thayer, S.S., Conn, E.E.: Plant Physiol.,63,1022–1028
(1979)
[3] Surkin Wurtele, E., Thayer, S.S., Conn, E.E.: Plant Physiol.,70,1732–1737 (1982)

1 NOMENCLATURE

EC number
2.4.1.86

Systematic name
UDPgalactose:N-acetyl-D-glucosaminyl-1,3-D-galactosyl-1,4-D-glucosyl-ceramide beta-D-galactosyltransferase

Recommended name
Glucosaminylgalactosylglucosylceramide beta-galactosyltransferase

Synonyms
Galactosyltransferase, uridine diphosphogalactose-acetyl-glucosaminylgalactosylglucosylceramide
GalT-4 [1]
Paragloboside synthase

CAS Reg. No.
9073-46-5

2 REACTION AND SPECIFICITY

Catalysed reaction
UDPgalactose + N-acetyl-D-glucosaminyl-1,3-beta-D-galactosyl-1,4-D-glucosylceramide →
→ UDP + 1,3-beta-D-galactosyl-N-acetyl-D-glucosaminyl-1,3-beta-D-galactosyl-1,4-D-glucosylceramide

Reaction type
Hexosyl group transfer

Natural substrates

Substrate spectrum
1 UDPgalactose + N-acetyl-D-glucosaminyl-1,3-beta-D-galactosyl-1,4-D-glucosylceramide [1, 2]
2 More (UDPglucose cannot substitute for UDPgalactose, Tay-Sachs ganglioside (GM2) is a poor galactose acceptor) [2]

Product spectrum
1 UDP + 1,3-beta-D-galactosyl-N-acetyl-D-glucosaminyl-1,3-beta-D-galactosyl-1,4-D-glucosylceramide [1, 2]
2 ?

Inhibitor(s)
  EDTA (complete inhibition) [2]

Cofactor(s)/prosthetic group(s)/activating agents
  Triton X-100 (5–12 mg/ml, highest activity) [2]

Metal compounds/salts
  Mn^{2+} (necessary for activity, cannot be replaced by Ni^{2+}, Cu^{2+}, Fe^{3+}, Zn^{2+}, Ca^{2+}) [2]; Mg^{2+} (6–7% of activity with Mn^{2+}) [2]; Cd^{2+} (6–7% of activity with Mn^{2+}) [2]; Co^{2+} (6–7% of activity with Mn^{2+}) [2]

Turnover number (min^{-1})

Specific activity (U/mg)

K_m-value (mM)
  0.12 (UDPgalactose) [2]; 0.5 (triglycosylceramide) [2]

pH-optimum
  6.4 [2]

pH-range

Temperature optimum (°C)
  37 (assay at) [2]

Temperature range (°C)

3 ENZYME STRUCTURE

Molecular weight

Subunits

Glycoprotein/Lipoprotein
  –

4 ISOLATION/PREPARATION

Source organism
  Rabbit [2]; Human (fetus with Tay-Sachs desease) [1]

Source tissue
  Bone marrow [2]; TSD cell culture (SV40-transformed glial cell culture derived from the cerebrum of a human fetus with Tay-Sachs desease) [1]

Localization in source

Purification
  Rabbit (partial) [2]; Human (TSD cell culture, partial) [1]

2

Crystallization

–

Cloned

–

Renatured

–

5 STABILITY

pH

Temperature (°C)

Oxidation

Organic solvent

General stability information
Purification has to be carried out at 0–5°C [2]

Storage

6 CROSSREFERENCES TO STRUCTURE DATABANKS

PIR/MIPS code

Brookhaven code

7 LITERATURE REFERENCES

[1] Basu, M., Presper, K.A., Basu, S., Hoffmann, L.M., Brooks, S.E.: Proc. Natl. Acad. Sci.
    USA,76,4270–4274 (1979)
[2] Basu, M., Basu, S.: J. Biol. Chem.,247,1489–1495 (1972)

1 NOMENCLATURE

EC number
2.4.1.87

Systematic name
UDPgalactose:beta-D-galactosyl-beta-1,4-N-acetyl-D-glucosaminyl-glyco-
peptide alpha-1,3-D-galactosyltransferase

Recommended name
beta-D-Galactosyl-N-acetylglucosaminylglycopeptide alpha-1,3-galactosyl-
transferase

Synonyms
alpha-Galactosyltransferase [1]
Galactosyltransferase, uridine diphosphogalactose-galactosylacetylglu-
cosaminylgalactosylglucosylceramide

CAS Reg. No.
62213-42-7

2 REACTION AND SPECIFICITY

Catalysed reaction
UDPgalactose + beta-D-galactosyl-beta-1,4-N-acetyl-D-glucosaminyl-glyco-
peptide →
→ UDP + alpha-D-galactosyl-1,3-D-galactosyl-beta-1,4-N-acetyl-
glucosaminyl-glycopeptide

Reaction type
Hexosyl group transfer

Natural substrates
UDPgalactose + galactosyl-1,3-N-acetylglucosaminyl-1,3-galactosyl-1,4-glu-
cosyl-1,1-ceramide (biosynthesis of blood group B specific penta-
glycosylceramide) [1]

Substrate spectrum
1 UDPgalactose + galactosyl-1,3-N-acetylglucosaminyl-1,3-galactosyl-1,4-
glucosyl-1,1-ceramide [1]

Product spectrum
1 UDP + O-alpha-galactosyl-1,3-galactosyl-1,3-N-acetylglucosaminyl-1,3-ga-
lactosyl-1,4-glucosyl-1,1-ceramide or O-alpha-galactosyl-1,4-galactosyl-1,3-
N-acetylglucosaminyl-1,3-galactosyl-1,4-glucosyl-1,1-ceramide [1]

Inhibitor(s)
  EDTA [1]

Cofactor(s)/prosthetic group(s)/activating agents

Metal compounds/salts
  $MnCl_2$ (10–20 mM, required) [1]

Turnover number (min^{-1})

Specific activity (U/mg)

K_m-value (mM)
  0.14 (UDPgalactose) [1]; 1.67 (galactosyl-1,3-N-acetylglucosaminyl-1,3-ga-
  lactosyl-1,4-glucosyl-1,1-ceramide) [1]

pH-optimum
  7.3 [1]

pH-range

Temperature optimum (°C)

Temperature range (°C)

3 ENZYME STRUCTURE

Molecular weight

Subunits

Glycoprotein/Lipoprotein
  –

4 ISOLATION/PREPARATION

Source organism
  Rabbit [1]

Source tissue
  Bone marrow [1]

Localization in source

Purification
  Rabbit (partial) [1]

Crystallization
  –

Cloned

–

Renatured

–

5 STABILITY

pH

Temperature (°C)

Oxidation

Organic solvent

General stability information

Storage

6 CROSSREFERENCES TO STRUCTURE DATABANKS

PIR/MIPS code

Brookhaven code

7 LITERATURE REFERENCES

[1] Basu, M., Basu, S.: J. Biol. Chem.,248,1700–1706 (1973)

1 NOMENCLATURE

EC number
2.4.1.88

Systematic name
UDP-N-acetyl-D-galactosamine:N-acetyl-D-galactosaminyl-1,3-D-galac-
tosyl-1,4-D-galactosyl-1,4-D-glucosylceramide alpha-N-acetyl-D-galac-
tosaminyltransferase

Recommended name
Globoside alpha-N-acetylgalactosaminyltransferase

Synonyms
Acetylgalactosaminyltransferase, uridine diphosphoacetylgalactosamine-
globoside alpha-
Forssman synthase
Globoside acetylgalactosaminyltransferase

CAS Reg. No.
52037-97-5

2 REACTION AND SPECIFICITY

Catalysed reaction
UDP-N-acetyl-D-galactosamine + N-acetyl-D-galactosaminyl-1,3-D-galac-
tosyl-1,4-D-galactosyl-1,4-D-glucosylceramide →
→ UDP + N-acetyl-D-galactosaminyl-N-acetyl-D-galactosaminyl-1,3-D-galac-
tosyl-1,4-D-galactosyl-1,4-D-glucosylceramide

Reaction type
Hexosyl group transfer

Natural substrates
UDP-N-acetyl-D-galactosamine + N-acetyl-D-galactosaminyl-1,3-D-galac-
tosyl-1,4-D-galactosyl-1,4-D-glucosylceramide (involved in biosynthesis of
carbohydrate moiety of sphingoglycolipids) [3]

Substrate spectrum
1 UDP-N-acetyl-D-galactosamine + N-acetyl-D-galactosaminyl-1,3-D-galac-
  tosyl-1,4-D-galactosyl-1,4-D-glucosylceramide (i.e. globoside, catalyzes
  the transfer of N-acetylgalactosamine in alpha-1,3-linkage to globoside
  [2], general structure of preferred substrates: N-acetylgalactosaminyl-
  beta-1,3-galactosyl-O-R [2], no substrates: UDP, UDPgalactose, 2'-fuco-
  syllactose, H blood group glycoprotein, bovine submaxillary deglycosyla-

ted mucin, GM1–3, glucosyl-beta-1,1-ceramide, galactosyl-beta-1,1-cer-
amide, galactosyl-beta-1,4-glucosylceramide [2], oligosaccharide of
blood type H substance, asialo-GM1, asialo-GM2 and GD1a, galac-
tosyl-alpha-1,4-galactosyl-beta-1,4-glucosylceramide [2]) [1–3]
2 UDP-N-acetyl-D-galactosamine + N-acetylgalactosaminyl-1,3-beta-galac-
tosyl-alpha-1,3-galactosyl-beta-1,4-glucosylceramide [2]
3 UDP-N-acetyl-D-galactosamine + oligosaccharide of globoside [2]

Product spectrum
1 UDP + N-acetyl-D-galactosaminyl-N-acetyl-D-galactosaminyl-1,3-D-galac-
tosyl-1,4-D-galactosyl-1,4-D-glucosylceramide (i.e. Forssman hapten)
[1–3]
2 ?
3 ?

Inhibitor(s)
EDTA (strong [3]) [1, 3]; NaN_3 (weak [3]) [1, 3]; PCMB (not [1, 4]) [3]; Co^{2+}
(in the presence of Mn^{2+}) [2]; Cu^{2+} (in the presence of Mn^{2+}) [2]; Ni^{2+} (in the
presence of Mn^{2+}) [2]; Zn^{2+} (in the presence of Mn^{2+}) [2]; Triton CF-54 [1];
Sodium deoxycholate [1]; Tween 20 and 80 [1]; Uridine (weak [3]) [3, 4];
UMP [4]; UDP (competitive to UDP-N-acetylgalactosamine, non-competitive
to globoside [2]) [2–4]; UDP-N-acetylglucosamine (strong [4], not [1]) [4];
ADP [3]; IDP [3]; GDP [3]; CDP [3]; More (no inhibitors: NaF [1, 3], GSH,
N-acetylgalactosamine, glucuronolactone [1]) [1, 3]

Cofactor(s)/prosthetic group(s)/activating agents
Triton X-100 (activation) [1]; Sodium taurocholate (activation) [1]; Phosphatidylserine (activation) [3]; GSH (slight stimulation) [3]; UDPglucose (slight
stimulation) [4]

Metal compounds/salts
Mn^{2+} (requirement, 10 mM [1], K_m-value 0.45 mM [2], Mg^{2+} and Ca^{2+} can
hardly replace Mn^{2+} [1]) [1–3]; Mg^{2+} (stimulation, only in the presence of
Mn^{2+} [2], not [1, 3]) [2]; Zn^{2+} (can replace Mn^{2+} with 83% efficiency) [2];
Fe^{2+} (can replace Mn^{2+} with 76% efficiency) [2]; Cu^{2+} (can replace Mn^{2+} with
31% efficiency, inhibits in the presence of Mn^{2+}) [2]; Co^{2+} (can replace Mn^{2+}
with 20% efficiency, inhibits in the presence of Mn^{2+}) [2]; More (no activation
by Ni^{2+}) [2]

Turnover number (min^{-1})

Specific activity (U/mg)
0.000059 [2]

K_m-value (mM)
0.00156 (globoside) [2]; 0.0094–0.1 (UDP-N-acetylgalactosamine) [1–3];
0.4 (globoside) [1]; 0.5 (globoside) [3]

pH-optimum
    6.6 [3]; 6.7 [1]; 6.7–6.9 [2]

pH-range

Temperature optimum (°C)
    37 (assay at) [2]

Temperature range (°C)

---

3 ENZYME STRUCTURE

Molecular weight
    120000 (dog, gel filtration) [2]

Subunits
    Dimer (1 × 56000 + 1 × 60000, dog, SDS-PAGE) [2]

Glycoprotein/Lipoprotein
    –

---

4 ISOLATION/PREPARATION

Source organism
    Guinea pig [1, 4]; Dog (mongrel dog, adult) [2, 3]

Source tissue
    Spleen [1–3]; Kidney [1, 4]; Small intestine [1]

Localization in source
    Microsomes [1, 4]; Membrane-bound [2, 3]

Purification
    Dog (affinity chromatography [2], partial [3]) [2, 3]

Crystallization
    –

Cloned
    –

Renatured
    –

---

5 STABILITY

pH

Temperature (°C)

Oxidation

Organic solvent

General stability information
  Repeated freeze-thawing inactivates, UDP and glycerol protect [2]; UDP,
  0.1 mM, stabilizes during storage [2]; Glycerol, 50%, stabilizes during storage [2]; Phosphatidylserine stabilizes [3]

Storage
  −80°C, 0.1 mg/ml enzyme in 25 mM sodium cacodylate buffer with 1% Triton
  X-100, at least 6 months [2]; −20°C, at least 2 weeks [3]; 4°C, 0.1 mg/ml
  enzyme in 25 mM sodium cacodylate buffer with 1% Triton X-100, 40% loss
  of activity within 3 months [2]

6 CROSSREFERENCES TO STRUCTURE DATABANKS

PIR/MIPS code

Brookhaven code

7 LITERATURE REFERENCES

[1] Kijimoto, S., Ishibashi, T., Makita, A.: Biochem. Biophys. Res. Commun.,56,177–184
    (1974)
[2] Taniguchi, N., Yokosawa, N., Gasa, S., Makita, A.: J. Biol. Chem.,257,10631–10637
    (1982)
[3] Ishibashi, T., Ohkubo, I., Makita, A.: Biochim. Biophys. Acta,484,24–34 (1977)
[4] Ishibashi, T., Atsuta, T., Makita, A.: Biochim. Biophys. Acta,429,759–767 (1976)

1 NOMENCLATURE

EC number
2.4.1.90

Systematic name
UDPgalactose:N-acetyl-D-glucosamine 4-beta-D-galactosyltransferase

Recommended name
N-Acetyllactosamine synthase

Synonyms
Galactosyltransferase, uridine diphosphogalactose-acetylglucosamine
beta-1,4-Galactosyltransferase
Acetyllactosamine synthetase
Lactosamine synthase
Lactosamine synthetase
Lactose synthetase A protein
N-Acetyllactosamine synthetase
UDP-galactose N-acetylglucosamine beta-4-galactosyltransferase
UDP-galactose-acetylglucosamine galactosyltransferase
UDP-galactose-N-acetylglucosamine beta-1,4-galactosyltransferase
UDP-galactose-N-acetylglucosamine galactosyltransferase
Uridine diphosphogalactose-acetylglucosamine galactosyltransferase
beta1–4-Galactosyltransferase [5, 17]
UDP-Gal:N-acetylglucosamine beta1–4-galactosyltransferase [5]
beta1–4GalT [5]
NAL synthetase [4]
UDP-beta-1,4-galactosyltransferase [7, 15, 16]
Gal-T [17]
UDP-galactose:N-acetylglucosaminide beta1–4-galactosyltransferase [18, 19]
UDPgalactose:N-acetylglucosaminyl(beta1–4)galactosyltransferase [21]
beta-N-Acetylglucosaminide beta1–4-galactosyltransferase [27]
UDPgalactose-N-acetylglucosamine beta-D-galactosyltransferase
EC 2.4.1.98 (formerly)
More (cf. EC 2.4.1.22, not distinguishable from EC 2.4.1.38)

CAS Reg. No.
9054-94-8

2 REACTION AND SPECIFICITY

Catalysed reaction

UDPgalactose + N-acetyl-D-glucosamine →
→ UDP + N-acetyllactosamine (not distinguishable from EC 2.4.1.38 which
has identical substrate specificities. EC 2.4.1.38/90 is identical with the A
protein of EC 2.4.1.22. The sequences of cDNA isolated from mammary and
F9 cell lines are identical, thus indicating that EC 2.4.1.38 and EC 2.4.1.90
are non-distinguishable [5], mechanism [1, 32]. In the presence of
alpha-lactalbumin the reaction of EC 2.4.1.22 is catalyzed)

Reaction type

Hexosyl group transfer

Natural substrates

UDPgalactose + N-acetylglucosaminyl at the non-reducing ends of protein-bound oligosaccharides (biosynthesis of glycoproteins [1], of plasma
glycoproteins [3], of carbohydrate moieties of glycoproteins and glycolipids,
role in intercellular recognition and adhesion [7, 8], biosynthesis of keratan
sulfate-like polysaccharides [35]) [1, 3, 7, 8, 23, 35]

Substrate spectrum

1 UDPgalactose + N-acetylglucosamine (the acceptor sugar N-acetylglucosamine may be the free monosaccharide or the non-reducing terminal
  monosaccharide of a carbohydrate side chain of a glycoprotein or glycolipid [15]. Whether or not this enzyme synthesizes all beta-1,4-galac-
  tosyl linkages found in oligosaccharide chains of glycoconjugates is
  uncertain [5]) [1–5, 15, 19, 20, 24, 27, 28, 30, 32–35, 40, 41, 42, 44]
2 UDPgalactose + beta-methylglucose [1]
3 UDPgalactose + beta-indoxylglucose [1]
4 UDPgalactose + cellobiose [1]
5 UDPgalactose + asialo-agalacto-transferrin [5]
6 UDPgalactose + ovomucoid [18]
7 UDPgalactose + ovalbumin [21, 29, 43]
8 UDPgalactose + desialated degalactosylated fetuin [31, 38, 42, 44]
9 UDPgalactose + alpha$_1$-acid glycoprotein [38]
10 UDPgalactose + N-acetylglucosaminyl-beta-1,2-man-
   nosyl-alpha-1,6-(N-acetylglucosaminyl-beta-1,2-man-
   nosyl-alpha-1,3-)mannosyl-beta-1,4-N-acetylglucosaminyl-beta-1,4-
   (fucosyl-alpha-1,6-)N-acetylglucosaminyl-asparagine [23]
11 UDPgalactose + N-acetylglucosaminyl-beta-1,2-mannosyl-alpha-1,6-
   (galactosyl-beta-1,4-N-acetylglucosaminyl-beta-1,2-man-
   nosyl-alpha-1,3-)mannosyl-beta-1,4-N-acetylglucosaminyl-beta-1,4-
   (fucosyl-alpha-1,6-)N-acetylglucosaminyl-asparagine [23]

12 UDPgalactose + galactosyl-beta-1,4-N-acetylglucosaminyl-beta-1,2-
    mannosyl-alpha-1,6-(N-acetylglucosaminyl-beta-1,2-man-
    nosyl-alpha-1,3-)mannosyl-beta-1,4-N-acetylglucosaminyl-beta-1,4-
    (fucosyl-alpha-1,6-)N-acetylglucosaminyl-asparagine [23]
13 UDPgalactose + N-acetylglucosaminyl-beta-1,3-galactose [27]
14 UDPgalactose + N-acetylglucosaminyl-beta-1,6-galactose [27]
15 UDPgalactose + N-acetylglucosaminyl-beta-1,3-(N-acetylglucosa-
    minyl-beta-1,6-)galactose [27]
16 UDPgalactose + N-acetylglucosaminyl-beta-1,3-(galac-
    tosyl-beta-1,4-N-acetylglucosaminyl-beta-1,6-)galactose [27]

Product spectrum
 1 UDP + N-acetyllactosamine [2, 19, 28, 35, 38, 40]
 2 ?
 3 ?
 4 ?
 5 ?
 6 UDP + ovomucoid with beta-1,4-bound galactose [18]
 7 ?
 8 UDP + fetuin containing beta-1,4-galactose linkages [38]
 9 ?
10 UDP + galactosyl-beta-1,4-N-acetylglucosaminyl-beta-1,2-mannosyl-
    alpha-1,6-(galactosyl-beta-1,4-N-acetylglucosaminyl-beta-1,2-man-
    nosyl-alpha-1,3-)mannosyl-beta-1,4-N-acetylglucosaminyl-beta-1,4-(fuco-
    syl-alpha-1,6-)N-acetylglucosaminyl-asparagine (galactose is transferred
    much faster to the N-acetylglucosaminyl-beta-1,2-mannosyl-alpha-1,3-
    branch than to the N-acetylglucosaminyl-beta-1,2-mannosyl-alpha-1,6-
    branch) [23]
11 UDP + galactosyl-beta-1,4-N-acetylglucosaminyl-beta-1,2-mannosyl-
    alpha-1,6-(galactosyl-beta-1,4-N-acetylglucosaminyl-beta-1,2-man-
    nosyl-alpha-1,3-)mannosyl-beta-1,4-N-acetylglucosaminyl-beta-1,4-(fuco-
    syl-alpha-1,6-)N-acetylglucosaminyl-asparagine [23]
12 UDP + galactosyl-beta-1,4-N-acetylglucosaminyl-beta-1,2-mannosyl-
    alpha-1,6-(galactosyl-beta-1,4-N-acetylglucosaminyl-beta-1,2-man-
    nosyl-alpha-1,3-)mannosyl-beta-1,4-N-acetylglucosaminyl-beta-1,4-(fuco-
    syl-alpha-1,6-)N-acetylglucosaminyl-asparagine [23]
13 ?
14 ?
15 ?
16 ?

Inhibitor(s)

UDP (noncompetitive to Mg^{2+} and N-acetylglucosamine, competitive to UDPgalactose) [1]; Phosphatidylserine [20]; Phosphatidic acid [20, 22]; Zn^{2+} [21]; Phosphatidylethanolamine [22]; Phosphatidylglycerol [22]; alpha-Lactalbumin (decreases the affinity for N-acetylglucosamine, increases the affinity for glucose and thus synthesis of lactose) [24, 34, 35]; Poly(L-Glu) [29], EDTA [30, 31, 39, 44]; N-Acetylglucosamine (above 10 mM) [38]; $alpha_1$-Acid glycoprotein (above 1.4 mM with respect to acceptor sites) [38]; Iodination by lactoperoxidase [41]; p-Hydroxymercuribenzoate [45]

Cofactor(s)/prosthetic group(s)/activating agents

Phosphatidylcholine (activation) [20, 22]; Phosphatidylethanolamine (activation) [20]; Lysophosphatidylcholine (activation) [20]; Phosphatidylglycerol (activation) [20]; Methylphosphatidylic acid (activation) [20]; Dimyristoylphosphatidylcholine (activation) [20, 22]; Poly(L-Arg) (activation) [29]; Poly(L-Lys) (activation) [29]; Histone (activation) [20]; Protamine sulfate (activation) [29]; Dioleoylphosphatidylcholine (activation) [22]; Dipalmitoylphosphatidylcholine (activation) [22]; Distearoylphosphatidylcholine (activation) [22]

Metal compounds/salts

Mn^{2+} (required [1, 2, 5, 21, 29–33, 35, 37–39, 42–44], optimal concentration 3–5 mM [24], 12.5 mM [42, 44], partially replaceable by Mg^{2+} [1, 2], Ca^{2+} [1, 2], Co^{2+} [35], Zn^{2+} [35], not replaceable by Mg^{2+} [39, 43, 44], Zn^{2+} [39, 44], Ca^{2+} [39, 43, 44], Co^{2+} [39, 44]) [1, 2, 5, 21, 24, 29–33, 35, 37–39, 42–44]; Zn^{2+} (activation, 9.2% of Mn^{2+} activity) [35]; Co^{2+} (activation, 14.9% of Mn^{2+} activity) [35]; Mg^{2+} (can partially replace Mn^{2+}) [1, 2]; Ca^{2+} (can partially replace Mn^{2+}) [1, 2]

Turnover number (min^{-1})

Specific activity (U/mg)

4700 [17]; 19.8 [19]; 0.291 [22]; More [21, 26, 31, 32, 35, 37, 38, 40]

K_m-value (mM)

0.024 (UDPgalactose) [5]; 0.028 (UDPgalactose, recombinant enzyme) [17]; 0.03 (UDPgalactose) [21]; 0.043 (UDPgalactose, bovine milk) [24]; 0.052 (UDPgalactose, rat serum) [24]; 0.056 (asialo-agalacto-transferrin) [5]; 0.065 (UDPgalactose, rat liver) [24]; 0.082 (UDPgalactose, human) [17]; 0.13 (N-acetylglucosaminyl-beta-1,2-mannosyl-alpha-1,6-(N-acetylglucosaminyl-beta-1,2-mannosyl-alpha-1,3-)mannosyl-beta-1,4-N-acetylglucosaminyl-beta-1,4-(fucosyl-alpha-1,6-)N-acetylglucosaminyl-asparagine) [23]; 0.25 (UDPgalactose) [2]; 0.33 (N-acetylglucosamine, rat liver) [24]; 0.43 (N-acetylglucosaminyl-beta-1,2-mannosyl-alpha-1,6-(galactosyl-beta-1,4-N-acetylglucosaminyl-beta-1,2-mannosyl-alpha-1,3-)mannosyl-beta-1,4-N-acetylglucosaminyl-beta-1,4-(fucosyl-alpha-1,6-)N-ace-

tylglucosaminyl-asparagine) [23]; 0.5 (N-acetylglucosamine, bovine milk)
[24]; 0.8 (N-acetylgalactosamine) [28]; 1.0 (N-acetylglucosamine) [2]; 1.49
(N-acetylglucosamine, bovine milk) [24]; 1.5 (N-acetylglucosaminyl-
beta-1,6-galactose, N-acetylglucosaminyl-beta-1,3-(N-acetylglucosaminyl-
beta-1,6-)galactose) [27]; 1.9 (N-acetylglucosamine [5], N-acetylglucos-
aminyl-beta-1,3-galactose [27]) [5, 27]; 2.3 (N-acetylglucosamine) [28]; 2.8
(N-acetylglucosamine, human) [17]; 3.4 (N-acetylglucosaminyl-beta-1,3-(ga-
lactosyl-beta-1,4-N-acetylglucosaminyl-beta-1,6-)galactose) [27]; 3.6 (N-acetylglucosamine, recombinant enzyme) [17]; 6.28 (galactosyl-beta-1,4-N-ace-
tylglucosaminyl-beta-1,2-mannosyl-alpha-1,6-(N-acetylglucosaminyl-beta-
1,2-mannosyl-alpha-1,3-)mannosyl-beta-1,4-N-acetylglucosaminyl-beta-
1,4-(fucosyl-alpha-1,6-)N-acetylglucosaminyl-asparagine) [23]; 8.2 (N-acetylglucosamine) [27]; More (effects of cationic polypeptides [29]) [29–31,
34, 38–40, 42–44]

pH-optimum
6.4–7.6 [31]; 6.5 [18, 30, 39]; 6.8 [40, 42]; 7.5 [38]; 7.5–10.5 [35]; 8.2 [2]

pH-range
5.5–8.0 (less than 50% of maximal activity above and below) [39]

Temperature optimum (°C)
30 [38]; 37 [31, 42]; 42 [17]

Temperature range (°C)
25–45 (less than 50% of maximal activity above and below) [42]

---

3 ENZYME STRUCTURE

Molecular weight
440000 (rat, gel chromatography, extremely protease-sensitive form) [19]
85000–90000 (human, gel filtration) [38]
70000 (pig, gel filtration) [31]
57000 (pig, sucrose gradient centrifugation) [40]
44876 (bovine, calculation from gene sequence, unglycosylated enzyme
[14], localization of UDPgalactose binding in primary structure [13]) [13, 14]
44416 (mouse, calculation from gene sequence, long form with NH_2-terminal
extension of 13 amino acids [7], only the long form is expressed in sperma-
togenic cells, other cells express both forms of the enzyme [9]) [7, 9]
42960 (mouse, calculation from gene sequence, short form, transmembrane
enzyme) [7]

Subunits
Multimer (x × 50000–51000, rat, SDS-PAGE [19], bovine, SDS-PAGE,
tendency to form aggregates [26]) [19, 26]
Monomer (1 × 74000, pig, SDS-PAGE) [31]

? (x × 42200, bovine, SDS-PAGE [35], x × 43000, rat, SDS-PAGE [44], x × 48000, human, glycosylated form [17], x × 47000, human, deglycosylated form, SDS-PAGE [17], x × 54000, bovine, SDS-PAGE [32], x × 65000–70000, rat, SDS-PAGE [39], x × 70000–80000, human, SDS-PAGE [38], x × 70000–75000, rat, SDS-PAGE [21]) [17, 21, 32, 35, 38, 39, 44]

Glycoprotein/Lipoprotein
Glycoprotein (structure of mucin-type sugar depends on blood group [12]) [1, 12, 17]

---

4 ISOLATION/PREPARATION

Source organism
Mammals [1]; Mouse [2, 5–7, 9, 15, 34]; Rat [3, 4, 18, 19, 21–24, 30, 36, 39, 43, 44]; Bovine [8, 10, 13, 14, 20, 25–27, 32, 33, 35, 41]; Human (expressed in Saccharomyces cerevisiae [17]) [11, 12, 16, 17, 28, 29, 38, 42, 45]; Pig [31, 40]; Sheep [37]

Source tissue
Mastocytoma cells [2]; Liver [3, 18, 19, 22–24, 30, 36, 39, 44]; Brain [4]; F9 embryonal carcinoma cells [5, 6]; Mammary cell line C127 [7]; MDBK cells (ATCC No. CCL22) [8]; Milk [12, 13, 20, 32, 33, 35, 41, 45]; Testis [9]; Mammary gland [10, 37]; HeLa cells [16, 17]; Intestine (regional distribution) [21]; Serum [24, 28, 43, 44]; Thymus [26]; Colostrum [27, 32]; Fibroblasts [29]; Thyroid gland [31]; Balb/c 3T12 cell line [34]; Eyes (cornea) [35]; Kidney [36]; Plasma [38]; Mesentary lymph nodes [40]; Amniotic fluid [42]

Localization in source
Golgi apparatus (membrane-bound) [1, 3, 4, 7, 14, 16, 18, 22–25, 31, 36, 37]; Membrane-bound [34, 40, 44]; Soluble [13, 18, 30, 35, 38, 40–45]

Purification
Mouse [2]; Rat [18, 19, 21, 22, 30, 36, 39, 44]; Bovine [26, 27, 35]; Pig [31, 40]; Human (expressed in Saccharomyces cerevisiae [17]) [17, 38]; Sheep [37]

Crystallization
–

Cloned
(expression in COS-1 cells [5], expression in Saccharomyces cerevisiae [17], expression of short and long form in CHO-cells [25], comparison of sequences of enzyme from placenta and HeLa cells [11]) [5–9, 11, 12, 14–17, 25]

Renatured
–

5 STABILITY

pH
   5–9 [2]

Temperature (°C)
   45 (up to) [4]

Oxidation

Organic solvent

General stability information
   Triton X-100 essential for stability during purification [27]; Ammonium sulfate
   stabilizes during storage [30]; Glycerol stabilizes during storage [30]; No
   stabilization by N-acetylglucosamine, dithiothreitol, 2-mercaptoethanol [44]

Storage
   –20°C, at least 1 month [17]; –20°C, partially purified enzyme stable for several weeks, purified enzyme stable for 1 week [21]; –20°C, bovine serum albumin, 10–20% loss of activity upon thawing [26]; –20°C, bovine serum albumin, up to 60 d [30]; –20°C, 1 mg/ml bovine serum albumin [32]; –20°C,
   0.02 M Tris/HCl buffer, pH 7.5, several months [40]; –20°C, 3 weeks [43];
   4°C, 0.1% bovine serum albumin, 3 months [37]; 4°C, concentrated enzyme
   [22]; 4°C, 4 weeks [24]

6 CROSSREFERENCES TO STRUCTURE DATABANKS

PIR/MIPS code

Brookhaven code

7 LITERATURE REFERENCES

[1] Hill, R., Brew, K. in "Adv. Enzymol. Relat. Areas Mol. Biol.",43,411–490 (1975)
    (Review)
[2] Helting, T., Erbing, B.: Biochim. Biophys. Acta,293,94–104 (1973)
[3] Schachter, H., Jabbal, I., Hudgin, R.L., Pinteric, L.: J. Biol. Chem.,245,1090–1100
    (1970)
[4] Deshmukh, D.S., Bear, W.D., Soifer, D.: Biochim. Biophys. Acta,542,284–295 (1978)
[5] Nakazawa, K., Furukawa, K., Kobata, A., Narimatsu, H.: Eur. J. Biochem.,196,
    363–368 (1991)
[6] Nakazawa, K., Ando, T., Kimura, T., Narimatsu, H.: J. Biochem.,104,165–168 (1988)
[7] Shaper, N.L., Hollis, G.F., Douglas, J.G., Kirsch, I.R., Shaper, J.H.: J. Biol. Chem.,
    263,10420–10428 (1988)
[8] Shaper, N.L., Shaper, J.H., Meuth, J.L., Fox, J.L., Chang, H., Kirsch, I.R., Hollis,
    G.F.: Proc. Natl. Acad. Sci. USA,83,1573–1577 (1986)
[9] Shaper, N.L., Wright, W.W., Shaper, J.H.: Proc. Natl. Acad. Sci. USA,87,791–795
    (1990)

[10] Narimatsu, H., Sinha, S., Brew, K., Okayama, H., Qasba, P.K.: Proc. Natl. Acad. Sci. USA,83,4720–4724 (1986)
[11] Watzele, G., Berger, E.G.: Nucleic Acids Res.,18,7174 (1990)
[12] Amano, J., Straehl, P., Berger, E.G., Kochibe, N., Kobata, A.: J. Biol. Chem.,266, 11461–11477 (1991)
[13] Yadav, S., Brew, K.: J. Biol. Chem.,265,14163–14169 (1990)
[14] D'Agostaro, G., Bendiak, B., Tropak, M.: Eur. J. Biochem.,183,211–217 (1989)
[15] Hollis, G.F., Douglas, J.G., Shaper, N.L., Shaper, J.H., Stafford-Hollis, J.M., Evans, R.J., Kirsch, I.R.: Biochem. Biophys. Res. Commun.,162,1069–1075 (1989)
[16] Mengle-Gaw, L., McCoy-Haman, M.F., Tiemeier, D.C.: Biochem. Biophys. Res. Commun.,176,1269–1276 (1991)
[17] Krezdorn, C.H., Watzele, G., Kleene, R.B., Ivanov, S.X., Berger, E.G.: Eur. J. Biochem.,212,113–120 (1993)
[18] Kawano, J.-i., Oinuma, T., Nakayama, T., Suganuma, T.: J. Biochem.,111,568–572 (1992)
[19] Bendiak, B., Ward, L.D., Simpson, R.J.: Eur. J. Biochem.,216,405–417 (1993)
[20] Mitranic, M.M., Moscarello, M.A.: Can. J. Biochem.,58,809–814 (1980)
[21] Weiser, M.M., Majumdar, S., Wilson, J.R., Luther, R.: Biochim. Biophys. Acta,924, 323–331 (1987)
[22] Clark, P.E., Moscarello, M.A.: Biochim. Biophys. Acta,859,143–150 (1986)
[23] Paquet, M.R., Narasimhan, S., Schachter, H., Moscarello, M.A.: J. Biol. Chem., 259,4716–4721 (1984)
[24] Paquet, M.R., Moscarello, M.A.: Biochem. J.,218,745–751 (1984)
[25] Russo, R.N., Shaper, N.L., Taatjes, D.J., Shaper, J.H.: J. Biol. Chem.,267,9241–9247 (1992)
[26] Blanken, W.M., van den Eijnden, D.H.: J. Biol. Chem.,260,12927–12934 (1985)
[27] Blanken, W.M., Hooghwinkel, G.J.M., van den Eijnden, D.H.: Eur. J. Biochem.,127, 547–552 (1982)
[28] Berger, E.G., Kozdrowski, I., Weiser, M.M., van den Eijnden, D.H., Schiphorst, W.E.C.M.: Eur. J. Biochem.,90,213–222 (1978)
[29] Rao, G.J.S., Chyatte, D., Nadler, H.L.: Biochim. Biophys. Acta,541,435–443 (1978)
[30] Fraser, I.H., Wadden, P., Mookerjea, S.: Can. J. Biochem.,58,878–884 (1980)
[31] Bouchilloux, S.: Biochim. Biophys. Acta,569,135–144 (1979)
[32] Tsopanakis, A.D., Herries, D.G.: Eur. J. Biochem.,83,179–188 (1978)
[33] Andree, P.J., Berliner, L.J.: Biochemistry,19,929–934 (1980)
[34] Cummings, R.D., Cebula, T.A., Roth, S.: J. Biol. Chem.,254,1233–1240 (1979)
[35] Christner, J.E., Distler, J.J., Jourdian, G.W.: Arch. Biochem. Biophys.,192,548–558 (1979)
[36] Fleischer, B., Smigel, M.: J. Biol. Chem.,253,1632–1638 (1978)
[37] Smith, C.A., Brew, K.: J. Biol. Chem.,252,7294–7299 (1977)
[38] Bella, A., Whitehead, J.S., Kim, Y.S.: Biochem. J.,167,621–628 (1977)
[39] Fraser, I.H., Mookerjea, S.: Biochem. J.,164,541–547 (1977)
[40] Rao, A.K., Garver, F., Mendicino, J.: Biochemistry,15,5001–5009 (1976)
[41] Chandler, D.K., Silvia, J.C., Ebner, K.E.: Biochim. Biophys. Acta,616,179–187 (1980)
[42] Nelson, J.D., Jato-Rodriguez, J.J., Mookerjea, S.: Can. J. Biochem.,52,42–50 (1974)
[43] Wagner, R.R., Cynkin, M.A.: Biochem. Biophys. Res. Commun.,45,57–62 (1971)
[44] Fraser, I.H., Mookerjea, S.: Biochem. J.,156,347–355 (1976)
[45] Kitchen, B.J., Andrews, P.: Biochem. J.,143,587–590 (1974)

1 NOMENCLATURE

EC number
2.4.1.91

Systematic name
UDPglucose:flavonol 3-O-D-glucosyltransferase

Recommended name
Flavonol 3-O-glucosyltransferase

Synonyms
Glucosyltransferase, uridine diphosphoglucose-flavonol 3-O-
UDP-glucose:flavonol 3-O-glucosyltransferase
UDPG:flavonoid-3-O-glucosyltransferase
GTI
More (different from EC 2.4.1.81)

CAS Reg. No.
50812-18-5

---

2 REACTION AND SPECIFICITY

Catalysed reaction
UDPglucose + a flavonol →
→ UDP + a flavonol 3-O-D-glucoside

Reaction type
Hexosyl group transfer

Natural substrates
UDP-D-glucose + quercetin (involved in flavonol glucoside biosynthesis) [8]

Substrate spectrum
1 UDP-D-glucose + quercetin (r [6], best substrate [1, 8], ADP-D-glucose
  or GDP-D-glucose cannot replace UDP-D-glucose [4]) [1–8]
2 UDP-D-glucose + dihydroquercetin (poor substrate [8], not [1, 2, 5])
  [7, 8]
3 UDP-D-glucose + kaempferol (r [6], best substrate [7], glucosylation at
  70% the rate of quercetin [1]) [1–3, 5–8]
4 UDP-D-glucose + dihydrokaempferol [7]
5 UDP-D-glucose + kaempferid (i.e. kaempferol-4'-O-methylether,
  glucosylation at 57% the rate of quercetin [1]) [1, 2]
6 UDP-D-glucose + fisetin (i.e. 5-deoxyquercetin, glucosylation at 57% the
  rate of quercetin [1]) [1, 5]

7 UDP-D-glucose + isorhamnetin (i.e. 3'-O-methylquercetin, best substrate
   [2], glucosylation at 28% the rate of quercetin [1]) [1–3, 5]
8 UDP-D-glucose + quercetin 7-O-glucoside [1]
9 UDP-D-glucose + 4,7-dihydroxyflavonol (glucosylation at 27% the rate of
   quercetin) [1]
10 UDP-D-glucose + isosakuranetin [7]
11 UDP-D-glucose + flavonolaglycones [2]
12 More (no acceptors are flavonol 3-O-glucosides, luteolin [2], catechin,
    malvidin, pelargonidin, apigenin [7, 8], naringenin [2, 5, 8], cyanidin
    [2, 8], 4,2',4'-trihydroxychalcone, daidzein, texasin, cinnamic acids [5],
    p-coumaric acid [2], some phenols [2, 5]) [2, 5, 7, 8]

Product spectrum
1 UDP + quercetin 3-O-glucoside [1, 2]
2 ?
3 UDP + kaempferol 3-O-glucoside [1]
4 ?
5 ?
6 ?
7 ?
8 UDP + quercetin 3,7-di-O-glucoside [1]
9 ?
10 ?
11 ?
12 ?

Inhibitor(s)
Mg^{2+} (above 5 mM [2], 1 mM: stimulation [4]) [2]; EDTA (weak) [2]; Zn^{2+}
(above 0.5 mM) [2]; Cu^{2+} (strong [3, 4], above 5 mM [2]) [2–4]; Mn^{2+} (strong
[3, 4], above 0.05 mM [2]) [2–4]; PCMB (not 2-mercaptoethanol, GSH or dithioerythritol) [2]

Cofactor(s)/prosthetic group(s)/activating agents
Dithioerythritol (slight activation) [2]; Sucrose (slight activation) [2]; GSH
(slight activation) [2]; 2-Mercaptoethanol (slight activation, 14 mM) [2]; More
(no cofactor requirement [5], no bovine serum albumin required [2]) [2, 5]

Metal compounds/salts
Mg^{2+} (stimulation, 1 mM [4], inhibition above 5 mM [2]) [4]; More (no NH_4^+
or Ca^{2+}-requirement) [2]

Turnover number (min^{-1})

Specific activity (U/mg)
0.0567 [7]; 3.396 [8]

K_m-value (mM)
    0.00069 (kaempferol) [3]; 0.0009 (quercetin, isorhamnetin) [2]; 0.001 (be-
    low, quercetin) [1]; 0.00121 (quercetin) [4]; 0.007 (quercetin) [8]; 0.0098
    (UDP-D-glucose) [4]; 0.012–0.0125 (kaempferol) [7, 8]; 0.07 (quercetin, cy-
    anidin) [7]; 0.1 (UDP-D-glucose) [2]; 0.126 (quercetin) [5]; 0.172 (kaempfe-
    rol) [5]; 0.18 (UDP-D-glucose) [8]; 0.2 (isorhamnetin) [5]; 0.27 (fisetin) [5];
    0.3 (UDP-D-glucose (+ quercetin)) [5]; 0.5 (UDP-D-glucose) [1]; 1.0
    (UDP-D-glucose) [7]; 1.67 (UDP-D-glucose) [3]

pH-optimum
    More (pI: 4.25–4.55 (possibly several isozymes) [8], pI: 5.6 [7], pI: 6.1 [3])
    [3, 7, 8]; 5.0 (kaempferol, shoulder with 69% of maximal activity at pH 8–8.5)
    [7]; 5.8–6.2 [8]; 7.5 [4]; 8.0 (broad [1], glycine-HCl buffer preferred [1]) [1,
    3]; 8.5 [5]; 8.5–9.0 [2]

pH-range
    4.2–8.3 (about half-maximal activity at pH 4.2 and 8.3) [8]

Temperature optimum (°C)
    30 (assay at) [1]

Temperature range (°C)

---

3 ENZYME STRUCTURE

Molecular weight
    40000 (Tulipa, gel filtration) [2]
    43000 (Vigna mungo, gel filtration) [3]
    49000 (Hippeastrum, gel filtration) [7]
    51000 (Prunus x yedoensis, gel filtration) [4]
    59000 (Brassica olercea, gel filtration) [8]

Subunits
    Dimer (2 × 24500, Hippeastrum, SDS-PAGE [7], 2 × 29500, Brassica oleracea,
    SDS-PAGE [8]) [7, 8]

Glycoprotein/Lipoprotein
    –

---

4 ISOLATION/PREPARATION

Source organism
    Petroselinum hortense (parsley) [1, 6]; Brassica oleracea (red cabbage, cv.
    Red Danish) [8]; Zea mays (maize) [9, 10]; Tulipa sp. (tulip, cv. Apeldoorn)
    [2]; Vigna mungo [3]; Glycine max (soy bean) [5]; Hippeastrum [7]; Prunus
    x yedoensis Matsum. (cherry tree) [4]

Source tissue
Cell suspension culture (illuminated for 24 h [1]) [1, 6]; Pollen [9, 10]; Anthers [2]; Seedlings (illuminated [8]) [3, 8]; Leaf [4]; Petals (post-anthesis stage) [7]

Localization in source
Tapetum [2]; Endoplasmic reticulum [8]

Purification
Tulipa [2]; Petroselinum hortense (partial) [1]; Vigna mungo [3]; Prunus x yedoensis [4]; Glycine max (partial) [5]; Hippeastrum [7]; Brassica oleracea [8]

Crystallization
–

Cloned
–

Renatured
–

---

5 STABILITY

pH

Temperature (°C)

Oxidation

Organic solvent

General stability information

Storage

---

6 CROSSREFERENCES TO STRUCTURE DATABANKS

PIR/MIPS code
PIR2:S01052 ((allele Bz-McC) maize); PIR2:S01037 ((allele Bz-W22) maize); PIR2:S08325 ((allele BzMcC2) maize); PIR1:XUBHFG (barley)

Brookhaven code

7 LITERATURE REFERENCES

[1] Sutter, A., Grisebach, H.: Biochim. Biophys. Acta,309,289–295 (1973)
[2] Kleinehollenhorst, G., Behrens, H., Pegels, G., Srunk, N., Wiermann, R.: Z. Naturforsch.,37c,587–599 (1982)
[3] Ishikura, N., Mato, M.: Plant Cell Physiol.,34,329–335 (1993)
[4] Ishikura, N., Kazumi, Y.: Plant Cell Physiol.,31,1109–1115 (1990)
[5] Poulton, J.E., Kauer, M.: Planta,136,53–59 (1977)
[6] Sutter, A., Grisebach, H.: Arch. Biochem. Biophys.,167,444–447 (1975)
[7] Hrazdina, G.: Biochim. Biophys. Acta,955,301–309 (1988)
[8] Sun, Y., Hrazdina, G.: Plant Physiol.,95,570–576 (1991)
[9] Larson, R.L.: Phytochemistry,10,3073–3076 (1971)
[10] Larson, R.L., Lonergan, C.M.: Planta,103,361–364 (1972)

1 NOMENCLATURE

EC number
   2.4.1.92

Systematic name
   UDP-N-acetyl-D-galactosamine:(N-acetylneuraminyl)-D-galactosyl-D-
   glucosylceramide N-acetyl-D-galactosaminyltransferase

Recommended name
   (N-Acetylneuraminyl)-galactosylglucosylceramide N-acetylgalactosaminyltransferase

Synonyms
   Acetylgalactosaminyltransferase, uridine diphosphoacetylgalactosamine-
   ganglioside GM3
   Ganglioside GM2 synthase
   Ganglioside GM3 acetylgalactosaminyltransferase
   GM2 synthase
   UDP acetylgalactosamine-(N-acetylneuraminyl)-D-galactosyl-D-glucosylcer-
   amide acetylgalactosaminyltransferase
   UDP-N-acetylgalactosamine GM3 N-acetylgalactosaminyltransferase
   Uridine diphosphoacetylgalactosamine-acetylneuraminylgalactosylglucosyl-
   ceramide acetylgalactosaminyltransferase
   Uridine diphosphoacetylgalactosamine-hematoside acetylgalactosaminyl-
   transferase
   GM2/GD2-synthase [7]

CAS Reg. No.
   67338-98-1

2 REACTION AND SPECIFICITY

Catalysed reaction
   UDP-N-acetyl-D-galactosamine + (N-acetylneuraminyl)-D-galactosyl-D-
   glucosylceramide →
   → UDP + N-acetyl-D-galactosaminyl-(N-acetylneuraminyl)-D-galactosyl-D-
   glucosylceramide (mechanism [7])

Reaction type
   Hexosyl group transfer

Natural substrates
   UDP-N-acetyl-D-galactosamine + (N-acetylneuraminyl)-D-galactosyl-D-
   glucosylceramide (involved in biosynthesis of gangliosides in brain) [2]

Substrate spectrum

1 UDP-N-acetyl-D-galactosamine + (N-acetylneuraminyl)-D-galactosyl-D-glucosylceramide (i.e. GM3(N-acetylneuraminic acid), best substrate [2], from human spleen or rat liver [3], 78.2% [10], 73% [6] as effective as GM3(N-glycolylneuraminic acid), not as effective: GM3(O-acetyl-(N-glycolyl)neuraminic acid) [2], poor substrates are sialyllactose [6], GM2-, GM1-, GD1b- and GT1-gangliosides and various ceramides [2], acceptor specificity [4], no substrates are N-acetylglucosamine [2], lactosylceramide, sialylparagloboside, GM1b(N-glycolylneuraminic acid), GD1a(N-acetylneuraminic acid), GT1b(N-acetylneuraminic acid), GQ1b(N-acetylneuraminic acid), GM1a(N-acetylneuraminic acid), GD1b(N-acetylneuraminic acid), globoside [6], blood type H glycoprotein, deglucosylated bovine submaxillary mucin, various ceramides [10]) [2–4, 6–10]

2 UDP-N-acetyl-D-galactosamine + GM3(N-glycolylneuraminic acid)-ganglioside (best substrate [6, 10], not as effective as GM3(N-acetylneuraminic acid) [2]) [2, 6, 10]

3 UDP-N-acetyl-D-galactosamine + lactosylceramide [7]

4 UDP-N-acetyl-D-galactosamine + GD3-ganglioside (GD3(N-glycolylneuraminic acid): 66% as effective as GM3(N-glycolylneuraminic acid), GD3(N-acetylneuraminic acid): 76% as effective as GM3(N-glycolylneuraminic acid) [6]) [6, 7]

5 UDP-N-acetyl-D-galactosamine + SM3-ganglioside (best substrate, no substrates are SM4-ganglioside, Gb3- and Gb4-ceramide) [9]

6 UDP-N-acetyl-D-galactosamine + N-acetylneuraminelactose (i.e. sialyllactose) [10]

Product spectrum

1 UDP + N-acetyl-D-galactosaminyl-(N-acetylneuraminyl)-D-galactosyl-D-glucosylceramide (i.e. GM2(N-acetylneuraminic acid)-ganglioside) [2–4, 6–10]

2 UDP + GM2(N-glycolylneuraminic acid)-ganglioside [10]

3 UDP + GA2-ganglioside [7]

4 UDP + GD2-ganglioside [7]

5 UDP + SM2-ganglioside [7]

6 ?

Inhibitor(s)

Chelex-100 [10]; CMP [3]; IMP [3]; Adenosine-5'-[alpha,beta-methylene]-triphosphate [3]; UDPgalactose [10]; GM3 (GD3 as substrate) [7]; GD3 (GM3 as substrate) [7]; Gangliosides (e.g. GD1a, GT1b, GQ1b, the latter not in detergent-solubilized enzyme assays, kinetics) [8]; EDTA [10]

Cofactor(s)/prosthetic group(s)/activating agents

Detergents (requirement, solubilized enzyme, e.g. Triton X-100 [3, 4, 6–8], CF-54, Tween 80 [2]) [2–4, 6–8]; Octylglucoside (activation) [3, 8]; Heptylthioglucoside (activation) [9]; Cardiolipin (activation) [3]; Phosphatidylglycerol (activation, detergent-free assay) [4, 8]; Endogen lipid factor (activation) [6]

Metal compounds/salts

Mn^{2+} (requirement, 2.5–10 mM [6], can replace Ni^{2+} to some extent [10]) [2, 3, 6, 9, 10]; Ni^{2+} (requirement, most active, 10–20 mM [10], can replace Mn^{2+} to some extent [3], 5% as effective as Mn^{2+} [6], not [2]) [3, 6, 10]; Co^{2+} (activation, can replace Mn^{2+} to some extent [3], 62% as effective as Mn^{2+} [6], not [10]) [3, 6]; Fe^{2+} (activation, can replace Mn^{2+} [3], Ni^{2+} to some extent, 17% as effective as Mn^{2+} [6]) [3, 6, 10]; Cd^{2+} (activation, can replace Mn^{2+} to some extent) [3]; Cu^{2+} (activation, can replace Ni^{2+} to some extent) [10]; Mg^{2+} (activation, 22% as effective as Mn^{2+} [6], not [2, 3, 9, 10]) [6]; Ca^{2+} (activation, 27% as effective as Mn^{2+} [6], not [2, 3, 9, 10]) [6]; More (no activation by K^+, Al^{3+} [2], Zn^{2+} [9, 10], Ba^{2+} [3]) [2, 3, 9, 10]

Turnover number (min^{-1})

Specific activity (U/mg)

0.000149 [10]; 0.0004–0.0005 [3]; 3.6 [6]

K_m-value (mM)

More (kinetic study) [7]; 0.007 (UDP-N-acetyl-D-galactosamine) [6]; 0.0166 (GM3) [2]; 0.017 (UDP-N-acetyl-D-galactosamine) [10]; 0.026 (UDP-N-acetyl-D-galactosamine (+ GM3)) [9]; 0.027 (GD3(N-glycolylneuraminic acid)) [6]; 0.035 (UDP-N-acetyl-D-galactosamine) [3]; 0.057 (UDP-N-acetyl-D-galactosamine) [2]; 0.082 (UDP-N-acetyl-D-galactosamine (+ SM3)) [9]; 0.1 (GM3) [3]; 0.16 (GM3(N-glycolylneuraminic acid)) [6]; 0.19 (GM3) [9]; 0.35 (GD3(N-acetylneuraminic acid)) [6]; 2.1 (GM3(N-acetylneuraminic acid)) [9]

pH-optimum

6.7–6.9 [10]; 6.8–7.2 [2]; 7.0–7.5 [9]; 7.2 [3]; 7.5–7.9 [6]

pH-range

6.2–8.5 (about half-maximal activity at pH 6.2 and about 85% of maximal activity at pH 8.5) [9]

Temperature optimum (°C)

37 (assay at) [2, 4, 6–8]

Temperature range (°C)

3 ENZYME STRUCTURE

Molecular weight
  120000 (rat, gel filtration) [10]

Subunits
  Dimer (2 × 64000, rat, SDS-PAGE [10], 2 × 65000, mouse, SDS-PAGE [6]) [6, 10]

Glycoprotein/Lipoprotein
  –

4 ISOLATION/PREPARATION

Source organism
  Rat [1–4, 7–10]; Human [5]; Mouse [6]

Source tissue
  NG 108–15 cells (cell suspension culture) [1]; Brain [1, 2, 9, 10]; Liver [3, 4, 6–8]; Melanoma cell line [5]; Neuroblastoma cell line [5]

Localization in source
  Mitochondria [2]; Microsomes [2]; Golgi apparatus (membrane bound) [3, 6–8]

Purification
  Mouse [6]; Rat (Triton X-100 solubilized, affinity chromatography on GM3-acid Sepharose) [10]

Crystallization
  –

Cloned
  (human cDNA clones by transfecting with polyoma T antigen, host recipient: mouse melanoma cell line B16) [5]

Renatured
  –

5 STABILITY

pH

Temperature (°C)

Oxidation

Organic solvent

General stability information
  Detergents stabilize [10]; Glycerol, 20% v/v, stabilizes [6]

Storage

−80°C, several months, crude Triton X-100 extract [9]; −80°C, stable in 20% v/v glycerol, 1% v/v Triton X-100 [10]; −70°C, 3 months, crude [6]; 4°C, purified enzyme, at least 6 days in 20% v/v glycerol, 1% w/v heptylthioglucoside, 1 M sucrose, 0.2 M NaCl and 1 mM EDTA in 0.03 M cacodylate buffer, pH 6.9 [6]

6 CROSSREFERENCES TO STRUCTURE DATABANKS

PIR/MIPS code

Brookhaven code

7 LITERATURE REFERENCES

[1] Scheideler, M.A., Dawson, G.: J. Neurochem.,46,1639–1643 (1986)
[2] Dicesare, J.L., Dain, J.A.: Biochim. Biophys. Acta,231,385–393 (1971)
[3] Senn, H.-J., Cooper, C., Warnke, P.C., Wagner, M., Decker, K.: Eur. J. Biochem., 120,59–67 (1981)
[4] Klein, D., Pohlentz, G., Schwartzmann, G., Sandhoff, K.: Eur. J. Biochem.,167, 417–424 (1987)
[5] Nagata, Y., Yamashiro, S., Yodoi, J., Lloyd, K.O., Shiku, H., Furukawa, K.: J. Biol. Chem.,267,12082–12089 (1992)
[6] Hashimoto, Y., Sekine, M., Iwasaki, K., Suzuki, A.: J. Biol. Chem.,268,25857–25864 (1993)
[7] Pohlentz, G., Klein, D., Schwartzmann, G., Schmitz, D., Sandhoff, K.: Proc. Natl. Acad. Sci. USA,85,7044–7048 (1988)
[8] Yusuf, H.K.M., Schwartzmann, G., Pohlentz, G., Sandhoff, K.: Biol. Chem. Hoppe-Seyler,368,455–462 (1987)
[9] Nagai, K.-I., Ishizuka, I.: J. Biochem.,101,1115–1127 (1987)
[10] Yanagisawa, K., Taniguchi, N., Makita, A.: Biochim. Biophys. Acta,919,213–220 (1987)

1 NOMENCLATURE

EC number
2.4.1.93

Systematic name
Inulin D-fructosyl-D-fructosyltransferase (1,2':2,3'-dianhydride-forming)

Recommended name
Inulin fructotransferase (depolymerizing, difructofuranose-1,2':2,3'-dianhydride-forming)

Synonyms
Inulin fructotransferase (depolymerizing)
Fructotransferase, inulin (depolymerizing)
Inulase II
Inulinase II
More (cf. EC 1.1.1.200)

CAS Reg. No.
50936-42-0

2 REACTION AND SPECIFICITY

Catalysed reaction
Removes successive terminal D-fructosyl-D-fructofuranosyl groups from inulin as the cyclic 1,2':2,3'-dianhydride, leaving a residual di- or trisaccharide

Reaction type
Hexosyl group transfer

Natural substrates
Inulin (inulin decomposition) [3, 4]

Substrate spectrum
1 Inulin (the enzyme attacks 2,1-beta-linked fructan molecules from the non-reducing fructose ends and requires the presence of at least 2 adjacent 2,1-beta-fructofuranosyl linkages [2]) [1, 2]
2 2,1-beta-Linked fructose oligosaccharides (with the exception of inulobiose) [2]
3 More (not: 2,6-beta-fructans [2], sucrose [5], 1-kestose [5], nystose [5], 1-F-fructofuranosyl-nystose [5], raffinose [5], melibiose [5], melezitose [5], stachyose [5]) [2, 5]

Product spectrum

1 Di-D-fructofuranose 1,2':2,3' dianhydride (+ oligosaccharides (small amount) [1, 2, 5], fructose-glucose oligosaccharides: O-beta-D-fructofuranosyl-2,1-O-beta-D-fructofuranosyl alpha-D-glucopyranoside (1-kestose) [2], O-beta-D-fructofuranosyl-[2,1-O-beta-D-fructofuranosyl]$_2$ alpha-D-glucopyranoside [2], O-beta-D-fructofuranosyl-[2,1-O-beta-D-fructofuranosyl]$_3$ alpha-D-glucopyranoside [2], after exhaustive digestion of inulin by this enzyme, nystose and 1F-fructofuranosyl-nystose are produced in addition to di-D-fructofuranose 1,2':2,3' dianhydride [5]) [1–5]

2 ?

3 ?

Inhibitor(s)

HgCl$_2$ [1]; CuCl$_2$ [1]; PbCl$_2$ [1]

Cofactor(s)/prosthetic group(s)/activating agents

Metal compounds/salts

Turnover number (min^{-1})

Specific activity (U/mg)

More [1]; 294 [3]; 853 [4]

K$_m$-value (mM)

0.8 (inulin) [5]

pH-optimum

5.0 [3]; 5.5 [4, 5]; 6.0 [1]

pH-range

4–8 (about 50% of activity maximum at pH 4 and 8) [5]

Temperature optimum (°C)

50 [1]; 60 [4, 5]

Temperature range (°C)

40–75 (about 50% of activity maximum at 40°C and 75°C) [5]

3 ENZYME STRUCTURE

Molecular weight

50000 (Arthrobacter globiformis, gel filtration [3], Arthrobacter ilicis, gel filtration [4]) [3, 4]
100000 (Arthrobacter sp. H65–7, gel filtration) [5]

Subunits

Monomer (1 × 45000, Arthrobacter globiformis, SDS-PAGE) [3]

Dimer (2 × 27000, Arthrobacter ilicis, SDS-PAGE [4], 2 × 49000, Arthrobacter sp. H65–7, SDS-PAGE [5]) [4, 5]

Glycoprotein/Lipoprotein

–

4 ISOLATION/PREPARATION

Source organism
Arthrobacter ilicis (OKU17B) [4]; Arthrobacter ureafaciens [1, 2]; Arthrobacter globiformis (C11–1) [3]; Arthrobacter sp. H65–7 [5]

Source tissue
Culture broth [4, 5]

Localization in source

Purification
Arthrobacter ilicis [4]; Arthrobacter ureafaciens [1]; Arthrobacter globiformis [3]; Arthrobacter sp. H65–7 [5]

Crystallization

–

Cloned

–

Renatured

–

5 STABILITY

pH
4.0–9.0 (25°C, 24 h, stable) [3]; 4–11 (stable) [1, 4]; 4.5–9.0 (stable) [5]

Temperature (°C)
50 (stable below) [1]; 70 (pH 7, 30 min, stable up to [4], stable up to [5]) [4, 5]; 75 (pH 5, 20 min, stable) [3]; 80 (pH 5, rapid inactivation) [3]

Oxidation

Organic solvent

General stability information

Storage
In refrigerator, pH 6.0, under toluene, stable for a few months [1]

6 CROSSREFERENCES TO STRUCTURE DATABANKS

PIR/MIPS code
  PIR2:PT0024 (Arthrobacter globiformis (fragment))

Brookhaven code

7 LITERATURE REFERENCES

[1] Uchiyama, T., Niwa, S., Tanaka, K.: Biochim. Biophys. Acta,315,412–420 (1973)
[2] Uchiyama, T.: Biochim. Biophys. Acta,397,153–163 (1975)
[3] Haraguchi, K., Kishimoto, M., Seki, K., Hayashi, K., Kobayashi, S., Kainuma, K.:
    Agric. Biol. Chem.,52,291–292 (1988)
[4] Kawamura, M., Takahashi, S., Uchiyama, T.: Agric. Biol. Chem.,52,3209–3210 (1988)
[5] Yokota, A., Enomoto, K., Tomita, F.: J. Ferment. Bioeng.,72,262–265 (1991)

1 NOMENCLATURE

EC number
2.4.1.94

Systematic name
UDP-N-acetyl-D-glucosamine:protein beta-N-acetyl-D-glucosaminyltrans-
ferase

Recommended name
Protein N-acetylglucosaminyltransferase

Synonyms
Acetylglucosaminyltransferase, uridine diphosphoacetylglucosamine-protein
Uridine diphospho-N-acetylglucosamine:polypeptide beta-N-acetylglu-
cosaminyltransferase
O-GlcNAc transferase [1]

CAS Reg. No.
72319-34-7

2 REACTION AND SPECIFICITY

Catalysed reaction
UDP-N-acetyl-D-glucosamine + protein →
→ UDP + 4-N-(N-acetyl-D-glucosaminyl)-protein

Reaction type
Hexosyl group transfer

Natural substrates

Substrate spectrum
1 UDP-N-acetylglucosamine + Tyr-Ser-Asp-Ser-Pro-Ser-Thr-Ser-Thr [1, 2]
2 UDP-N-acetylglucosamine + pancreatic ribonuclease A [3–5]
3 Nuclear pore protein p62 + UDP-N-acetylglucosamine [6]
4 More (no substrates: Tyr-Ser-Asp-Ser-Gly-Ser-Thr-Ser-Thr, Tyr-Ser-Asp-
Ser-Pro) [2]

Product spectrum
1 UDP + N-acetylglucosamine with O-linkage to peptide [2]
2 UDP + pancreatic rionuclease A with N-acetylglucosamine linked to
Asp-34 [3–5]
3 UDP + nuclear pore protein with O-linked N-acetylglucosamine [6]
4 ?

Inhibitor(s)
UDP [2]; EDTA (5 mM, slight) [2]; Tunicamycin (not inhibitory [2, 3]) [4];
4-Thiouridine diphosphate (photoinactivation) [1]

Cofactor(s)/prosthetic group(s)/activating agents

Metal compounds/salts
Co^{2+} (activation) [3, 5]; Mn^{2+} (activation) [3, 5]; Ca^{2+} (activation) [3]; Ni^{2+} (slight activation) [5]

Turnover number (min^{-1})

Specific activity (U/mg)
0.00111 [1]; More [3]

K_m-value (mM)
0.0046 (UDP-N-acetylglucosamine) [3]

pH-optimum
6.4 [3]

pH-range

Temperature optimum (°C)

Temperature range (°C)

3 ENZYME STRUCTURE

Molecular weight
340000 (rat, gel filtration) [1]

Subunits
Heterotrimer ($1 \times 78000 + 2 \times 110000$, rat, SDS-PAGE) [1]

Glycoprotein/Lipoprotein
–

4 ISOLATION/PREPARATION

Source organism
Rat [1–3]; Rabbit [4–6]; Saccharomyces cerevisiae [4]

Source tissue
Liver (rabbit [4]) [1–5]; Reticulocytes (commercial product) [6]

Localization in source
Cytosol (85% of overall activity [2]) [1, 2]; Membrane-bound (15% of overall activity [2]) [2–4]; Rough endoplasmic reticulum [5]

Purification
    Rat (partial [3]) [1, 3]

Crystallization
    –

Cloned
    –

Renatured
    –

---

5 STABILITY

pH

Temperature (°C)

Oxidation

Organic solvent

General stability information

Storage
    –20°C, 20 mM Tris/HCl buffer, pH 7.5, 40% ethylene glycol plus cytochrome
    c or bovine serum albumin [1]

---

6 CROSSREFERENCES TO STRUCTURE DATABANKS

PIR/MIPS code

Brookhaven code

---

7 LITERATURE REFERENCES

[1] Haltiwanger, R.S., Blomberg, M.A., Hart, G.W.: J. Biol. Chem.,267,9005–9013 (1992)
[2] Haltiwanger, R.S., Holt, G.D., Hart, G.W.: J. Biol. Chem.,265,2563–2568 (1990)
[3] Arakawa, H., Mookerjea, S.: Eur. J. Biochem.,140,297–302 (1984)
[4] Khalkhali, Z., Marshall, R.D., Reuvers, F., Habets-Willems, C., Boer, P.: Biochem. J.,
    160,37–41 (1976)
[5] Khalkhali, Z., Marshall, R.D.: Biochem. J.,146,299–307 (1975)
[6] Starr, C.M., Hanover, J.A.: J. Biol. Chem.,265,6868–6873 (1990)

.

1 NOMENCLATURE

EC number
2.4.1.95

Systematic name
Bilirubin-glucuronoside:bilirubin-glucuronoside D-glucuronosyltransferase

Recommended name
Bilirubin-glucuronoside glucuronosyltransferase

Synonyms
Glucuronosyltransferase, bilirubin glucuronoside
Bilirubin glucuronoside glucuronosyltransferase
Bilirubin monoglucuronide transglucuronidase

CAS Reg. No.
71822-22-5

2 REACTION AND SPECIFICITY

Catalysed reaction
2 Bilirubin-glucuronoside →
→ bilirubin + bilirubin-bisglucuronoside

Reaction type
Hexosyl group transfer
Transglucuronidation

Natural substrates

Substrate spectrum
1 Bilirubin monoglucuronide + bilirubin monoglucuronide [1–3]

Product spectrum
1 Bilirubin diglucuronide + bilirubin [1]

Inhibitor(s)

Cofactor(s)/prosthetic group(s)/activating agents

Metal compounds/salts

Turnover number (min^{-1})

Specific activity (U/mg)
More [2, 3]

K_m-value (mM)
    0.032–0.034 (bilirubin monoglucuronide) [2, 3]

pH-optimum
    6.6 [1]

pH-range

Temperature optimum (°C)

Temperature range (°C)

3 ENZYME STRUCTURE

Molecular weight
    160000 (rat, gel filtration) [2, 3]

Subunits
    Oligomer (x × 28000, rat, SDS-PAGE with and without 2-mercaptoethanol, subunits not linked by disulfide bonds) [3]

Glycoprotein/Lipoprotein
    –

4 ISOLATION/PREPARATION

Source organism
    Rat (Wistar) [1–3]

Source tissue
    Liver [1–3]

Localization in source
    Plasma membrane [1–3]; Microsomes [2]

Purification
    Rat [2, 3]

Crystallization
    –

Cloned
    –

Renatured
    –

5 STABILITY

pH

Temperature (°C)

Oxidation

Organic solvent

General stability information

Storage

6 CROSSREFERENCES TO STRUCTURE DATABANKS

PIR/MIPS code

Brookhaven code

7 LITERATURE REFERENCES

[1] Jansen, P.L.M., Chowdhury, J.R., Fischberg, E.B., Arias, I.M.: J. Biol. Chem.,252, 2710–2716 (1977)
[2] Chowdhury, J.R., Arias, I.M.: Methods Enzymol.,77,192–197 (1981)
[3] Chowdhury, J.R, Chowdhury, N.R., Bhargava, M.M., Arias, I.M.: J. Biol. Chem.,254, 8336–8339 (1979)

1 NOMENCLATURE

EC number
2.4.1.96

Systematic name
UDPgalactose:sn-glycerol-3-phosphate 1-alpha-D-galactosyltransferase

Recommended name
sn-Glycerol-3-phosphate 1-galactosyltransferase

Synonyms
Isofloridoside-phosphate synthase
UDP-Gal:sn-glycero-3-phosphoric acid 1-alpha-galactosyl-transferase
UDPgalactose:sn-glycerol-3-phosphate alpha-D-galactosyltransferase [3]
Galactosyltransferse, uridine diphosphogalactose-glycerol phosphate
Galactosyltransferase, glycerol 3-phosphate 1alpha-
More (cf. EC 2.4.1.137)

CAS Reg. No.
9076-70-4

---

2 REACTION AND SPECIFICITY

Catalysed reaction
UDPgalactose + sn-glycerol 3-phosphate →
→ UDP + alpha-D-galactosyl-(1,1')-sn-glycerol 3-phosphate

Reaction type
Hexosyl group transfer

Natural substrates

Substrate spectrum
1 UDPgalactose + sn-glycerol 3-phosphate [1–5]

Product spectrum
1 UDP + alpha-D-galactosyl-sn-glycerol 3-phosphate (the product is hydrolyzed by a phosphatase to isofloridoside involved in osmoregulation [1, 2,
4, 5]) [1–5]

Inhibitor(s)

Cofactor(s)/prosthetic group(s)/activating agents
More (enzyme appears to exist as an inactive proenzyme which can be activated by incubation of crude cell extracts with endogenous or exogenous
proteases) [5]

Metal compounds/salts
More (no requirement for Ca^{2+}, Mn^{2+} or Mg^{2+}) [3]

Turnover number (min^{-1})

Specific activity (U/mg)
198 [1]

K_m-value (mM)
0.01 (UDPgalactose) [3]; 0.02 (sn-glycerol 3-phosphate) [3]

pH-optimum
7–8 [3]

pH-range
6–9 (about 50% of activity maximum at pH 6 and 9) [3]

Temperature optimum (°C)
25 (assay at) [1]

Temperature range (°C)

3 ENZYME STRUCTURE

Molecular weight
68000 (Poterioochromonas malhamensis, gel filtration) [3]

Subunits
Monomer (1 × 70000, Poterioochromonas malhamensis, SDS-PAGE) [3]

Glycoprotein/Lipoprotein
–

4 ISOLATION/PREPARATION

Source organism
Ochromonas malhamensis [1, 2, 5]; Poterioochromonas malhamensis
(Peterfi [4]) [3, 4]

Source tissue

Localization in source

Purification
Poterioochromonas malhamensis [3]

Crystallization
–

Cloned

–

Renatured

–

---

5 STABILITY

pH

Temperature (°C)

Oxidation

Organic solvent

General stability information
Unstable to repeated freezing and thawing [3]; Cetyltrimethylammonium bromide, optimal concentration for stabilization, 0.075% w/v at pH 8.5, 0.1% w/v at pH 7.2, 0.2 w/v at pH 6.2 [3]; Glycerol, 30% v/v, stabilizes the enzyme at pH 7.8 but not at lower or higher pH values [3]

Storage

---

6 CROSSREFERENCES TO STRUCTURE DATABANKS

PIR/MIPS code

Brookhaven code

---

7 LITERATURE REFERENCES

[1] Kauss, H., Schobert, B.: FEBS Lett.,19,131–135 (1971)
[2] Kauss, H., Quader, H.: Plant Physiol.,58,295–298 (1976)
[3] Thomson, K.-S.: Biochim. Biophys. Acta,759,154–159 (1983)
[4] Kauss, H., Thomson, K.S., Thomson, M., Jeblick, W.: Plant Physiol.,63,455–459 (1979)
[5] Kauss, H., Thomson, K.S., Tetour, M., Jeblick, W.: Plant Physiol.,61,35–37 (1978)

1 NOMENCLATURE

EC number
   2.4.1.97

Systematic name
   1,3-beta-D-Glucan:orthophosphate alpha-D-glucosyltransferase

Recommended name
   1,3-beta-D-Glucan phosphorylase

Synonyms
   Laminarin phosphorylase
   Phosphorylase, 1,3-beta-glucan
   1,3-beta-D-Glucan:orthophosphate glucosyltransferase [2]
   Laminarin phosphoryltransferase
   More (different from EC 2.4.1.30 and EC 2.4.1.31)

CAS Reg. No.
   37340-31-1

---

2 REACTION AND SPECIFICITY

Catalysed reaction
   (1,3-beta-D-Glucosyl)$_n$ + phosphate $\rightarrow$
   $\rightarrow$ (1,3-beta-D-glucosyl)$_{n-1}$ + alpha-D-glucose 1-phosphate

Reaction type
   Hexosyl group transfer

Natural substrates
   (1,3-beta-D-Glucosyl)$_n$ + phosphate (enzyme may be a key factor in the
   regulation of cell-wall extension and build up by switching hydrolase and
   synthase activity in a balance dependent manner) [1]

Substrate spectrum
   1 Laminarin + phosphate (r [2], i.e. oligosaccharide with beta-1,3-glucosi-
      dic linkages [1, 2], reaction leads to oligosaccharide synthesis in experi-
      ments performed in imidazole buffer containing 5 mM cysteine, it lies in
      the direction of phosphorolytic activity in citrate-phosphate buffer [2]) [1, 2]
   2 Laminaribiose + phosphate [2]
   3 Laminaribiose + glucose 1-phosphate [2]
   4 Laminarin + glucose 1-phosphate (specific for glucose 1-phosphate) [1]

Product spectrum
1 (1,3-beta-D-Glucosyl)$_{n-1}$ + alpha-D-glucose 1-phosphate [1, 2]
2 Glucose + glucose 1-phosphate [2]
3 ?
4 ?

Inhibitor(s)
AMP (low concentration: stimulation, K_m: 0.04 mM, high concentration: inhibition) [1]

Cofactor(s)/prosthetic group(s)/activating agents
AMP (low concentration: stimulation, K_m: 0.04 mM, high concentration: inhibition) [1]

Metal compounds/salts

Turnover number (min^{-1})

Specific activity (U/mg)
More [1]

K_m-value (mM)
0.25 (glucose (+ glucose 1-phosphate)) [2]; 0.34 (laminaribiose (+ phosphate)) [2]; 0.47 (laminarin (+ phosphate)) [2]; 1 (laminarin) [1]; 1.45 (laminaribiose (+ glucose 1-phosphate)) [2]; 1.7 (laminarin (+ glucose 1-phosphate)) [2]; 12 (glucose 1-phosphate) [1]; 25 (phosphate) [1]

pH-optimum
5.5 [1]; 6.5 (assay at) [2]

pH-range
5.0–5.5 (5.0: 50% of activity maximum, 5.5: activity maximum) [1]

Temperature optimum (°C)
22.5 [1]; 40 (assay at) [2]

Temperature range (°C)
16–26.5 (about 50% of activity maximum at 16°C and 26.5°C) [1]

3 ENZYME STRUCTURE

Molecular weight

Subunits

Glycoprotein/Lipoprotein
–

4 ISOLATION/PREPARATION

Source organism
Ochromonas malhamensis [1]; Acacia verek [2]

Source tissue
Cells [1]; Cultured cells [2]

Localization in source
Cell wall (bound) [2]

Purification
Ochromonas malhamensis [1]

Crystallization
[1]

Cloned
–

Renatured
–

5 STABILITY

pH

Temperature (°C)

Oxidation

Organic solvent

General stability information
Freezing and thawing, loss of activity [1]

Storage
0–4°C, stable for several weeks [1]

6 CROSSREFERENCES TO STRUCTURE DATABANKS

PIR/MIPS code

Brookhaven code

7 LITERATURE REFERENCES

[1] Albrecht, G.J., Kauss, H.: Phytochemistry,10,1293–1298 (1971)
[2] Lienart, Y., Comtat, J., Barnoud, F.: Plant Sci.,58,165–170 (1988)

1 NOMENCLATURE

EC number
2.4.1.99

Systematic name
Sucrose:sucrose 1^F-beta-D-fructosyltransferase

Recommended name
Sucrose 1^F-fructosyltransferase

Synonyms
SST [3]
Sucrose:sucrose 1-fructosyltransferase [3]
Fructosyltransferase, sucrose 1^F-
Sucrose-sucrose 1-fructosyltransferase
Sucrose:sucrose fructosyltransferase

CAS Reg. No.
73379-56-3

2 REACTION AND SPECIFICITY

Catalysed reaction
Sucrose + sucrose →
→ D-glucose + 1^F-beta-D-fructosylsucrose

Reaction type
Hexosyl group transfer
Transfructosylation [4]

Natural substrates
Sucrose + sucrose (first enzyme in biosynthetic pathway of most fructans [6], initiates fructan synthesis [1], key enzyme in inulin metabolism [4]) [1, 4, 6]

Substrate spectrum
1 Sucrose + sucrose (r [3]) [1–7]
2 Sucrose + raffinose [5]
3 Sucrose + stachyose [5]
4 More (fructosyltransfer from sucrose to the fructose moiety of neokestose and its homologous $6^G(1$-beta-fructofuranosyl$)_n$-sucrose, not: fructosyl-transfer to $1^F(1$-beta-fructofuranosyl$)_n$-sucrose and $1^F(1$-beta-fructofura-nosyl$)_m$-$6^G(1$-fructofuranosyl$)_n$-sucrose (except for m is 0)) [3]

Product spectrum
1 D-Glucose + 1^F-beta-D-fructosylsucrose (i.e. 1-kestose) [1–7]
2 ?
3 ?
4 ?

Inhibitor(s)
Cu^{2+} [4, 6]; Co^{2+} [4]; Mn^{2+} [3–5]; KCl (no inhibition of enzymes from Allium cepa bulbs and Asparagus officinalis) [6]; Hg^{2+} [3–5]; High ionic strength [6]; p-Chloromercuribenzoate [3–5]; Ag^+ [3, 5]; $AgNO_3$ [4]; Zn^{2+} [6]

Cofactor(s)/prosthetic group(s)/activating agents

Metal compounds/salts

Turnover number (min^{-1})

Specific activity (U/mg)
2.97 [3]; 0.126 [4]; More [5]

K_m-value (mM)
42 (sucrose) [4]; 83 (sucrose) [5]; 110 (sucrose) [3]; 120 (sucrose) [2]; 230 (sucrose) [1]; 266 (sucrose) [6]

pH-optimum
5.0 [3]; 5.2 (phosphate-citrate buffer) [1]; 5.4 [4, 5]; 5.5 [2, 6]

pH-range
3–7 (less than 10% of activity maximum at pH 3 and 7) [1]; 3.5–8.0 (3.5: about 70% of activity maximum, 8.0: about 50% of activity maximum) [1]

Temperature optimum (°C)
30 (assay at) [3]; 34 [4]

Temperature range (°C)

3 ENZYME STRUCTURE

Molecular weight
65000 (Asparagus officinalis, gel filtration) [3]
67000 (Helianthus tuberosus, gel filtration) [4]
68000 (Lolium rigidum [2], Allium cepa, gel filtration [5]) [2, 5]
69000 (Allium cepa, gel filtration) [1]

Subunits

Glycoprotein/Lipoprotein
–

4 ISOLATION/PREPARATION

Source organism
    Dactylis glomerata (orchard grass) [6]; Triticum aestivum [6]; Hordeum vulgare [6]; Lolium rigidum (Gaudin) [2]; Allium cepa (onion) [1, 5]; Asparagus officinalis [3, 5]; Helianthus tuberosus [4]; Chiocrium intybus (chicory) [7]

Source tissue
    Bulbs [6]; Inner leaf bases [1]; Root [3]; Tuber [4]; Seed [5]; Stem [6]; Leaf [6]

Localization in source

Purification
    Allium cepa [1, 5]; Hordeum vulgare [6]; Lolium rigidum (partial) [2]; Asparagus officinalis [3]

Crystallization
–

Cloned
–

Renatured
–

5 STABILITY

pH
    4.0–8.0 (30°C, 30 min, stable) [3]; 4.0 (10 min, 45°C, 63% loss of activity) [5]; 4.6 (10 min, 45°C, 48% loss of activity) [5]; 5.4 (10 min, 45°C, 48% loss of activity) [5]; 6.0 (10 min, 45°C, 65% loss of activity) [5]; 7.0 (10 min, 45°C, 95% loss of activity) [5]

Temperature (°C)
    20–37 (10 min, stable) [5]; 30 (30 min, pH 4.0–8.0, stable) [3]; 45 (30 min, pH 5.0–6.5, 50% loss of activity) [3]; 50 (10 min, about 20% loss of activity) [3]; 50–60 (10 min, inactivation) [5]; 60 (10 min, 80% loss of activity) [3]

Oxidation

Organic solvent

General stability information

Storage

6 CROSSREFERENCES TO STRUCTURE DATABANKS

PIR/MIPS code

Brookhaven code

7 LITERATURE REFERENCES

[1] Henry, R.J., Darbyshire, B.: Phytochemistry,19,1017–1020 (1980)
[2] St. John, J.A., Bonnett, G.D., Simpson, R.J.: New Phytol.,123,705–715 (1993)
[3] Shiomi, N., Izawa, M.: Agric. Biol. Chem.,44,603–614 (1980)
[4] Praznik, W., Beck, R.H.F., Spies, T.: Agric. Biol. Chem.,54,2429–2431 (1990)
[5] Shiomi, N., Kido, H., Kiriyama, S.: Phytochemistry,24,695–698 (1985)
[6] Chevalier, P.M., Rupp, R.A.: Plant Physiol.,101,589–594 (1993)
[7] Singh, R., Bhatia, I.S.: Phytochemistry,10,2037–2039 (1971)

1 NOMENCLATURE

EC number
2.4.1.100

Systematic name
1,2-beta-D-Fructan:1,2-beta-D-fructan 1^F-beta-D-fructosyltransferase

Recommended name
1,2-beta-Fructan 1^F-fructosyltransferase

Synonyms
Fructosyltransferase, 1,2-beta-D-fructan 1^F-
Fructan:fructan fructosyl transferase [3]
FFT [3]

CAS Reg. No.
73379-55-2

---

2 REACTION AND SPECIFICITY

Catalysed reaction
$(1,2\text{-beta-D-Fructosyl})_m + (1,2\text{-beta-D-fructosyl})_n \rightarrow$
$\rightarrow (1,2\text{-beta-D-fructosyl})_{m-1} + (1,2\text{-beta-D-fructosyl})_{n+1}$

Reaction type
Hexosyl group transfer

Natural substrates
$(1,2\text{-beta-D-Fructosyl})_m + (1,2\text{-beta-D-fructosyl})_n$ (role of enzyme in synthesis of fructan) [4]

Substrate spectrum
1 $(1,2\text{-beta-D-Fructosyl})_m + (1,2\text{-beta-D-fructosyl})_n$ (active on different oligofructans of the inulin series [3], 1-kestose-dependent nystose production [3], transfers fructosyl groups from oligofructans (degree of polymerization: 3–8) of the inulin series [2], 1-kestose is an efficient donor of fructosyl units to sucrose [2], enzyme is specific for fructosyl transfer from beta-2,1-linked 1-kestose or fructan to sucrose and beta-2,1-fructosyl transfer to other fructans [4]) [1–4]
2 6-Kestose + 1-kestose + sucrose [1]
3 1-Kestose + 6-kestose [4]
4 1-Kestose + 1-kestose [4]
5 Sucrose + 1-kestose [4]

---

6 More (no glycosyl transfer with 6-kestose [2, 3], neokestose [2, 3], maltose [2], raffinose [2], maltotriose [2], melezitose [3], maltopentaose [3] or sucrose [3] as the sole substrate) [2, 3]

Product spectrum
1 $(1,2\text{-beta-D-Fructosyl})_{m-1} + (1,2\text{-beta-D-fructosyl})_{n+1}$ (the enzyme produces tetrasaccharides and higher polymers from trisaccharide [1])
2 Mixture of tetrasachharides of beta-2,6- and beta-2,1-linked fructans [1]
3 1,6-Nystose + 6,1-nystose [4]
4 1,1-Nystose + 1,1,1-logose (inulin-type tetra- and pentasaccharides) [4]
5 1,1-Nystose [4]
6 ?

Inhibitor(s)

Cofactor(s)/prosthetic group(s)/activating agents

Metal compounds/salts

Turnover number (min^{-1})

Specific activity (U/mg)

K_m-value (mM)
0.2 (sucrose) [2]

pH-optimum
6.5 [2–4]

pH-range
5–8 (5: about 70% of activity maximum, 8: about 65% of activity maximum) [4]

Temperature optimum (°C)
30 [4]

Temperature range (°C)

---

3 ENZYME STRUCTURE

Molecular weight
72800 (Jerusalem artichoke, preparative gel electrophoresis) [2]

Subunits
? ($x \times 49000$, Taraxacum officinale, SDS-PAGE) [3]

Glycoprotein/Lipoprotein
Glycoprotein [3]

4 ISOLATION/PREPARATION

Source organism
Allium cepa [1]; Jerusalem artichoke [2]; Taraxacum officinale (Weber) [3];
Triticum aestivum [4]

Source tissue
Inner leaf bases [1]; Tuber [2]

Localization in source

Purification
Allium cepa [1]; Taraxacum officinale (Weber) [3]; Triticum aestivum [4]

Crystallization
–

Cloned
–

Renatured
–

5 STABILITY

pH

Temperature (°C)
30 (1 h, stable) [3]; 40 (1 h, up to 20% loss of activity at temperatures up to
40°C) [2]

Oxidation

Organic solvent

General stability information

Storage

6 CROSSREFERENCES TO STRUCTURE DATABANKS

PIR/MIPS code

Brookhaven code

7 LITERATURE REFERENCES

[1] Henry, R.J., Darbyshire, B.: Phytochemistry,19,1017–1020 (1980)
[2] Luescher, M., Frehner, M., Noesberger, J.: New Phytol.,123,717–724 (1993)
[3] Luescher, M., Frehner, M., Noesberger, J.: New Phytol.,123,437–442 (1993)
[4] Jeong, B.-R., Housley, T.L.: Plant Physiol.,100,199–204 (1992)

1 NOMENCLATURE

EC number
2.4.1.101

Systematic name
UDP-N-acetyl-D-glucosamine:glycoprotein (N-acetyl-D-glucosamine to alpha-D-mannosyl-1,3-(R$_1$)-beta-D-mannosyl-R$_2$) beta-1,2-N-acetyl-D-glucosaminyl-transferase

Recommended name
alpha-1,3-Mannosyl-glycoprotein beta-1,2-N-acetylglucosaminyltransferase

Synonyms
N-Glycosyl-oligosaccharide-glycoprotein N-acetylglucosaminyltransferase I
N-Acetylglucosaminyltransferase I
Acetylglucosaminyltransferase, uridine diphosphoacetylglucosamine-alpha-1,3-mannosylglycoprotein beta-1,2-N-
UDP-N-acetylglucosaminyl: alpha-1,3-D-mannoside-beta-1,2-N-acetylglu-cosaminyltransferase I
UDP-N-acetylglucosaminyl:alpha-3-D-mannoside beta-1,2-N-acetyl-glucosaminyltransferase I
More (cf. EC 2.4.1.143, EC 2.4.1.144, EC 2.4.1.145 and EC 2.4.1.155)

CAS Reg. No.
102576-81-8

2 REACTION AND SPECIFICITY

Catalysed reaction
UDP-N-acetyl-D-glucosamine + alpha-D-mannosyl-1,3-(R$_1$)-beta-D-man-nosyl-R$_2$ →
→ UDP + N-acetyl-beta-D-glucosaminyl-1,2-alpha-D-mannosyl-1,3-(R$_1$)-beta-D-mannosyl-R$_2$ (mechanism [10])

Reaction type
Hexosyl group transfer

Natural substrates
UDP-N-acetyl-D-glucosamine + alpha-D-mannosyl-1,6-(alpha-D-man-nosyl-1,3)-alpha-D-mannosyl-1,6-(alpha-D-mannosyl-1,3)-beta-D-man-nosyl-1,4-N-acetyl-D-glucosaminyl-1,4-N-acetyl-D-glucosaminyl-Asn-peptide (key enzyme of biosynthesis of complex and hybrid N-glycans [1, 10], stimulates galactosyltransferase of bovine milk [8]) [1, 8, 10]
UDP-N-acetyl-D-glucosamine + ovalbumin glycopeptide V (physiological substrate) [4]

Substrate spectrum

1 UDP-N-acetyl-D-glucosamine + alpha-D-mannosyl-1,6-(alpha-D-man-
nosyl-1,3)-alpha-D-mannosyl-1,6-(alpha-D-mannosyl-1,3)-beta-D-man-
nosyl-1,4-N-acetyl-D-glucosaminyl-R (R: H [2, 4, 6], 1,4-N-ace-
tyl-D-glucosaminyl-Asn [1, 4, 5, 8, 10], (fucose-1,6)-1,4-N-ace-
tyl-D-glucosaminyl-Asn [4], 1,4-N-acetyl-D-glucosaminyl-Asn-peptide [4] or
pyridinylamine [9], transfers an N-acetylglucosamine in beta-1,2-linkage
to mannosyl-1,3-beta-mannosyl-terminus [1], can act on alpha-man-
nosyl-1,6-beta-mannosyl-terminus if alpha-mannosyl-1,3-beta-man-
nosyl-terminus is not available (appreciable higher K_m-value) [4]. Essential for activity: unsubstituted equatorial hydroxyl on C-4 of beta-linked
mannosyl-residue [10], substrate specificity [1, 10], acceptor substrates
are dehexoso-orosomucoid [2], ovalbumin [5], beta-galactosidase- and
beta-N-acetylhexosamidase-treated asialofetuin or asialotransferrin [5],
desialo-degalacto-dehexosamine-orosomucoid [8], free and protein-matrix-bound glycans [7], does not act on alpha-mannosyl-1,2-alpha-man-
nosyl-, alpha-mannosyl-1,3-alpha-mannosyl-, alpha-mannosyl-1,6-alpha-
mannosyl-termini [4]. No acceptor substrates are N-acetyl-D-glucosamine,
N-acetyl-D-glucosaminyl-1,2-alpha-D-mannosyl-1,3-(N-acetyl-D-
glucosaminyl-1,2-alpha-D-mannosyl-1,6)-beta-D-mannosyl-1,4-N-ace-
tyl-D-glucosaminyl-1,4-N-acetyl-D-glucosaminyl-Asn [10], ovalbumin
glycopeptides III A-C and IV [4], asialofetuin, beta-galactosidase-,
beta-N-acetylhexosamidase- and alpha-mannosidase-treated asialofetuin,
dolichol phosphate, asialo-bovine submaxillary mucin or galactosylated
asialo-bovine submaxillary mucin [5], synthetic substrate analogues [10])
[1–12]

2 UDP-N-acetyl-D-glucosamine + alpha-D-mannosyl-1,3-(alpha-D-man-
nosyl-1,6)-beta-D-mannosyl-1,4-N-acetyl-D-glucosaminyl-R (R: H [2, 4, 6,
10], pyridylamine [9], 1,4-N-acetyl-D-glucosaminyl-R [12], or 1,4-N-ace-
tyl-D-glucosaminyl-Asn [10], synthetic-beta-D-mannosyl-(1,6-anhydro)-de-
rivative acts as substrate, too [10]) [2, 4, 6, 9, 10, 12]

3 UDP-N-acetyl-D-glucosamine + alpha-D-mannosyl-1,3-beta-D-man-
nosyl-1,4-N-acetyl-D-glucosamine [2, 6]

4 UDP-N-acetyl-D-glucosamine + N-acetyl-D-glucosaminyl-1,2-alpha-D-man-
nosyl-1,3-(alpha-D-mannosyl-1,6)-beta-D-mannosyl-1,4-N-acetylglucosa-
mine-1,4-(fucose-1,6)-N-acetyl-glucosamine-Asn-peptide (poor substrate,
not [2, 6]) [4]

5 UDP-N-acetyl-D-glucosamine + ovalbumin glycopeptide V (best substrate) [4, 5]

Product spectrum

1 UDP + alpha-D-mannosyl-1,6-(alpha-D-mannosyl-1,3)-alpha-D-man-
nosyl-1,6-(N-acetyl-D-glucosaminyl-1,2-alpha-D-mannosyl-1,3)-beta-D-man-
nosyl-1,4-N-acetyl-D-glucosaminyl-R (R: H [2, 4, 6], 1,4-N-ace-
tyl-D-glucosaminyl-Asn [1, 4, 5, 8, 10], or (fucose-1,6)-1,4-N-ace-
tyl-D-glucosaminyl-Asn [4] or pyridinylamine [9]) [1, 2, 4–6, 8–10, 12]

2 UDP + N-acetyl-D-glucosaminyl-1,2-alpha-D-mannosyl-1,3-(alpha-D-man-
nosyl-1,6)-beta-D-mannosyl-1,4-N-acetyl-D-glucosaminyl-R (R: H [2, 10], R:
pyridylamine [9]) [2, 9, 10, 12]
3 UDP + N-acetyl-D-glucosaminyl-1,2-alpha-D-mannosyl-1,3-beta-D-man-
nosyl-1,4-N-acetyl-D-glucosamine [2]
4 UDP + N-acetyl-D-glucosaminyl-1,2-alpha-D-mannosyl-1,3-(N-ace-
tyl-D-glucosaminyl-1,2-alpha-D-mannosyl-1,6)-beta-D-mannosyl-1,4-N-ace-
tylglucosamine-1,4-(fucose-1,6)-N-acetyl-glucosamine-Asn-peptide [4]
5 UDP + ovalbumin glycopeptide IIIA [4]

Inhibitor(s)
alpha-D-Mannosyl-1,6-(N-acetyl-D-glucosaminyl-1,2-alpha-D-man-
nosyl-1,3)-beta-D-mannosyl-1,4-N-acetyl-D-glucosaminyl-R [10];
alpha-D-Mannosyl-1,3-(alpha-D-mannosyl-1,6)-4-O-methyl-beta-D-man-
nosyl-1,4-N-acetyl-D-glucosamine (weak) [1]; EDTA (3 mM) [2, 4]; High ionic
strength (e.g. 0.1 M NaCl) [2]; UDP [2, 10]; Antibodies to pig liver N-acetylglucosaminyltransferase I [6]; UDPglucose (weak) [10]; UDP-N-acetylgalactosamine (weak) [10]; 5-Mercury-UDP [10]; 5-Bromo-UTP [10]; TDP
[10]; 2'-Deoxy-UDP (weak) [10]; 2'-O-Methyl-UDP (weak) [10]; UTP (weak)
[10]; UMP (weak) [10]; More (no inhibitors are CDP, UDPhexanolamine,
AMP) [10]

Cofactor(s)/prosthetic group(s)/activating agents
Triton X-100 (slight activation, rabbit [12]) [5, 10, 12]; More (no activation by
2-mercaptoethanol, rabbit) [10, 12]

Metal compounds/salts
Mn^{2+} (requirement [2–10, 12], 18–26 mM [4], 2 mM [5], 25–100 mM, rabbit
[12]) [2–10, 12]; Co^{2+} (activation [2, 10, 12], can substitute for Mn^{2+} with
80% [10] or 91% [2] efficiency) [2, 10, 12]; Mg^{2+} (activation [2, 10, 12], can
substitute for Mn^{2+} with 40% [10] or 61% [2] efficiency) [2, 10, 12]; Cd^{2+} (activation [2, 10, 12], can substitute for Mn^{2+} with 20% [10] or 78% [2] efficiency) [2, 10, 12]; Ni^{2+} (activation [2, 10, 12], can substitute for Mn^{2+} with 20%
[10] or 34% [2] efficiency) [2, 10, 12]; Zn^{2+} (activation [2, 10, 12], can substitute for Mn^{2+} with 8% [10] or 19% [2] efficiency) [2, 10, 12]; Fe^{2+} (activation [2, 10, 12], can substitute for Mn^{2+} with 65% efficiency [10], slight [2])
[2, 10, 12]; Ca^{2+} (slight activation) [2, 10, 12]; Ba^{2+} (slight activation [10, 12],
not [2]) [10, 12]; Sn^{2+} (slight activation) [10, 12]; More (no activation by Hg^{2+}
[2], Sr^{2+}, Pb^{2+} [10, 12] or Cu^{2+} [2, 10, 12]) [2, 10, 12]

Turnover number (min^{-1})

Specific activity (U/mg)
0.000015 (transferase A, liver) [5]; 0.00065 (transferase B, hepatoma) [5];
0.079 (alpha-D-mannosyl-1,3-(alpha-D-mannosyl-1,6)-beta-D-man-
nosyl-1,4-N-acetyl-D-glucosaminyl-R immunoglobulin G glycopeptide) [4];

0.106 (alpha$_1$-acid glycoprotein treated with mild acid, beta-galactosidase
and beta-N-acetylglucosamidase) [4]; 2.51 [2]; 4.8 (liver) [6]; 19.8 (low-molecular weight form) [10]

K_m-value (mM)
More (kinetic data of free and protein-matrix-bound acceptor substrates [7],
kinetic study [10]) [7, 10]; 0.0384 (UDP-N-acetylglucosamine (+ alpha-D-
mannosyl-1,3-(alpha-D-mannosyl-1,6)-beta-D-mannosyl-1,4-N-acetyl-D-
glucosaminyl-R)) [10]; 0.078 (UDP-N-acetylglucosamine (+ alpha-D-man-
nosyl-1,3-(alpha-D-mannosyl-1,6)-alpha-D-mannosyl-1,3-(alpha-D-man-
nosyl-1,6)-beta-D-mannosyl-1,4-N-acetyl-D-glucosamine), rabbit) [12]; 0.1
(UDP-N-acetylglucosamine) [4]; 0.12 (ovalbumin glycopeptide V) [4]; 0.2
(alpha-D-mannosyl-1,3-(alpha-D-mannosyl-1,6)-beta-D-mannosyl-1,4-N-ace-
tyl-D-glucosaminyl-R, immunoglobulin G glycopeptide) [4]; 0.25 (alpha-D-
mannosyl-1,3-(alpha-D-mannosyl-1,6)-alpha-D-mannosyl-1,3-(alpha-D-man-
nosyl-1,6)-beta-D-mannosyl-1,4-N-acetyl-D-glucosaminyl-1,4-N-acetylglu-
cosamine, rabbit) [12]; 0.33 (glycopeptide V, transferase A) [5]; 0.39
(alpha-D-mannosyl-1,6-(alpha-D-mannosyl-1,3)-alpha-D-mannosyl-1,6-
(alpha-D-mannosyl-1,3)-beta-D-mannosyl-1,4-N-acetyl-D-glucosamine) [6];
0.43 (N-acetyl-D-glucosaminyl-1,2-alpha-D-mannosyl-1,3-(alpha-D-man-
nosyl-1,6)-beta-D-mannosyl-1,4-N-acetyl-D-glucosaminyl-pyridinylamine) [9];
0.44 (ovalbumin, transferase A) [5]; 0.45 (alpha-D-mannosyl-1,6-(alpha-
D-mannosyl-1,3)-alpha-D-mannosyl-1,6-(alpha-D-mannosyl-1,3)-beta-D-man-
nosyl-1,4-N-acetyl-D-glucosamine) [2]; 2.03 (alpha-D-mannosyl-1,3-
(alpha-D-mannosyl-1,6)-beta-D-mannosyl-1,4-N-acetyl-D-glucosamine) [10];
4.5 (ovalbumin, transferase B) [5]; 7.4 (alpha-mannosyl-1,3-beta-man-
nosyl-1,4-N-acetylglucosamine) [4]; 10 (N-acetyl-D-glucosaminyl-1,2-
alpha-D-mannosyl-1,3-(alpha-D-mannosyl-1,6)-beta-D-mannosyl-1,4-N-ace-
tyl-D-glucosaminyl-R, immunoglobulin G glycopeptide) [4]

pH-optimum
5.6 (rabbit) [10, 12]; 6 [4, 9]; 6.3 [2]; 7–7.3 [5]

pH-range
5.1–6.9 (about half-maximal activity at pH 5.1 and about 70% of maximal activity at pH 6.9) [10]; 5.8–6.7 [2]

Temperature optimum (°C)
37 (assay at) [2, 4–10]

Temperature range (°C)

---

3 ENZYME STRUCTURE

Molecular weight
More (two MW-species of pig liver enzyme: major MW 52000 and minor MW
59000, SDS-PAGE [6], rabbit: two MW-species [2, 10, 12], separable by gel
filtration [10, 12]) [2, 6, 10, 12]

Subunits

? (x × 58000 + x × 46000, rabbit, SDS-PAGE [2], x × 45000 + x × 50000 +
x × 54000, rabbit low molecular weight form, SDS-PAGE, major band: MW
45000 [10, 12]) [2, 10, 12]

Glycoprotein/Lipoprotein

–

---

4 ISOLATION/PREPARATION

Source organism

Rat (male Donryu [5]) [1, 5, 7]; Rabbit [2, 10, 12]; Pig [3, 6, 12]; Bovine [4,
8, 12]; Human [11]; Chicken (hen) [12]; Hamster [12]; Acer pseudoplatanus
(sycamore) [9]

Source tissue

Liver (rabbit [12]) [1, 2, 5–7, 10, 12]; Trachea mucosa (pig) [3, 6, 12]; Colostrum (bovine) [4, 8, 12]; Hepatoma (diethylnitrosamine- or dimethyl-
aminoazobenzene-induced hepatoma, Morris 5123D hepatoma or AH-109A
(solid or ascitic)) [5]; Oviduct (hen) [12]; Cell suspension culture [9]

Localization in source

Membrane-bound (transferase A [5], not [4, 8]) [5, 7, 9–11]; Golgi apparatus
(resident type II transmembrane protein [11]) [7, 9, 11]; Microsomes [3, 6];
Soluble (transferase B, hepatoma [5]) [4, 5, 8]

Purification

Rat (partial [5], two forms: transferase A and B, the latter only in hepatoma
[5]) [1, 5]; Rabbit (two molecular weight forms separable by gel filtration,
purification of low-molecular weight form [10]) [2, 10, 12]; Bovine (partial)
[4]; Pig (partial [3], liver [6]) [3, 6]

Crystallization

–

Cloned

(human cDNA clone, constructed type-II-surface membrane-protein/human
transferase chimeras inserted into mammalian expression vector pRSN and
transfected into Madin-Darby canine kidney cells) [11]

Renatured

–

---

5 STABILITY

pH

Temperature (°C)

Oxidation

Organic solvent

General stability information
Repeated freeze-thawing, stable to [2, 6]; Freeze-thawing, solubilized enzyme, unstable to [3]; Triton X-100, 0.1%, and albumin, 0.01%, stabilize solubilized enzyme [3]

Storage
−20°C, at least 6 months [6]; −20°C, 20% glycerol v/v, up to a year [2]; 3°C, at least a month [3]; 4°C, in dilute solution, $t_{1/2}$: 1 day [4]; 4°C, in the presence of albumin, at least a month [4]; 4°C, in 20% glycerol, 0.1% Triton X-100, 25 mM MES-buffer, pH 6.5, 10 mM $MnCl_2$, 1 mM PMSF, 0.1 mM 6-aminocaproic acid and 0.02% NaN_3, several months [12]

6 CROSSREFERENCES TO STRUCTURE DATABANKS

PIR/MIPS code

Brookhaven code

7 LITERATURE REFERENCES

[1] Möller, G., Reck, F., Paulsen, H., Kaur, K.J., Sarkar, M., Schachter, H., Brockhausen, I.: Glycoconjugate J.,9,180–190 (1992)
[2] Oppenheimer, C.L., Hill, R.L.: J. Biol. Chem.,256,799–804 (1981)
[3] Mendicino, J., Chandrasekaran, E.V., Rao Anumula, K., Davila, M.: Biochemistry,20, 967–976 (1981)
[4] Harpaz, N., Schachter, H.: J. Biol. Chem.,255,4885–4893 (1980)
[5] Miyagi, T., Tsuiki, S.: Biochim. Biophys. Acta,661,148–157 (1981)
[6] Oppenheimer, C.L., Eckhardt, A.E., Hill, R.L.: J. Biol. Chem.,256,11477–11482 (1981)
[7] Shao, M.-C., Wold, F.: J. Biol. Chem.,263,5771–5774 (1988)
[8] Moscarello, M.A., Mitranic, M.M., Vella, G.: Biochim. Biophys. Acta,831,192–200 (1985)
[9] Tezuka, K., Hayashi, M., Ishihara, H., Akazawa, T., Takahashi, N.: Eur. J. Biochem., 203,401–413 (1992)
[10] Nishikawa, Y., Pegg, W., Paulsen, H., Schachter, H.: J. Biol. Chem.,263,8270–8281 (1988)
[11] Tang, B.L., Wong, S.H., Low, S.H., Hong, W.: J. Biol. Chem.,267,10122–10126 (1992)
[12] Schachter, H., Brockhausen, I., Hull, E.: Methods Enzymol.,179,351–397 (1989) (Review)

1 NOMENCLATURE

EC number
2.4.1.102

Systematic name
UDP-N-acetyl-D-glucosamine:O-glycosyl-glycoprotein (N-acetyl-D-glucosamine to N-acetyl-D-galactosamine of beta-D-galactosyl-1,3-N-acetyl-D-galactosaminyl-R) beta-1,6-N-acetyl-D-glucosaminyltransferase

Recommended name
beta-1,3-Galactosyl-O-glycosyl-glycoprotein beta-1,6-N-acetylglucosaminyltransferase

Synonyms
O-Glycosyl-oligosaccharide-glycoprotein N-acetylglucosaminyltransferase I
beta6-N-Acetylglucosaminyltransferase
Acetylglucosaminyltransferase, uridine diphosphoacetylglucosamine-mucin beta-(1→6)-
Core 2 acetylglucosaminyltransferase
Acetylglucosaminyltransferase, uridine diphosphoacetylglucosamine-mucin beta-(1→6)-, A
Core 6-beta-GlcNAc-transferase A [2]
More (cf. EC 2.4.1.146–148)

CAS Reg. No.
95978-15-7; 87927-97-7

2 REACTION AND SPECIFICITY

Catalysed reaction
UDP-N-acetyl-D-glucosamine + beta-D-galactosyl-1,3-N-acetyl-D-galactosaminyl-R →
→ UDP + beta-D-galactosyl-1,3-(N-acetyl-beta-D-glucosaminyl-1,6)-N-acetyl-D-galactosaminyl-R

Reaction type
Hexosyl group transfer

Natural substrates
UDP-N-acetyl-D-glucosamine + beta-D-galactosyl-1,3-N-acetyl-D-galactosaminyl-R (involved in mucin oligosaccharide biosynthesis) [1, 2]

Substrate spectrum

1 UDP-N-acetyl-D-glucosamine + beta-D-galactosyl-1,3-N-acetyl-D-galac-
tosaminyl-R (i.e. core class 1, R: polypeptide [1], porcine [4, 5] or ovine
submaxillary mucin polypeptide [4], antifreeze glycoprotein polypeptide
[5], benzyl [1, 2, 4, 5], p-nitrophenyl (R can be beta-linked [1]) [1, 5], o-ni-
trophenyl [1, 2, 5], phenyl [1, 4], methyl or H [1, 5], substrate specificity
[5]. Less effective acceptors are asialofetuin [5] or fucosyl-alpha-1,2-ga-
lactosyl-beta-1,3-N-acetyl-D-galactosaminyl-R [1]. No substrates: galac-
tosyl-beta-1,3-N-acetyl-D-galactosamine [1, 5], N-acetylgalactosaminyl-
Ser(Thr)-mucin [1], asialo-ovine submaxillary mucin, asialo-alpha$_1$ acid
glycoprotein, galactosyl-beta-1,3-N-acetyl-D-glucosaminyl-beta-p-nitro-
phenyl, galactosyl-beta-1,3-N-acetyl-D-glucosaminyl-beta-methyl, galac-
tosyl-beta-1,3-N-acetyl-D-galactosaminitol, D-fucosyl-beta-1,3-N-ace-
tyl-D-galactosaminyl-alpha-benzyl, fetuin [5]) [1–6]

Product spectrum

1 UDP + beta-D-galactosyl-1,3-(N-acetyl-beta-D-glucosaminyl-1,6)-N-ace-
tyl-D-galactosaminyl-R (i.e. core class 2) [1, 2, 4, 5]

Inhibitor(s)

Mn^{2+} (slight stimulation at 5 mM, inhibition at higher concentrations [4])
[1, 4]; Triton X-100 (activation at 0.1% v/v, inhibition at higher concentrations, pig stomach) [1]

Cofactor(s)/prosthetic group(s)/activating agents

Triton X-100 (activation, 0.1% v/v [1], 0.125% v/v [4], inactivates at higher
concentrations, pig stomach [1]) [1, 4]; Empigen BB (activation, detergent)
[4]

Metal compounds/salts

Mn^{2+} (slight stimulation, 5 mM, inhibits at higher concentration) [1, 4]; Mg^{2+}
(slight stimulation, 5 mM) [4]

Turnover number (min^{-1})

Specific activity (U/mg)

K$_m$-value (mM)

0.52 (galactosyl-beta-1,3-N-acetyl-D-galactosaminyl-alpha-p-nitrophenyl) [1,
5]; 0.77 (galactosyl-beta-1,3-N-acetyl-D-galactosaminyl-alpha-benzyl) [1, 5];
0.86 (galactosyl-beta-1,3-N-acetyl-D-galactosaminyl-alpha-o-nitrophenyl) [1,
5]; 0.92 (galactosyl-beta-1,3-N-acetyl-D-galactosaminyl-beta-p-nitrophenyl)
[1, 5]; 1 (UDP-N-acetyl-D-glucosamine) [1]; 1.2 (galactosyl-beta-1,3-N-ace-
tyl-D-galactosaminyl-alpha-R, R: phenyl [1] or H [5]) [1, 5]; 4.2 (galactosyl-
beta-1,3-N-acetyl-D-galactosaminyl-alpha-methyl) [1, 5]; 5.2 (porcine submaxillary mucin) [1]

pH-optimum
 7 [1, 4]

pH-range

Temperature optimum (°C)
 25 (assay at) [4]; 30 (assay at) [1]

Temperature range (°C)

---

3 ENZYME STRUCTURE

Molecular weight

Subunits

Glycoprotein/Lipoprotein
 –

---

4 ISOLATION/PREPARATION

Source organism
 Dog [1–5]; Pig [1–3]; Rat [1–3]; Human [1, 6]; Rabbit [1]; Bovine [1]; Monkey [1]; Sheep [1]

Source tissue
 Submaxillary gland (dog, rat [1, 3], not pig [1, 3]) [1–5]; Trachea (bovine) [1]; Stomach (sheep, pig, rat, monkey [1]) [1–3]; Colon (human, pig, rat, monkey [1]) [1–3]; Intestine (rabbit) [1]; Small intestine (rat) [1]; T-Lymphocytes [6]; Platelets [6]; B-cell lines (Epstein-Barr virus immortalized) [6]; More (not in rat liver) [1]

Localization in source
 Microsomes [1, 3–5]

Purification

Crystallization
 –

Cloned
 –

Renatured
 –

5 STABILITY

pH

Temperature (°C)
37 (decrease of activity, in the presence of Triton X-100 [1]) [1, 4]

Oxidation

Organic solvent

General stability information
Freeze-thawing, after 2 cycles 13% loss of activity [4]

Storage
−70°C, microsomal preparation, detergent-free 0.25 M sucrose suspension, several years [1]; Frozen in liquid nitrogen, several months [4]

---

6 CROSSREFERENCES TO STRUCTURE DATABANKS

PIR/MIPS code

Brookhaven code

---

7 LITERATURE REFERENCES

[1] Schachter, H., Brockhausen, I., Hull, E.: Methods Enzymol.,179,351–397 (1989) (Review)
[2] Brockhausen, I., Rachaman, E.S., Matta, K.L., Schachter, H.: Carbohydr. Res.,120, 3–16 (1983)
[3] Brockhausen, I., Matta, K.L., Orr, J., Schachter, H.: Biochemistry,24,1866–1874 (1985)
[4] Williams, D., Schachter, H.: J. Biol. Chem.,255,11247–11252 (1980)
[5] Williams, D., Longmore, G., Matta, K.L., Schachter, H.: J. Biol. Chem.,255, 11253–11261 (1980)
[6] Higgins, E.A., Siminovitch, K.A., Zhuang, D., Brockhausen, I., Dennis, J.W.: J. Biol. Chem.,266,6280–6290 (1991)

1 NOMENCLATURE

EC number
2.4.1.103

Systematic name
UDPglucose:1,2-dihydroxy-9,10-anthraquinone 2-O-beta-D-glucosyltransferase

Recommended name
Alizarin 2-beta-glucosyltransferase

Synonyms
Glucosyltransferase, uridine diphosphoglucose-alizarin

CAS Reg. No.
74506-41-5

---

2 REACTION AND SPECIFICITY

Catalysed reaction
UDPglucose + alizarin →
→ UDP + 1-hydroxy-2-(beta-D-glucosyloxy)-9,10-anthraquinone (mechanism [1])

Reaction type
Hexosyl group transfer

Natural substrates
UDPglucose + alizarin [2]

Substrate spectrum
1 UDPglucose + alizarin (i.e. 1,2-dihydroxy-9,10-anthraquinone, poor substrates are 1-hydroxy-3-methoxy-9,10-anthraquinone, 1,4-, 1,5- and 1,8-dihydroxy-9,10-anthraquinone [2]) [1, 2]
2 UDPglucose + 1,3-dihydroxy-9,10-anthraquinone [2]
3 UDPglucose + 3-hydroxy-1-methoxy-9,10-anthraquinone [2]
4 UDPglucose + 1-hydroxy-9,10-anthraquinone [2]
5 UDPglucose + 2-hydroxy-9,10-anthraquinone [2]

Product spectrum
1 UDP + 1-hydroxy-2-(beta-D-glucosyloxy)-9,10-anthraquinone [1, 2]
2 ?
3 ?
4 ?
5 ?

Inhibitor(s)

Cofactor(s)/prosthetic group(s)/activating agents

Metal compounds/salts

Turnover number (min^{-1})

Specific activity (U/mg)
   976.74 [1]

K_m-value (mM)
   0.0108 (UDPglucose) [1]; 0.11 (alizarin) [1]

pH-optimum
   7.1 [1]

pH-range
   6.7–7.6 (about half-maximal activity at pH 6.7 and 7.6) [1]

Temperature optimum (°C)
   30 (assay at) [1, 2]

Temperature range (°C)

---

3 ENZYME STRUCTURE

Molecular weight

Subunits

Glycoprotein/Lipoprotein
   –

---

4 ISOLATION/PREPARATION

Source organism
   Streptomyces aureofaciens (mutant strains B96 [1, 2], Bg, 84–25, NMG-2
   and UV61 [2]) [1, 2]

Source tissue
   Mycelium [1, 2]

Localization in source
   Cytoplasm (soluble [2]) [1, 2]

Purification
   Streptomyces aureofaciens (partial) [1]

Crystallization
   –

Cloned

–

Renatured

–

5 STABILITY

pH

Temperature (°C)

Oxidation

Organic solvent

General stability information

Storage
 –20°C, 2 months [1]

6 CROSSREFERENCES TO STRUCTURE DATABANKS

PIR/MIPS code

Brookhaven code

7 LITERATURE REFERENCES

[1] Mateju, J., Cudlín, J., Steinerová, N., Blumauerová, M., Vanek, Z.: Folia Microbiol.,24, 205–210 (1979)
[2] Mateju, J., Nohynek, M.: Folia Microbiol.,36,314–316 (1991)

1 NOMENCLATURE

EC number
2.4.1.104

Systematic name
UDPglucose:7,8-dihydroxycoumarin 7-O-beta-D-glucosyltransferase

Recommended name
o-Dihydroxycoumarin 7-O-glucosyltransferase

Synonyms
Glucosyltransferase, uridine diphosphoglucose-o-dihydroxycoumarin 7-O-
UDP-glucose:o-dihydroxycoumarin glucosyltransferase [2]

CAS Reg. No.
74114-37-7

2 REACTION AND SPECIFICITY

Catalysed reaction
UDPglucose + 7,8-dihydroxycoumarin →
→ UDP + daphnin

Reaction type
Hexosyl group transfer

Natural substrates
More (physiological role: glycosylation of daphnetin, esculetin and possibly scopoletin to their corresponding 7-O-glucosides: daphnin, chicoriin and scopolin) [1]

Substrate spectrum
1 UDPglucose + 7,8-dihydroxycoumarin (i.e. daphnetin [1]) [1, 2]
2 UDPglucose + esculetin [1, 2]
3 UDPglucose + umbelliferone [1]
4 UDPglucose + hydrangetin [1]
5 UDPglucose + scopoletin [1]
6 UDPglucose + caffeic acid [1]
7 UDPglucose + protocatechuic acid (weak activity) [1]
8 UDPglucose + vanillic acid (weak activity) [1]
9 UDPglucose + syringic acid [1]
10 More (requires an intact coumarin ring system with o-dihydroxy groups for highest activity) [1]

Product spectrum
    1 UDP + daphnin [1, 2]
    2 UDP + cichoriin [1, 2]
    3 UDP + skimmin [1]
    4 UDP + hydrangin [1]
    5 UDP + scopolin [1]
    6 ?
    7 ?
    8 ?
    9 ?
   10 ?

Inhibitor(s)

Cofactor(s)/prosthetic group(s)/activating agents

Metal compounds/salts

Turnover number (min^{-1})

Specific activity (U/mg)
    More [1]

K_m-value (mM)
    0.050 (UDPglucose (+ esculetin)) [1]; 0.095 (daphnetin) [1, 2]; 0.11 (esculetin) [1, 2]; 1.43 (scopoletin) [1]

pH-optimum

pH-range

Temperature optimum (°C)
    30 (assay at) [1]

Temperature range (°C)

3 ENZYME STRUCTURE

Molecular weight

Subunits

Glycoprotein/Lipoprotein
    –

4 ISOLATION/PREPARATION

Source organism
    Nicotiana tabacum (L. cv. Wisconsin No. 38 [1]) [1, 2]

Source tissue
 Suspension culture [1, 2]

Localization in source

Purification
 Nicotiana tabacum (partial) [1, 2]

Crystallization
 –

Cloned
 –

Renatured
 –

---

5 STABILITY

pH

Temperature (°C)

Oxidation

Organic solvent

General stability information

Storage

---

6 CROSSREFERENCES TO STRUCTURE DATABANKS

PIR/MIPS code

Brookhaven code

---

7 LITERATURE REFERENCES

[1] Ibrahim, R.K., Boulay, B.: Plant Sci. Lett.,18,177–184 (1980)
[2] Ibrahim, R.K.: Phytochemistry,19,2459–2460 (1980)

1 NOMENCLATURE

EC number
2.4.1.105

Systematic name
UDPglucose:vitexin 2''-O-beta-D-glucosyltransferase

Recommended name
Vitexin beta-glucosyltransferase

Synonyms
Glucosyltransferase, uridine diphosphoglucose-vitexin 2''-

CAS Reg. No.
76828-68-7

2 REACTION AND SPECIFICITY

Catalysed reaction
UDPglucose + vitexin →
→ UDP + vitexin 2''-O-beta-D-glucoside

Reaction type
Hexosyl group transfer

Natural substrates
UDPglucose + vitexin (pathway in flavone glycoside metabolism) [1]

Substrate spectrum
1 UDPglucose + vitexin [1]

Product spectrum
1 UDP + vitexin 2''-O-beta-D-glucoside [1]

Inhibitor(s)
Mn^{2+} (stimulates at 2 mM, inhibits above 8 mM) [1]; Co^{2+} (stimulates at 2 mM, inhibits above 8 mM) [1]

Cofactor(s)/prosthetic group(s)/activating agents

Metal compounds/salts
Ca^{2+} (stimulation, 2 mM) [1]; Co^{2+} (stimulates at 2 mM, inhibits above 8 mM) [1]; Mg^{2+} (stimulation) [1]; Mn^{2+} (stimulates at 2 mM, inhibits above 8 mM) [1]

Turnover number (min^{-1})

Specific activity (U/mg)

K$_m$-value (mM)
    0.01 (vitexin) [1]; 0.2 (UDPglucose) [1]

pH-optimum
    7.5 [1]

pH-range

Temperature optimum (°C)

Temperature range (°C)

---

3 ENZYME STRUCTURE

Molecular weight

Subunits

Glycoprotein/Lipoprotein
    –

---

4 ISOLATION/PREPARATION

Source organism
    Silene alba (Armenian population, Caryophyllaceae) [1]

Source tissue
    Petals [1]

Localization in source

Purification

Crystallization
    –

Cloned
    –

Renatured
    –

5 STABILITY

pH

Temperature (°C)

Oxidation

Organic solvent

General stability information

Storage

---

6 CROSSREFERENCES TO STRUCTURE DATABANKS

PIR/MIPS code

Brookhaven code

---

7 LITERATURE REFERENCES

[1] Heinsbroek, R., van Brederode, J., van Nigtevecht, G., Maas, J., Kamsteeg, J., Besson, E., Chopin, J.: Phytochemistry,19,1935–1937 (1980)

3

1 NOMENCLATURE

EC number
2.4.1.106

Systematic name
UDPglucose:isovitexin 2''-O-beta-D-glucosyltransferase

Recommended name
Isovitexin beta-glucosyltransferase

Synonyms
Glucosyltransferase, uridine diphosphoglucose-isovitexin 2''-

CAS Reg. No.
72102-99-9

2 REACTION AND SPECIFICITY

Catalysed reaction
UDPglucose + isovitexin →
→ UDP + isovitexin 2''-O-beta-D-glucoside

Reaction type
Hexosyl group transfer

Natural substrates
UDPglucose + isovitexin (pathway in flavone glycoside metabolism) [1]

Substrate spectrum
1 UDPglucose + isovitexin (ionisation grade of substrates influences their binding to the enzyme [2]) [1, 2]
2 UDPglucose + isoorientin [2]

Product spectrum
1 UDP + isovitexin 2''-O-beta-D-glucoside [1, 2]
2 UDP + isoorientin 2''-O-beta-D-glucoside [2]

Inhibitor(s)
Mn^{2+} (inhibits above 25 mM, stimulates at lower concentrations) [1]

Cofactor(s)/prosthetic group(s)/activating agents

Metal compounds/salts
Ca^{2+} (stimulation) [1]; Co^{2+} (stimulation) [1]; Mg^{2+} (stimulation) [1]; Mn^{2+} (stimulates, inhibits above 25 mM) [1]

Turnover number (min^{-1})

Specific activity (U/mg)

K_m-value (mM)
  0.09 (isovitexin) [1, 2]; 0.3 (UDPglucose (+ isovitexin)) [1, 2]; 0.45 (isoorientin) [2]; 0.75 (UDPglucose (+ isoorientin))

pH-optimum
  7.5 (isoorientin) [2]; 8.5 (isovitexin) [1, 2]

pH-range
  6.5–8.2 (about half-maximal activity at pH 6.5 and 8.2, isoorientin) [2];
  7.2–9.5 (about half-maximal activity at pH 7.2 and 9.5, isovitexin) [2]

Temperature optimum (°C)

Temperature range (°C)

3 ENZYME STRUCTURE

Molecular weight

Subunits

Glycoprotein/Lipoprotein
  –

4 ISOLATION/PREPARATION

Source organism
  Silene alba (Caryophyllaceae) [1, 2]

Source tissue
  Petals [1, 2]; Leaf [2]

Localization in source

Purification

Crystallization
  –

Cloned
  –

Renatured
  –

5 STABILITY

pH

Temperature (°C)

Oxidation

Organic solvent

General stability information

Storage

---

6 CROSSREFERENCES TO STRUCTURE DATABANKS

PIR/MIPS code

Brookhaven code

---

7 LITERATURE REFERENCES

[1] Heinsbroek, R., van Brederode, J., van Nigtevecht, G., Maas, J., Kamsteeg, J.,
    Besson, E., Chopin, J.: Phytochemistry,19,1935–1937 (1980)
[2] van Brederode, J., Chopin, J., Kamsteeg, J., van Nigtevecht, G., Heinsbroek,
    R.: Phytochemistry,18,655–656 (1979)

1 NOMENCLATURE

EC number
2.4.1.109

Systematic name
Dolichyl-phosphate-D-mannose:protein O-D-mannosyltransferase

Recommended name
Dolichyl-phosphate-mannose-protein mannosyltransferase

Synonyms
Mannosyltransferase, dolichol phosphomannose-protein
Protein O-D-mannosyltransferase [3]

CAS Reg. No.
74315-99-4

2 REACTION AND SPECIFICITY

Catalysed reaction
Dolichyl phosphate D-mannose + protein →
→ dolichyl phosphate + O-D-mannosylprotein

Reaction type
Hexosyl group transfer

Natural substrates
Dolichyl phosphate D-mannose + protein (production of cell-wall manno-
proteins) [1–5]

Substrate spectrum
1 Dolichyl phosphate D-mannose + protein (Tyr-Asn-Pro-Thr-Ser-Val [3, 4],
Ac-Tyr-Asn-Pro-Thr-Ser-Val-NH_2 [3], Tyr-Ala-Thr-Ala-Val [3], Ac-Ala-Thr-
Ala-NH_2 [3], dinitrophenylated tetrapeptide Asn-Ala-Thr-Val [2], Tyr-Pro-
Thr-Ala-Val [3], Tyr-Leu-Thr-Ala-Val [3], Ac-Tyr-Ala-Thr-Ala-Val-NH_2 [3], bio-
tin-Tyr-Thr-Ala-Val-NH_2 [3], biotin-Tyr-Ala-Thr-Ala-Val-NH_2 [3], Tyr-Asn-Leu-
Thr-Ser-Val [4], Lys-Pro-Thr-Gly-Tyr [4], Pro-Tyr-Thr-Val [4], Pro-Thr-Val [4],
Lys-Pro-Thr-Pro-Tyr [4], Lys-Pro-Ser-Gly-Tyr [4], human-granulocyte-macro-
phage colony-stimulating-factor-derived peptide(4–11) [5], preferred
chain length: $C_{100} \gg C_{80} \gg C_{55} > C_{35}$ [2], acidic amino acids strongly
inhibit acceptor activity, as do glycine and proline residues as amino-
terminal and carboxy-terminal neighbours [4], the enzyme transfers
mannosyl residues to the hydroxyl of serine or threonine residues) [1–5]

Product spectrum
1 Dolichyl phosphate + O-D-mannosylprotein [1–5]

Inhibitor(s)
Phosphatidylinositol [3]; Mg^{2+} (above 5 mM) [3]

Cofactor(s)/prosthetic group(s)/activating agents
Deoxycholate (0.025%, stimulates slightly) [1]; Phosphatidylcholine (stimulates) [3]

Metal compounds/salts
Mg^{2+} (Mg^{2+} or Mn^{2+} stimulates [1], not required [3]) [1, 3]; Mn^{2+} (Mg^{2+} or Mn^{2+} stimulates) [1]

Turnover number (min^{-1})

Specific activity (U/mg)
More [4]

K_m-value (mM)
0.075 (biotin-Tyr-Thr-Ala-Val-NH_2) [3]; 0.1 (biotin-Tyr-Leu-Ala-Val-NH_2) [3]; 0.25 (Ac-Tyr-Ala-Thr-Ala-Val-NH_2) [3]; 0.4 (dolichyl phosphate D-mannose) [4]; 0.85 (biotin-Tyr-Pro-Thr-Ala-Val-NH_2) [3]; 2.2 (Tyr-Ala-Thr-Ala-Val) [3]; 2.5 (Tyr-Leu-Thr-Ala-Val) [3]; 3.3 (Tyr-Asn-Pro-Thr-Ser-Val) [4]; 4.3 (Ac-Tyr-Asn-Pro-Thr-Ser-Val-NH_2) [3]; 6.7 (Ac-Ala-Thr-Ala-NH_2) [3]; 7.3 (Tyr-Pro-Thr-Ala-Val) [3]

pH-optimum
5.7 [1]; 7.5 [4]; 7.5–8.0 (the best buffers are bicine pH 7.7, tricine pH 8.0 and HEPES pH 7.5) [3]

pH-range

Temperature optimum (°C)
25 (assay at) [3]

Temperature range (°C)

3 ENZYME STRUCTURE

Molecular weight

Subunits
? (x × 92000, Saccharomyces cerevisiae, SDS-PAGE) [4]

Glycoprotein/Lipoprotein
Glycoprotein (may contain 4 carbohydrate chains) [4]

4 ISOLATION/PREPARATION

Source organism
Saccharomyces cerevisiae [1, 2, 4, 5]; Candida albicans (2005 E) [3]

Source tissue
Microsomes [3]

Localization in source
Membrane (bound [1]) [1, 3, 4]; Endoplasmic reticulum [4]

Purification
Saccharomyces cerevisiae [1, 4]

Crystallization

Cloned
–

Renatured
–

---

5 STABILITY

pH

Temperature (°C)

Oxidation

Organic solvent

General stability information

Storage

---

6 CROSSREFERENCES TO STRUCTURE DATABANKS

PIR/MIPS code

Brookhaven code

---

7 LITERATURE REFERENCES

[1] Babczinski, P., Haselbeck, A., Tanner, W.: Eur. J. Biochem.,105,509–515 (1980)
[2] Palamarczyk, G., Lehle, L., Mankowski, T., Chojnacki, T., Tanner, W.: Eur. J. Biochem., 105,517–523 (1980)
[3] Weston, A., Nassau, P.M., Henley, C., Marriott, M.S.: Eur. J. Biochem.,215,845–849 (1993)
[4] Strahl-Bolsinger, S., Tanner, W.: Eur. J. Biochem.,196,185–190 (1991)
[5] Lorenz, C., Strahl-Bolsinger, S., Ernst, J.F.: Eur. J. Biochem.,205,1163–1167 (1992)

1 NOMENCLATURE

EC number
2.4.1.110

Systematic name
GDPmannose:tRNA^{Asp}-queuosine O-5''-beta-D-mannosyltransferase

Recommended name
tRNA-queuosine beta-mannosyltransferase

Synonyms

CAS Reg. No.

---

2 REACTION AND SPECIFICITY

Catalysed reaction
GDPmannose + tRNA^{Asp}-queuosine →
→ GDP + tRNA^{Asp}-O-5''-beta-D-mannosylqueuosine

Reaction type
Hexosyl group transfer

Natural substrates

Substrate spectrum
1 GDPmannose + tRNA^{Asp}-queuosine (strict specificity for E. coli tRNA^{Asp}, transfer with concomitant inversion of anomeric configuration at sugar C-1-carbon, no substrates are tRNA^{Tyr}, tRNA^{His} or tRNA^{Asn}) [1]

Product spectrum
1 GDP + tRNA^{Asp}-O-5''-beta-D-mannosylqueuosine [1]

Inhibitor(s)
EDTA [1]

Cofactor(s)/prosthetic group(s)/activating agents

Metal compounds/salts
Mg^{2+} (requirement) [1]; NaCl (activation) [1]

Turnover number (min^{-1})

Specific activity (U/mg)

K$_m$-value (mM)

---

pH-optimum
   7.5 [1]

pH-range

Temperature optimum (°C)
   37 (assay at) [1]

Temperature range (°C)

---

3 ENZYME STRUCTURE

Molecular weight

Subunits

Glycoprotein/Lipoprotein
   –

---

4 ISOLATION/PREPARATION

Source organism
   Rat (male Donryu) [1]

Source tissue
   Liver [1]

Localization in source
   Cytoplasm (soluble) [1]

Purification
   Rat (partial) [1]

Crystallization
   –

Cloned
   –

Renatured
   –

---

5 STABILITY

pH

Temperature (°C)

Oxidation

Organic solvent

General stability information

Storage

---

6 CROSSREFERENCES TO STRUCTURE DATABANKS

PIR/MIPS code

Brookhaven code

---

7 LITERATURE REFERENCES

[1] Okada, N., Nishimura, S.: Nucleic Acids Res.,4,2931–2937 (1977)

Enzyme Handbook © Springer-Verlag Berlin Heidelberg 1996
Duplication, reproduction and storage in data banks are only
allowed with the prior permission of the publishers   3

1 NOMENCLATURE

EC number
2.4.1.111

Systematic name
UDPglucose:coniferyl-alcohol 4'-beta-D-glucosyltransferase

Recommended name
Coniferyl-alcohol glucosyltransferase

Synonyms
Glucosyltransferase, uridine diphosphoglucose-coniferyl alcohol
UDP-glucose coniferyl alcohol glucosyltransferase

CAS Reg. No.
61116-23-2

---

2 REACTION AND SPECIFICITY

Catalysed reaction
UDPglucose + coniferyl alcohol →
→ UDP + coniferin (mechanism [3])

Reaction type
Hexosyl group transfer

Natural substrates
UDPglucose + coniferyl alcohol (pathway in lignin biosynthesis, glucosylation of coniferyl and possibly sinapyl alcohol [1], participates in lignification of spruce [2, 3]) [1–3]

Substrate spectrum
1 UDPglucose + coniferyl alcohol (r [3], pronounced specificity for UDPglucose, TDPglucose can replace it with 17% efficiency [3], substrates: scopoletin, isorhamnetin [1, 4], caffeic acid (not [3]), quercetin [1], poor substrates: homoeriodyctiol [1, 3], chrysoeriol (not [3]) [1], p-coumaric aldehyde, 5-hydroxyferulic acid, naringenin [3], no substrates are CDPglucose, ADPglucose, GDPglucose [3], p-coumaryl alcohol [1, 3, 4], p-coumaric acid, p-hydroxybenzoic acid, luteolin [1, 3], eriodyctiol, apigenin [3], isoferulic acid, syringic acid [1]) [1–4]
2 UDPglucose + sinapyl alcohol (glucosylated at 14% [3], 82% [4] the rate of coniferyl alcohol glucosylation, can be replaced by sinapic acid (not [3]) [1, 4]) [1, 3, 4]

3 UDPglucose + coniferyl aldehyde (glucosylated at 63% the rate of coniferyl alcohol glucosylation [3]) [1, 3]
4 UDPglucose + sinapaldehyde (glucosylated at 48% [3], 82% [4] the rate of coniferyl alcohol glucosylation) [3, 4]
5 UDPglucose + ferulic acid (glucosylated at 18% the rate of coniferyl alcohol glucosylation [1, 4], not [3]) [1, 4]
6 UDPglucose + vanillic acid (glucosylated at 14% the rate of coniferyl alcohol glucosylation [3], not [1]) [3]
7 UDPglucose + vanillin (glucosylated at 17% the rate of coniferyl alcohol glucosylation) [4]

Product spectrum
1 UDP + coniferin (i.e. coniferyl alcohol 7-O-beta-D-glucopyranoside) [1–4]
2 ?
3 ?
4 ?
5 ?
6 ?
7 ?

Inhibitor(s)
Coniferyl alcohol (above 0.08 mM, Paul's scarlet rose) [1]; UDP (non-competitive, product inhibition) [3]; Coniferin (non-competitive, product inhibition) [3]; EDTA (Paul's scarlet rose [1], not [3]) [1]; PCMB (Paul's scarlet rose [1]) [1, 3, 4]; More (Paul's scarlet rose: no inhibitors are iodoacetate, p-diazobenzene sulfonate) [1]

Cofactor(s)/prosthetic group(s)/activating agents

Metal compounds/salts
More (no activation by Mg^{2+}, Ca^{2+} [1, 3, 4], Mn^{2+} [3, 4], Cu^{2+}, Zn^{2+} [3], 1 mM each [3]) [1, 3, 4]

Turnover number (min^{-1})

Specific activity (U/mg)
0.036 (Paul's scarlet rose) [1]; 2.436 [3]

K_m-value (mM)
0.0033 (coniferyl alcohol, Paul's scarlet rose) [1]; 0.0037 (coniferyl alcohol) [4]; 0.0056 (sinapyl alcohol, Paul's scarlet rose) [1]; 0.02 (UDPglucose, Paul's scarlet rose) [1]; 0.025 (UDPglucose) [4]; 0.065 (coniferyl aldehyde, Paul's scarlet rose) [1]; 0.22 (UDPglucose) [2, 3]; 0.25 (coniferyl alcohol) [2, 3]; 0.58 (sinapyl alcohol) [3]; 0.64 (vanillin) [3]

pH-optimum
7.5 (Tris-HCl buffer [1, 4], Tris-HCl (0.1–0.2 M) or phosphate buffer [3]) [1, 3, 4]; 8.0 (phosphate buffer) [1, 4]

pH-range
5.5–8.2 (about half-maximal activity at pH 5.5 and about 70% of maximal
activity at pH 8.2, phosphate buffer) [3]; 6.5–8.4 (about half-maximal activity
at pH 6.5 and 8.4, Tris buffer) [1]; 7–9.2 (about half-maximal activity at pH 7
and 9.2, phosphate buffer) [1]; 6.5–8.2 (about half-maximal activity at pH 6.5
and 83% of maximal activity at pH 8.2, Tris buffer) [3]

Temperature optimum (°C)
36 [3]

Temperature range (°C)

---

3 ENZYME STRUCTURE

Molecular weight
44000 (Picea abies, gel filtration) [3]
52000 (Paul's scarlet rose, gel filtration) [1]

Subunits
Monomer (1 × 50000, Picea abies, SDS-PAGE) [3]

Glycoprotein/Lipoprotein
–

---

4 ISOLATION/PREPARATION

Source organism
Paul's scarlet rose [1]; Picea abies (spruce, 80–130 years old) [2, 3]; Forsythia orata [4]; Linum usitatissimum (flax) [1]; Glycine max v. Kanrich (soy
bean) [1]; Petroselinum hortense (parsley) [1]; Parthenocissus tricuspidata
[1]; Nicotiana tabacum (tobacco) [1]; Cicer arietum (chickpea) [1]; More
(distribution in higher plants) [4]

Source tissue
Cell suspension culture [1]; Callus tissue (Linum usitatissimum, Glycine
max, Parthenocissus tricuspidata, Nicotiana tabacum) [1]; Hypocotyls (epidermal and subepidermal layers, vascular bundles) [2]; Cambial sap [3];
Stem segments [4]

Localization in source
Cytoplasm (parietal cytoplasmic layer, localisation by immunofluorescence
technique) [2]

Purification
Paul's scarlet rose [1]; Picea abies [3]; Forsythia orata (partial) [4]

Enzyme Handbook © Springer-Verlag Berlin Heidelberg 1996
Duplication, reproduction and storage in data banks are only
allowed with the prior permission of the publishers

Crystallization

–

Cloned

–

Renatured

–

5 STABILITY

pH

Temperature (°C)

Oxidation

Organic solvent

General stability information
Glycerol or ethylene glycol, 10–20%, stabilizes dilute enzyme solutions [3];
Buffer solutions of high molarity, e.g. 0.2 M Tris-HCl, stabilize [3]; Ethylene
glycol, 10% v/v, stabilizes during purification and storage, Paul's scarlet
rose [1]; 2-Mercaptoethanol stabilizes [1, 3]

Storage
–20°C, about 20% loss of activity within several weeks, Paul's scarlet rose
[1]; –20°C, several months [3]; 0–2°C, several weeks [3]; Glycerol, 15%,
stabilizes during storage [3]

6 CROSSREFERENCES TO STRUCTURE DATABANKS

PIR/MIPS code

Brookhaven code

7 LITERATURE REFERENCES

[1] Ibrahim, R.K., Grisebach, H.: Arch. Biochem. Biophys.,176,700–708 (1976)
[2] Schmid, G., Hammer, D.K., Ritterbusch, A., Grisebach, H.: Planta,156,207–212
    (1982)
[3] Schmid, G., Grisebach, H.: Eur. J. Biochem.,123,363–370 (1982)
[4] Ibrahim, R.K.: Z. Pflanzenphysiol.,85,253–262 (1977)

1 NOMENCLATURE

EC number
2.4.1.112

Systematic name
UDPglucose:protein 4-alpha-glucosyltransferase

Recommended name
alpha-1,4-Glucan-protein synthase (UDP-forming)

Synonyms
UDPglucose:protein glucosyltransferase [1]
Glycogen initiator synthase [2]
UDPGlc:protein transglucosylase [5]
UPTG [5]
Uridine diphosphoglucose protein transglucosylase I [6]
Proglycogen synthase
Glucosyltransferase, uridine diphosphoglucose-protein 4-alpha-
Glucosyltransferase, uridine diphosphoglucose-protein
UDP-glucose protein transglucosylase
UDP-glucose-protein glucosyltransferase
Uridine diphosphate glucose-protein transglucosylase I (different from glucose-1-phosphate dependent UDP-glucose-protein transglucosylase, different enzymes or a single enzyme that occurs in 2 conformational states) [4]

CAS Reg. No.
152478-54-1; 39369-27-2

---

2 REACTION AND SPECIFICITY

Catalysed reaction
UDPglucose + protein →
→ UDP + alpha-D-glucosyl-protein

Reaction type
Hexosyl group transfer

Natural substrates
UDPglucose + protein (the enzyme builds up alpha-1,4-glucan chains covalently bound to protein, thus acting as an initiator of glycogen synthesis [1, 5], when the saccharide linked to the protein has reached a certain size it is almost exclusively enlarged by an ADPglucose-dependent enzyme [1], complex product of the reaction acts as a precursor for the synthesis of glycogen by EC 2.4.1.11 [2]) [1, 2, 5]

Substrate spectrum
1 UDPglucose + protein (high specificity for UDPglucose [4], purified
   UPTG undergoes selfglucosylation in an UDPglucose and Mn^{2+}-depen-
   dent reaction, UPTG is the enzyme and at the same time the priming protein required for the biogenesis of protein bound alpha-glucan in potato
   tuber [5], specific for glucosyl donor and for endogenous acceptor, a
   macromolecular component (MW 380000) which cannot be separated
   from the enzyme by ion-exchange chromatography, affinity chromatography, gel filtration and sucrose density gradient centrifugation [6]) [1–6]

Product spectrum
1 UDP + alpha-D-glucosyl-protein (alpha-1,4-glucan protein [2], acceptor
   protein-Glc [3]) [1–6]

Inhibitor(s)
Brij-58 (weak) [1]; p-Chloromercuribenzoate [4]

Cofactor(s)/prosthetic group(s)/activating agents

Metal compounds/salts
NaEDTA (enzyme requires presence of some salts e.g. NaEDTA at high concentrations) [2]; Mn^{2+} (5fold stimulation by 5 mM [5], stimulation by Mg^{2+} or
Mn^{2+}, 3fold activation by 4 mM [4]) [4, 5]; Mg^{2+} (stimulation by Mg^{2+} or
Mn^{2+}, 3fold activation by 4 mM) [4]

Turnover number (min^{-1})

Specific activity (U/mg)

K_m-value (mM)
More [4]

pH-optimum
5–9 [4]; 7.4 (assay at) [5]

pH-range

Temperature optimum (°C)
30 (assay at) [5]; 37 (reaction very rapid) [4]

Temperature range (°C)

3 ENZYME STRUCTURE

Molecular weight

Subunits
    ? (x × 38000, Solanum tuberosum, SDS-PAGE) [5]

Glycoprotein/Lipoprotein
    –

---

4 ISOLATION/PREPARATION

Source organism
    E. coli [1]; Rat [2]; Solanum tuberosum (potato) [3–6]

Source tissue
    Liver [2]; Tuber [3–6]

Localization in source

Purification
    Solanum tuberosum (potato) [3, 5]

Crystallization
    –

Cloned
    –

Renatured
    –

---

5 STABILITY

pH

Temperature (°C)
    45 (2 min, about 20% loss of activity) [4]

Oxidation

Organic solvent

General stability information

Storage
    0–4°C, stable for 2–3 weeks [4]; –10°C, rapidly destroyed [4]

6 CROSSREFERENCES TO STRUCTURE DATABANKS

PIR/MIPS code
    PIR2:B26206 (rabbit (fragment))

Brookhaven code

7 LITERATURE REFERENCES

[1] Barengo, R., Krisman, C.R.: Biochim. Biophys. Acta,540,190–196 (1978)
[2] Krisman, C.R., Barengo, R.: Eur. J. Biochem.,52,117–123 (1975)
[3] Moreno, S., Tandecarz, J.S.: FEBS Lett.,139,313–316 (1982)
[4] Lavintman, N., Tandecarz, J.S., Carceller, M., Mendiara, S., Cardini,
    C.E.: Eur. J. Biochem.,50,145–155 (1974)
[5] Ardila, F.J., Tandecarz, J.S.: Plant Physiol.,99,1342–1347 (1992)
[6] Moreno, S., Cardini, C.E., Tandecarz, J.S.: Eur. J. Biochem.,157,539–545 (1986)

1 NOMENCLATURE

EC number
2.4.1.113

Systematic name
ADPglucose:protein 4-alpha-D-glucosyltransferase

Recommended name
alpha-1,4-Glucan-protein synthase (ADP-forming)

Synonyms
ADPglucose:protein glucosyltransferase [1]
Glucosyltransferase, adenosine diphosphoglucose-protein
ADP-glucose-protein glucosyltransferase

CAS Reg. No.
67053-99-0

---

2 REACTION AND SPECIFICITY

Catalysed reaction
ADPglucose + protein →
→ ADP + alpha-D-glucosyl-protein

Reaction type
Hexosyl group transfer

Natural substrates
ADPglucose + protein (the enzyme builds up alpha-1,4-glucan chains covalently bound to protein thus acting as an initiator of glycogen synthesis,
once the saccharide linked to the protein has reached a certain size it is almost exclusively enlarged by another ADPglucose-dependent enzyme) [1]

Substrate spectrum
1 ADPglucose + protein [1]

Product spectrum
1 ADP + alpha-D-glucosyl-protein [1]

Inhibitor(s)

Cofactor(s)/prosthetic group(s)/activating agents

Metal compounds/salts

Turnover number (min^{-1})

Specific activity (U/mg)

K_m-value (mM)

pH-optimum

pH-range

Temperature optimum (°C)

Temperature range (°C)

3 ENZYME STRUCTURE

Molecular weight

Subunits

Glycoprotein/Lipoprotein

–

4 ISOLATION/PREPARATION

Source organism
  E. coli [1]

Source tissue

Localization in source

Purification

Crystallization

–

Cloned

–

Renatured

–

5 STABILITY

pH

Temperature (°C)

Oxidation

Organic solvent

General stability information

Storage

6 CROSSREFERENCES TO STRUCTURE DATABANKS

PIR/MIPS code

Brookhaven code

7 LITERATURE REFERENCES

[1] Barengo, R., Krisman, C.R.: Biochim. Biophys. Acta,540,190–196 (1978)

1 NOMENCLATURE

EC number
2.4.1.114

Systematic name
UDPglucose:trans-2-hydroxycinnamate O-beta-D-glucosyltransferase

Recommended name
2-Coumarate O-beta-glucosyltransferase

Synonyms
Glucosyltransferase, uridine diphosphoglucose-o-coumarate
UDPG:o-coumaric acid O-glucosyltransferase [2]

CAS Reg. No.
73665-97-1

---

2 REACTION AND SPECIFICITY

Catalysed reaction
UDPglucose + trans-2-hydroxycinnamate →
→ UDP + trans-beta-D-glucosyl-2-hydroxycinnamate

Reaction type
Hexosyl group transfer

Natural substrates
More (involved in the biosynthesis of coumarin) [1, 2]

Substrate spectrum
1 UDPglucose + 2-hydroxycinnamate (i. e. o-coumaric acid) [1, 2]
2 More (coumarinate is no acceptor) [1]

Product spectrum
1 UDP + trans-beta-D-glucosyl-o-hydroxycinnamic acid [1, 2]
2 ?

Inhibitor(s)
Cu^{2+} (1 mM, final concentration: 63% inhibition) [1]; Zn^{2+} (1 mM, final concentration: 77% inhibition) [1]; Sodium citrate (4 mM, final concentration: 37% inhibition) [1]; More (not inhibitory: 4 mM EDTA, 0.8 mM 2-coumaric acid glucoside) [1]

Cofactor(s)/prosthetic group(s)/activating agents
Sulfhydryl compound (absolute requirement, ascorbic acid or glutathione have no effect) [1]; 2-Mercaptoethanol (stimulation) [1]; Cysteine (stimulation, almost as effective as 2-mercaptoethanol) [1]

Metal compounds/salts

Turnover number (min^{-1})

Specific activity (U/mg)

K_m-value (mM)

pH-optimum
7.6–8.3 [1]

pH-range
6–9 (25% of maximal activity at pH 6, 10% of maximal activity at pH 9) [1]

Temperature optimum (°C)
30 (assay at) [1]; 35 (assay at) [2]

Temperature range (°C)

3 ENZYME STRUCTURE

Molecular weight

Subunits

Glycoprotein/Lipoprotein
–

4 ISOLATION/PREPARATION

Source organism
Melilotus alba (cucubb genotyp [1], var. White Blossom [2]) [1, 2]

Source tissue
Leaf [1, 2]

Localization in source
More (not associated with chloroplasts) [2]

Purification
Melilotus alba (partial) [1]

Crystallization
–

Cloned

–

Renatured

–

---

5 STABILITY

pH

Temperature (°C)

Oxidation

Organic solvent

General stability information
Freezing results in a complete loss of activity [1]

Storage
4°C, complete loss of activity within 1–2 days [1]

---

6 CROSSREFERENCES TO STRUCTURE DATABANKS

PIR/MIPS code

Brookhaven code

---

7 LITERATURE REFERENCES

[1] Kleinhofs, A., Haskins, F.A., Gorz, H.J.: Phytochemistry,6,1313–1318 (1967)
[2] Poulton, J.E., McRee, D.E., Conn, E.E.: Plant Physiol.,65,171–175 (1980)

1 NOMENCLATURE

EC number
   2.4.1.115

Systematic name
   UDPglucose:anthocyanidin 3-O-D-glucosyltransferase

Recommended name
   Anthocyanidin 3-O-glucosyltransferase

Synonyms
   Glucosyltransferase, uridine diphosphoglucose-anthocyanidin 3-O-
   UDP-glucose:anthocyanidin/flavonol 3-O-glucosyltransferase
   UDP-glucose:cyanidin-3-O-glucosyltransferase [1]
   More (cf. EC 2.4.1.91)

CAS Reg. No.
   65607-32-1

---

2 REACTION AND SPECIFICITY

Catalysed reaction
   UDPglucose + anthocyanidin →
   → UDP + anthocyanidin-3-O-glucoside

Reaction type
   Hexosyl group transfer

Natural substrates
   UDPglucose + anthocyanidin (important pathway in anthocyanin biosynthesis [1, 4], stabilizes highly unstable anthocyanidines [4]) [1, 4]

Substrate spectrum
   1 UDPglucose + cyanidin (best substrate [1], transfers glucosyl moiety to
     3-hydroxyl-group of cyanidin, TDPglucose can replace UDPglucose with
     75% efficiency [4], not ADPglucose [1, 4], no glucosylation of 5-hydroxyl-
     group of anthocyans, no acceptors are cyanidin 3-glucoside or cyanidin
     3-rhamnosylglucoside [1]) [1, 3, 4]
   2 UDPglucose + malvidin [2]
   3 UDPglucose + pelargonidin (more slowly than cyanidin [1]) [1]
   4 UDPglucose + delphinidin (more slowly than cyanidin [1]) [1]
   5 UDPglucose + paeonidin [4]
   6 UDPglucose + kaempferol (not [1]) [3, 4]
   7 UDPglucose + quercetin (not [1]) [3, 4]
   8 UDPglucose + myricetin [4]

Product spectrum
1 UDP + cyanidin 3-O-glucoside [1, 3]
2 UDP + malvidin 3-O-glucoside [2]
3 UDP + pelargonidin 3-O-glucoside [1]
4 UDP + delphinidin 3-O-glucoside [1]
5 ?
6 ?
7 ?
8 ?

Inhibitor(s)
PCMB (strong [1], 2-mercaptoethanol or cysteine restores activity [1]) [1, 4]; HgCl$_2$ [1]; N-Ethylmaleimide (weak, 2-mercaptoethanol or cysteine restores activity) [1]; Increasing ionic strength (above 0.01 M) [3]; Antiserum against UDPglucose:flavonol 3-O-glucosyltransferase [3]; Pelargonidin (strong, substrate inhibition) [4]; Quercetin (strong, substrate inhibition) [4]; Cu^{2+} (strong) [4]; Fe^{2+} [4]; Zn^{2+} [4]; Diethyldicarbonate [4]; More (no inhibition by diethyldithiocarbamate) [4]

Cofactor(s)/prosthetic group(s)/activating agents
2-Mercaptoethanol (activation) [1]; Cysteine (activation, can replace 2-mercaptoethanol) [1]; Detergents (activation, e.g. Tween 20 or 80, cetrimide, 2-methoxyethanol or Triton X-100) [1]

Metal compounds/salts
Ca^{2+} (slight stimulation [4], not [1]) [4]; Co^{2+} (slight stimulation [4], not [1]) [4]; Mg^{2+} (slight stimulation [4], not [1]) [4]; More (no divalent cations required, e.g. Mn^{2+} or Zn^{2+}) [1]

Turnover number (min^{-1})

Specific activity (U/mg)
More [3]; 0.0252 [1]

K$_m$-value (mM)
0.04 (cyanidin) [1]; 0.13 (pelargonidin, delphinidin) [1]; 0.41 (UDPglucose) [1]

pH-optimum
7.2 (rapid anthocyanidin degradation above pH 7.7) [3]; 7.5 [1]; 8.5 (pelargonidin) [4]; 9.5 (quercetin) [4]

pH-range

Temperature optimum (°C)
37 [4]

Temperature range (°C)
0–37 (about 20% of maximal activity at 0°C and maximal activity at 37°C) [4]

3 ENZYME STRUCTURE

Molecular weight
125000 (Silene dioica, gel filtration) [1]

Subunits
Monomer (1 × 60000, can exist as monomer or dimer, Silene dioica, SDS-PAGE) [1]
Dimer (2 × 60000, can exist as monomer or dimer, Silene dioica, SDS-PAGE) [1]

Glycoprotein/Lipoprotein
–

4 ISOLATION/PREPARATION

Source organism
Silene dioica (red campion) [1]; Hippeastrum (cv. Dutch Red Hybrid) [2]; Petunia hybrida hort. (various inbred lines: R27, V13 (wild-type), W22, W42) [3]; Tulipa cultivar (tulip, var. Most Miles) [2]; Matthiola incana [4]

Source tissue
Petals (highest activity in petals of opening flowers of young plants [1]) [1, 2, 4]; Rosette leaf [1]; Calyces [1]; Leaf (tulip) [2]

Localization in source
Cytosol (not: vacuole) [2]; Soluble [2]; More (subcellular distribution) [2]

Purification
Silene dioica (partial) [1]; Petunia hybrida hort. (partial) [3]

Crystallization
–

Cloned
–

Renatured
–

5 STABILITY

pH

Temperature (°C)
30–40 (crude, at least 1 h stable) [4]; 40 (crude, initial activation decreases to normal level after 30 min) [4]; 50 (crude, $t_{1/2}$: 10 min) [4]; 60 (crude, $t_{1/2}$: 4 min) [4]; 70 (crude, $t_{1/2}$: 2 min) [4]

Oxidation

Organic solvent

General stability information
Repeated freeze-thawing inactivates [1]; Very unstable at increasing degree of purity [3]

Storage
−20°C, crude, several months [4]; −17°C, about 40% loss of activity within 3 months [1]; 4°C, about 40% loss of activity within 1 week [1]

6 CROSSREFERENCES TO STRUCTURE DATABANKS

PIR/MIPS code

Brookhaven code

7 LITERATURE REFERENCES

[1] Kamsteeg, J., van Brederode, J., van Nigtevecht, G.: Biochem. Genet.,16, 1045–1058 (1978)
[2] Hrazdina, G., Wagner, G.J., Siegelmann, H.W.: Phytochemistry,17,53–56 (1978)
[3] Jonsson, L.M.V., Aarsman, M.E.G., Bastiaannet, J., Donker-Koopman, W.E., Gerats, A.G.M., Schram, A.W.: Z. Naturforsch.,39c,559–567 (1984)
[4] Teusch, M., Forkmann, G., Seyffert, W.: Z. Naturforsch.,41c,699–706 (1986)

1 NOMENCLATURE

EC number
2.4.1.116

Systematic name
UDPglucose:cyanidin-3-O-D-rhamnosyl-1,6-D-glucoside 5-O-D-glucosyl-
transferase

Recommended name
Cyanidin-3-rhamnosylglucoside 5-O-glucosyltransferase

Synonyms
Glucosyltransferase, uridine diphosphoglucose-cyanidin 3-rhamnosylgluco-
side 5-O-

CAS Reg. No.
70248-66-7

---

2 REACTION AND SPECIFICITY

Catalysed reaction
UDPglucose + cyanidin-3-O-D-rhamnosyl-1,6-D-glucoside →
→ UDP + cyanidin-3-O-[D-rhamnosyl-1,6-D-glucoside]-5-O-D-glucoside

Reaction type
Hexosyl group transfer

Natural substrates
UDPglucose + cyanidin-3-O-D-rhamnosyl-1,6-D-glucoside (involved in anthocyan biosynthesis) [1]

Substrate spectrum
1 UDPglucose + cyanidin-3-O-D-rhamnosyl-1,6-D-glucoside (not ADPglucose [1]) [1]
2 UDPglucose + pelargonidin-3-O-rhamnosylglucoside [1]
3 UDPglucose + cyanidin-3-O-glucoside [1, 2]
4 More (affinity and specificity for anthocyan acceptor vary with pH-value)
[2]

Product spectrum
1 UDP + cyanidin-3-O-[D-rhamnosyl-1,6-D-glucoside]-5-O-D-glucoside [1]
2 UDP + ? [1]
3 UDP + cyanidin-3,5-diglucoside [2]
4 ?

Inhibitor(s)

EDTA (does not inhibit glucosylation of cyanidin-3-glucoside [2]) [1, 2]; Hg^{2+} (strong) [1, 2]; Zn^{2+} [2]; PCMB (weak, 2-mercpatoethanol or cysteine does not protect) [1]; N-Ethylmaleimide (weak, 2-mercpatoethanol or cysteine does not protect) [1]

Cofactor(s)/prosthetic group(s)/activating agents

Metal compounds/salts

Ca^{2+} (activation, 1 mM) [1]; Co^{2+} (activation, 1 mM, can replace Ca^{2+}) [1]; Mg^{2+} (activation, 1 mM, can replace Ca^{2+}) [1]; Mn^{2+} (activation, 1 mM, can replace Ca^{2+}) [1]; More (diglucoside formation is not stimulated by divalent cations) [2]

Turnover number (min^{-1})

Specific activity (U/mg)

0.00953 [1]

K_m-value (mM)

0.5 (UDPglucose (+ cyanidin-3-rhamnosylglucoside), pH 7.4) [1, 2]; 0.59 (UDPglucose (+ cyanidin-3-rhamnosylglucoside), pH 6.5) [2]; 0.67 (UDPglucose (+ cyanidin-3-glucoside), pH 6.5) [2]; 3.6 (cyanidin-3-rhamnosylglucoside) [1, 2]; 23.4 (cyanidin-3-glucoside) [2]

pH-optimum

6.5 (cyanidin-3-glucoside) [2]; 7.4 (cyanidin-3-rhamnosylglucoside [2]) [1, 2]

pH-range

6.1–7.2 (about 80% of maximal activity at pH 6.1 and about half-maximal activity at pH 7.2, cyanidin-3-glucoside) [2]; 6.9–8.8 (about half-maximal activity at pH 6.9 and 8.8) [1]; 7–8.5 (about half-maximal activity at pH 7 and 8.5, cyanidin-3-rhamnosylglucoside) [2]

Temperature optimum (°C)

30 (assay at) [1, 2]

Temperature range (°C)

3 ENZYME STRUCTURE

Molecular weight

55000 (Silene dioica, gel filtration) [1]

Subunits

Glycoprotein/Lipoprotein

–

4 ISOLATION/PREPARATION

Source organism
Silene dioica (red campion) [1, 2]

Source tissue
Petals (not in green parts of the plant [1]) [1, 2]

Localization in source

Purification
Silene dioica (partial) [1]

Crystallization
–

Cloned
–

Renatured
–

5 STABILITY

pH
7.6 (stable below) [2]

Temperature (°C)

Oxidation

Organic solvent

General stability information
Repeated freeze-thawing inactivates [1]

Storage
–20°C, about 60–80% loss of activity within 3 months [1]

6 CROSSREFERENCES TO STRUCTURE DATABANKS

PIR/MIPS code

Brookhaven code

7 LITERATURE REFERENCES

[1] Kamsteeg, J., van Brederode, J., van Nigtevecht, G.: Biochem. Genet.,16, 1059–1071 (1978)
[2] Kamsteeg, J., van Brederode, J., van Nigtevecht, G.: Z. Pflanzenphysiol.,96,87–93 (1980)

1 NOMENCLATURE

EC number
2.4.1.117

Systematic name
UDPglucose:dolichyl-phosphate beta-D-glucosyltransferase

Recommended name
Dolichyl-phosphate beta-glucosyltransferase

Synonyms
Polyprenyl phosphate:UDP-D-glucose glucosyltransferase [1]
UDP-glucose dolichyl-phosphate glucosyltransferase [5]
Glucosyltransferase, uridine diphosphoglucose-dolichol
UDP-glucose:dolichol phosphate glucosyltransferase
UDP-glucose:dolicholphosphoryl glucosyltransferase
UDP-glucose:dolichyl monophosphate glucosyltransferase
UDP-glucose:dolichyl phosphate glucosyltransferase
UDP-glucose:dolichylphosphate glucosyltransferase

CAS Reg. No.
71061-42-2

---

2 REACTION AND SPECIFICITY

Catalysed reaction
UDPglucose + dolichyl phosphate →
→ UDP + dolichyl beta-D-glucosyl phosphate (sequential mechanism [6])

Reaction type
Hexosyl group transfer

Natural substrates
More (synthesis of dolichyl D-glucosyl phosphate, a glucosyl donor in the
formation of lipid-linked core oligosaccharides [1], enzyme is involved in
the biosynthesis of the lipid-linked oligosaccharides that are precursors of
N-linked glycoproteins [7]) [1, 7]

Substrate spectrum
1 UDPglucose + dolichyl phosphate (r [1]) [1–5]
2 UDPglucose + solanesyl phosphate [1]
3 UDPglucose + ficaprenyl phosphate [1]
4 More (not: GDP-D-glucose, UDP-D-glucuronic acid, UDP-N-acetyl-D-glucosamine, UDP-D-xylose) [1]

Product spectrum

1 UDP + dolichyl beta-D-glucosyl phosphate [1–4]
2 ?
3 ?
4 ?

Inhibitor(s)

$MnCl_2$ (stimulates at 1 mM, inhibits at higher concentration) [7]; UMP (reversible) [1]; UDP (competitive) [1]; EDTA (inhibition reversed by $MgCl_2$ [1]) [1, 7]

Cofactor(s)/prosthetic group(s)/activating agents

Reducing agent (required) [1]; Detergent (required for maximum activity, optimum for Triton X-100: 0.015%) [4, 8]; Phospholipid (required, phosphatidylethanolamine most effective) [5]

Metal compounds/salts

Ca^{2+} (divalent cation required, maximum stimulation in presence of Ca^{2+}. Co^{2+}, Mg^{2+} or Mn^{2+} can replace Ca^{2+} [6], divalent cation required: Mg^{2+} > Mn^{2+} >> Ca^{2+} > Co^{2+} > Zn^{2+} [4], $CaCl_2$ cannot replace $MgCl_2$ or $MnCl_2$ [7]) [4, 6]; Co^{2+} (divalent cation required, maximum stimulation in presence of Ca^{2+}. Co^{2+}, Mg^{2+} or Mn^{2+} can replace Ca^{2+} [6], divalent cation required: Mg^{2+} > Mn^{2+} >> Ca^{2+} > Co^{2+} > Zn^{2+} [4]) [4, 6]; Zn^{2+} (divalent cation required: Mg^{2+} > Mn^{2+} >> Ca^{2+} > Co^{2+} > Zn^{2+}) [4]; Mg^{2+} (divalent cation required, maximum stimulation in presence of Ca^{2+}. Co^{2+}, Mg^{2+} or Mn^{2+} can replace Ca^{2+} [6], divalent cation required: Mg^{2+} > Mn^{2+} >> Ca^{2+} > Co^{2+} > Zn^{2+} [4], K_m: 0.7 mM [4, 8], divalent cation required, either Mg^{2+} or Mn^{2+}, Mn^{2+} stimulates the enzyme to a greater extent than Mg^{2+} at low concentrations, Mg^{2+} stimulates to a greater extent than Mn^{2+} at high concentration, but each is maximally effective at a concentration around 8 mM [1]) [1, 4, 6, 8]; Mn^{2+} (divalent cation required, maximum stimulation in presence of Ca^{2+}. Co^{2+}, Mg^{2+} or Mn^{2+} can replace Ca^{2+} [6], divalent cation required: Mg^{2+} > Mn^{2+} >> Ca^{2+} > Co^{2+} > Zn^{2+} [4], divalent cation required, either Mg^{2+} or Mn^{2+}, Mn^{2+} stimulates the enzyme to a greater extent than Mg^{2+} at low concentrations, Mg^{2+} stimulates to a greater extent than Mn^{2+} at high concentration, but each is maximally effective at a concentration around 8 mM [1], $MnCl_2$ stimulates at 1 mM, inhibits at higher concentration [7]) [1, 4, 6, 7]

Turnover number (min^{-1})

Specific activity (U/mg)

More [5, 7]

K_m-value (mM)

0.00015 (UDPglucose, pH 7.2) [6]; 0.00033 (UDPglucose, pH 5.3) [6]; 0.002 (dolichyl phosphate) [7]; 0.0045 (dolichyl phosphate) [1]; 0.0091 (UDPglucose) [1]; 0.027 (UDPglucose) [7]; More [4, 8]

pH-optimum
6.0–7.0 [7]; 6.5 [4]; 7.0 [1]; 7.5 [8]

pH-range
5.6–8.8 (5.6: 25% of activity maximum, 8.8: about 40% of activity maximum)
[1]

Temperature optimum (°C)
30 [1]; 37 (assay at) [7]

Temperature range (°C)

---

3 ENZYME STRUCTURE

Molecular weight

Subunits
? (x x 36000, human, SDS-PAGE [5], x x 39000, Phaseolus aureus,
SDS-PAGE after photoaffinity labeling, catalytic subunit [7]) [5, 7]

Glycoprotein/Lipoprotein
–

---

4 ISOLATION/PREPARATION

Source organism
Acanthamoeba castellani [1]; Bovine (calf) [2]; Pig [3]; Zea mays (L. inbred
A 636) [4, 8]; Human [5, 6]; Phaseolus aureus (mung bean) [7]

Source tissue
Cysts [1]; Endosperm (culture [4]) [4, 8]; Pancreas [2]; Liver [3, 5, 6]

Localization in source
Soluble [1]; Microsomes (membrane [4]) [2–8]

Purification
Human [5]; Phaseolus aureus (partial) [7]; Zea mays (partial) [8]

Crystallization
–

Cloned

Renatured
–

5 STABILITY

pH

Temperature (°C)
30 (2 h, 50% loss of activity) [1]; 52 (10 min, complete loss of activity) [1]

Oxidation

Organic solvent

General stability information
Uridine or UDPglucuronic acid, 5 mM, protects solubilized enzyme against rapid inactivation [5]

Storage
−20°C, indefinitely stable below [1]; 4°C, stable for several days [1]; −70°C, up to 4 weeks [8]

6 CROSSREFERENCES TO STRUCTURE DATABANKS

PIR/MIPS code
PIR2:S48136 (yeast (Saccharomyces cerevisiae))

Brookhaven code

7 LITERATURE REFERENCES

[1] Villemez, C.L., Carlo, P.L.: J. Biol. Chem.,254,4814–4819 (1979)
[2] Herscovics, A., Bugge, B., Jeanloz, R.W.: J. Biol. Chem.,252,2271–2277 (1977)
[3] Behrens, N.H., Leloir, L.F.: Proc. Natl. Acad. Sci. USA,66,153–159 (1970)
[4] Riedell, W.E., Miernyk, J.A.: Plant Physiol.,87,420–426 (1988)
[5] Matern, H., Bolz, R., Matern, S.: Eur. J. Biochem.,190,99–105 (1990)
[6] Matern, H., Matern, S.: Biochim. Biophys. Acta,1004,67–72 (1989)
[7] Drake, R.R., Kaushal, G.P., Pastuszak, I., Elbein, A.D.: Plant Physiol.,97,396–401 (1991)
[8] Miernyk, J.A., Riedell, W.E.: Phytochemistry,30,2865–2867 (1991)

1 NOMENCLATURE

EC number
2.4.1.118

Systematic name
UDPglucose:zeatin 7-glucosyltransferase

Recommended name
Cytokinin 7-beta-glucosyltransferase

Synonyms
Glucosyltransferase, uridine diphosphoglucose-zeatin 7-
Cytokinin 7-glucosyltransferase [3]

CAS Reg. No.
72103-03-8

---

2 REACTION AND SPECIFICITY

Catalysed reaction
UDPglucose + N^6-alkylaminopurine →
→ UDP + N^6-alkylaminopurine-7-beta-D-glucoside

Reaction type
Hexosyl group transfer

Natural substrates
UDPglucose + N^6-alkylaminopurine [1]

Substrate spectrum
1 UDPglucose + N^6-alkylaminopurine (acts on a range of N^6-substituted adenines, including zeatin (i.e. N^6-(4-hydroxy-3-methylbut-trans-2-enylamino)purine) and N^6-benzylaminopurine [1, 2], 2 separate enzymes for the formation of the 7- and 9-glucopyranosyl derivatives of cytokinin 6-benzylaminopurine [1], with some acceptors 9-beta-D-glucosides are also formed [2], specificity overview: adenine derivatives with an alkyl side chain at least three carbon atoms in length at position N^6 are preferentially glucosylated [2]) [1, 2]

Product spectrum
1 UDP + N^6-alkylaminopurine-7-beta-D-glucoside [1]

Inhibitor(s)
6-Benzylamino-2-(2-hydroxyethylamino)-7-methylpurine [3]; 6-Benzylamino-2-(2-hydroxyethylamino)-9-methylpurine [3]; 6-Benzylamino-9-(3-hydroxyprolyl)purine (weak) [3]; 2-Chloro-6-(5-hydroxypentylamino)-7-methylpurine (weak) [3]; 2-Chloro-6-(5-hydroxypentylamino)-9-methyl-purine [3]; 6-(5-Hydroxypentylamino)-9-methylpurine (weak) [3]; 3-Methyl-7-pentylaminopyrazolo[4,3-d]pyrimidine [3]; 4-Furfurylaminopyra-zolo[3,4-d]pyrimidine (weak) [3]; 4-Cyclopentylamino-2-methylthiopyr-rolo[2,3-d]pyrimidine [3]; Theophylline (weak) [3]; N-(3-Chlorophenyl)-N'-phenylurea [3]; N-Benzyl-N'-phenylurea [3]

Cofactor(s)/prosthetic group(s)/activating agents

Metal compounds/salts
More (divalent metal ions, Mg^{2+} or Mn^{2+}, have negligible influence on the re-action in the range from 1–10 mM) [2]

Turnover number (min^{-1})

Specific activity (U/mg)
More [2]

K_m-value (mM)
0.15 (zeatin) [2]; 0.19 (UDPglucose) [2]

pH-optimum
7.3–7.4 [2]; 7.5–7.6 (assay at) [1]

pH-range

Temperature optimum (°C)
25 (assay at) [1]; 35 (assay at) [2]

Temperature range (°C)

3 ENZYME STRUCTURE

Molecular weight
46500 (Raphanus sativus, gel filtration) [2]

Subunits

Glycoprotein/Lipoprotein
–

4 ISOLATION/PREPARATION

Source organism
Raphanus sativus (radish) [1–3]

2

Source tissue
Cotyledons (expanded [1]) [1–3]

Localization in source
Soluble [1]

Purification
Raphanus sativus (partial) [1, 2]

Crystallization
–

Cloned
–

Renatured
–

---

5 STABILITY

pH

Temperature (°C)

Oxidation

Organic solvent

General stability information
Highest stability in potassium phosphate buffer, lowest in Tris-chloride [2]

Storage

---

6 CROSSREFERENCES TO STRUCTURE DATABANKS

PIR/MIPS code

Brookhaven code

---

7 LITERATURE REFERENCES

[1] Entsch, B., Letham, D.S.: Plant Sci. Lett.,14,205–212 (1979)
[2] Entsch, B., Parker, C.W., Letham, D.S., Summons, R.E.: Biochim. Biophys. Acta,570, 124–139 (1979)
[3] Parker, C.W., Entsch, B., Letham, D.S.: Phytochemistry,25,303–310 (1986)

1 NOMENCLATURE

EC number
2.4.1.119

Systematic name
Dolichyl-diphosphooligosaccharide:protein-L-asparagine oligopolysaccharidotransferase

Recommended name
Dolichyl-diphosphooligosaccharide-protein glycotransferase

Synonyms
Glycosyltransferase, dolichyldiphosphooligosaccharide-protein
Asparagine N-glycosyltransferase
Dolichyldiphosphooligosaccharide-protein oligosaccharyltransferase
Glycosyltransferase, dolichylpyrophosphodiacetylchitobiose-protein
Oligomannosyltransferase
Oligosaccharide transferase
Oligosaccharyltransferase, dolichyldiphosphoryloligosaccharide-protein

CAS Reg. No.
75302-32-8

---

2 REACTION AND SPECIFICITY

Catalysed reaction
Dolichyl diphosphooligosaccharide + protein L-asparagine →
→ dolichyl diphosphate + a glycoprotein with the oligosaccharide chain
attached by N-glycosyl-linkage to protein L-asparagine

Reaction type
Hexosyl group transfer

Natural substrates
Lipid-linked oligosaccharide + unfolded nascent polypeptide chain (central
enzyme in synthesis of N-linked glycoproteins) [7, 8]

Substrate spectrum
1 Dolichyl diphosphooligosaccharide + protein L-asparagine (glycosyl donors for trypanosomatid protozoa enzyme: dolichyl diphospho-chitobiose-(mannosyl)$_9$-(glucosyl)$_{1-3}$ and dolichyl diphospho-chitobiose-(mannosyl)$_{7-9}$ [5], glycosyl donor specificity [12], peptide acceptor specificity [2, 4, 9–11], catalyzes reaction between beta-amido nitrogen of Asn-residue and oligosaccharyl-diphosphodolichol [2], Asn-Xaa-Thr/Ser is a necessary and sufficient prerequisite for N-glycosylation [4, 10]) [1–5, 9–12]

---

   2 Dolichyl diphosphooligosaccharide + synthetic tripeptides (e.g. Asn-Tyr-Thr [8], Asn-Leu-Thr and its N-terminal acetyl-, benzoyl-, octanoyl- or t-butoxycarbonyl-derivatives [10], peptide hydrophobicity increases its acceptor activity [10], no substrates are Asn-Leu-Thr-derivatives containing asparagine modifications or substitution [10]) [3, 8–10]

   3 Dolichyl diphosphate-di-N-acetylchitobiose + synthetic hexapeptides [4]

   4 Dolichyl diphosphooligosaccharide + Tyr-Asn-Leu-Thr-Ser-Val [4]

   5 Dolichyl diphosphochitobiose-(mannosyl)$_9$-(glucosyl)$_3$ + synthetic hexapeptide (e.g. Tyr-Asn-Leu-Thr-Ser-Val [12], rat liver, Saccharomyces [5], glycosyl-donors are also dolichyl diphosphochitobiose or dolichyl diphospho-chitobiose-mannose, no substrates are dolichyl diphosphochitobiose-(mannosyl)$_9$ or dolichyl diphospho-N-acetyl-D-glucosamine [12]) [5, 12]

   6 Dolichyl-diphosphochitobiose + Arg-Asn-Gly-Thr-Ala-Val-methylester [6]

Product spectrum

   1 Dolichyl diphosphate + a glycoprotein with the oligosaccharide chain attached by N-glycosyl-linkage to protein L-asparagine [1–5, 9–12]

   2 ?

   3 ?

   4 ?

   5 ?

   6 ?

Inhibitor(s)

EDTA [1, 10]; Mg^{2+} (above 3 mM [1], not [10]) [1]; Ca^{2+} [10]; Zn^{2+} [10]; Cu^{2+} [10]; Octylglucoside [12]; Arg-Asn-Gly-epoxyethylglycine-Ala-Val-methylester (acceptor peptide Arg-Asn-Gly-Thr-Ala-Val-methylester protects partially) [6]; N-Acyl-derivatives of Asn-Leu-Thr (benzoyl- and octanoyl-derivatives most effective, alternative substrates, dose-dependent inhibition) [10]; More (no inhibitors: peptides containing Asn-modifications or substitutions, DTT, 5% v/v DMSO [10], Nonidet P-40 (up to 0.1% v/v [10]) [10, 12]) [10, 12]

Cofactor(s)/prosthetic group(s)/activating agents

More (no activation by detergents) [10]

Metal compounds/salts

Mn^{2+} (requirement [1, 7, 10, 12], 3 mM [1], 1 mM [12]) [1, 7, 10, 12]; Mg^{2+} (activation [1, 12], 10 mM [12], 30% as effective as Mn^{2+} [1, 12], inhibition above 3 mM [1]) [1, 12]; More (no activation by Ca^{2+} or K$^+$) [12]

Turnover number (min^{-1})

0.22 (tripeptide) [8]

Specific activity (U/mg)

More [12]; 0.000985 [8]

K$_m$-value (mM)
More (kinetic study with intact microsomal membrane) [10]; 0.0005 (dolichyl diphosphochitobiose-(mannosyl)$_9$-(glucosyl)$_3$) [12]; 0.0012 (dolichyl diphosphochitobiose) [12]; 0.05 (Tyr-Asn-Leu-Thr-Ser-Val (+ dolichyl diphosphochitobiose-(mannosyl)$_9$-(glucosyl)$_3$)) [12]; 0.08 (Tyr-Gln-Ser-Asn-Ser-Thr-Met, acetyl-Asn-Ala-Thr, membrane-bound enzyme) [11]; 0.127 (Tyr-Gln-Ser-Asn-Ser-Thr-Met, solubilized enzyme) [11]; 0.143 (acetyl-Asn-Ala-Thr, solubilized enzyme) [11]; 0.25 (Asn-Leu-Thr) [9]; 0.278 (acetyl-Asn-Lys-Thr) [11]; 0.29 (Tyr-Asn-Leu-Thr-Ser-Val) [4]; 0.3–0.358 (Ala-Leu-Gln-Asn-Ala-Thr-Arg) [11]; 0.56 (Asn-Ala-Thr, solubilized enzyme) [11]; 0.6 (Tyr-Asn-Leu-Thr-Ser-Val (+ dolichyl diphosphochitobiose)) [12]; 2.09 (Asn-Ala-Thr, membrane-bound enzyme) [11]; 3.3 (Asn-Asp-Thr) [9]

pH-optimum
6.5–7.7 [12]; 7.0–7.5 [1]

pH-range
6–8 (30% [1], 50% [12] of maximal activity at pH 6 and 20% [1], 50% [12] of maximal activity at pH 8) [1, 12]

Temperature optimum (°C)
20 (assay at) [6]; 25 (assay at) [8]; 30 (assay at) [10, 11]

Temperature range (°C)

---

3 ENZYME STRUCTURE

Molecular weight

Subunits
Oligomer (x × 66000 + x × 63000 + x × 48000, dog, SDS-PAGE) [8]
? (x × 60000, chicken, SDS-PAGE [2, 7], x × 80000, Saccharomyces cerevisiae, SDS-PAGE [2]) [2, 7]
More (oligosaccharyltransferase activity is mediated by a protein complex composed of ribophorins I and II and a MW 48000 protein) [8]

Glycoprotein/Lipoprotein
Glycoprotein (N-linked) [7]

---

4 ISOLATION/PREPARATION

Source organism
Chicken (hen) [1, 2, 7, 10, 11]; Mouse [2]; Rat [2, 4, 5]; Dog [2, 8]; Rabbit [2]; Pig [9]; Bovine (calf) [6]; Trypanosoma cruzi [5]; Leptomonas samueli [5]; Crithidia fasciculata [5]; Blastocrithidia culicis [5]; [5]; Drosophila melanogaster [2]; Saccharomyces cerevisiae (yeast, strain X-2180–1B [5], wild-type strain X-2180–1A [12]) [2, 3, 5, 12]; More (no activity in radish root and E. coli) [2]

Source tissue
   Oviduct [1, 2, 7, 10]; Thyroidea [11]; Pancreas (dog) [2, 8]; Kidney (hen) [2];
   Intestine (hen) [2]; Liver (mouse [2]) [2, 4–6, 9]; Brain (hen) [2]; Hepatoma
   cell line FAZA (rat) [2]; Cervix (rabbit) [2]; Cell (yeast [2], epimastigotes (Trypanosoma cruzi) [5]) [2, 3, 5, 12]

Localization in source
   Rough endoplasmic reticulum (integral membrane protein, tightly associated to luminal side of membrane [2], structure [2]) [1, 2, 7, 8, 11]; Microsomes (luminal face of membrane [10]) [3, 4, 6, 7, 10, 11]; Membrane-
   bound [3, 5, 6, 12]

Purification
   Chicken (slight, unstable to purification, solubilized with Nonidet P-40 [1],
   sodium deoxycholate [11], photoaffinity labeling technique [2]) [1, 2, 11];
   Dog (partial, solubilized with digitonin) [8]; Saccharomyces cerevisiae (partial, solubilized with 0.5% Nonidet P-40 at protein/detergent ratio of 2) [12]

Crystallization
   –

Cloned
   –

Renatured
   –

---

5 STABILITY

pH

Temperature (°C)
   50 (stable below) [7]; 60 (inactivation) [7]

Oxidation

Organic solvent

General stability information
   Solubilization with detergents inactivates [7]; Phosphatidylcholine stabilizes
   detergent-solubilized enzyme [8]; Glycerol, 25%, stabilizes [12]

Storage
   –85°C, membrane preparation, 30% glycerol, several months [9]; –70°C,
   Nonidet-solubilized crude enzyme, 25% glycerol, 0.01% 2-mercaptoethanol,
   retains 90% of original activity after 1 month and 55% after 5 months [12];
   –20°C, Nonidet-solubilized crude enzyme, 25% glycerol, 0.01% 2-mercapto-
   ethanol, 18% loss of activity after 1 month and 87% after 5 months [12]; 4°C,
   less than 20% loss of activity within 1 week [1]

6 CROSSREFERENCES TO STRUCTURE DATABANKS

PIR/MIPS code

Brookhaven code

7 LITERATURE REFERENCES

[1] Das, R.C., Heath, E.C.: Proc. Natl. Acad. Sci. USA,77,3811–3815 (1980)
[2] Kaplan, H.A., Welply, J.K., Lennarz, W.J.: Biochim. Biophys. Acta,906,161–173 (1987) (Review)
[3] Lee, J., Coward, J.K.: Biochemistry,32,6794–6801 (1993)
[4] Bause, E.: FEBS Lett.,103,296–299 (1979)
[5] Bosch, M., Trombetta, S., Engstrom, U., Parodi, A.J.: J. Biol. Chem.,263, 17360–17365 (1988)
[6] Bause, E.: Biochem. J.,209,323–330 (1983)
[7] Welply, J.K., Shenbagamurthi, P., Naider, F., Park, H.R., Lennarz, W.J.: J. Biol. Chem., 260,6459–6465 (1985)
[8] Kelleher, D.J., Kreibich, G., Gilmore, R.: Cell,69,55–65 (1992)
[9] Imperiali, B., Shannon, K.L.: Biochemistry,30,4374–4380 (1991)
[10] Welply, J.K., Shenbagamurthi, P., Lennarz, W.J., Naider, F.: J. Biol. Chem.,258, 11856–11863 (1983)
[11] Ronin, C., Granier, C., Caseti, C., Bouchilloux, S., Van Rietschoten, J.: Eur. J. Biochem.,118,159–164 (1981)
[12] Sharma, C.B., Lehle, L., Tanner, W.: Eur. J. Biochem.,116,101–108 (1981)

1 NOMENCLATURE

EC number
2.4.1.120

Systematic name
UDPglucose:sinapate D-glucosyltransferase

Recommended name
Sinapate 1-glucosyltransferase

Synonyms
Glucosyltransferase, uridine diphosphoglucose-sinapate
UDPglucose:sinapic acid glucosyltransferase [2]
Uridine 5'-diphosphoglucose-hydroxycinnamic acid acylglucosyltransferase
More (cf. EC 2.4.1.126)

CAS Reg. No.
74082-53-4

---

2 REACTION AND SPECIFICITY

Catalysed reaction
UDPglucose + sinapate →
→ UDP + 1-sinapoyl-D-glucose

Reaction type
Hexosyl group transfer

Natural substrates
UDPglucose + sinapic acid (detoxification of sinapic acid and activation of sinapic acid for the subsequent sinapoyltransferase reaction leading to sin-apoylmalate) [1, 2]

Substrate spectrum
1 UDPglucose + sinapic acid (r) [1, 2]
2 UDPglucose + p-coumaric acid (40% of activity compared to sinapic acid) [1]
3 UDPglucose + ferulic acid (38% of activity compared to sinapic acid) [1]
4 UDPglucose + caffeic acid (38% of activity compared to sinapic acid) [1]
5 UDPglucose + benzoic acid (39% of activity compared to sinapic acid) [1]
6 UDPglucose + vanillic acid (36% of activity compared to sinapic acid) [1]
7 UDPglucose + syringic acid (35% of activity compared to sinapic acid) [1]
8 UDPglucose + anthranilic acid (31% of activity compared to sinapic acid) [1]

9 UDPglucose + 4-hydroxybenzoic acid (31% of activity compared to sinapic acid) [1]
10 UDPglucose + 3,5-dihydroxybenzoic acid (5% of activity compared to
   sinapic acid) [1]
11 UDPglucose + cinnamic acid (16% of activity compared to sinapic acid)
   [1]
12 TDPglucose + sinapic acid [2]
13 ?

Product spectrum
  1 UDP + 1-sinapoylglucose [1, 2]
  2 UDP + p-coumaroylglucose [1]
  3 UDP + feruloylglucose [1]
  4 UDP + caffeoylglucose [1]
  5 ?
  6 ?
  7 ?
  8 ?
  9 ?
  10 ?
  11 ?
  12 ?
  13 More (only glucose esters are formed, no glucosides) [1]

Inhibitor(s)
  PCMB (1 mM: 100% inhibition, activity can be restored up to 50% by addition of 10 mM dithiothreitol, 0.1 mM: 24% inhibition, activity can be completely restored by 10 mM dithiothreitol) [1]; UDP (strong) [1]

Cofactor(s)/prosthetic group(s)/activating agents
  Dithiothreitol (10 mM: 3-fold stimulation) [1]; 2-Mercaptoethanol (stimulation)
  [1]; More (SH-group required) [1, 2]

Metal compounds/salts
  More (no requirement for Mg^{2+} or Ca^{2+}) [1]

Turnover number (min^{-1})

Specific activity (U/mg)
  0.0094 (sinapic acid) [2]

K_m-value (mM)
  0.03 (sinapic acid) [1, 2]; 0.055 (UDPglucose) [1, 2]

pH-optimum
  5.8–6.0 (depending on buffer) [2]; 7.0 [1]

pH-range
    5–7 (85% of maximal activity at pH 5, 63% of maximal activity at pH 7) [2]

Temperature optimum (°C)
    30 (assay at) [1, 2]; 40 [1]

Temperature range (°C)
    25–45 (50% of maximal activity at 25°C, 80% of maximal activity at 45°C) [1]

---

3 ENZYME STRUCTURE

Molecular weight

Subunits

Glycoprotein/Lipoprotein
    –

---

4 ISOLATION/PREPARATION

Source organism
    Rhaphanus sativus [1, 2]

Source tissue
    Seedling (highest content in cotyledons) [1, 2]

Localization in source

Purification
    Rhaphanus sativus (partial) [1, 2]

Crystallization
    –

Cloned
    –

Renatured
    –

---

5 STABILITY

pH

Temperature (°C)

Oxidation

Organic solvent

General stability information
   Enzyme is less stable at –20°C than at 4°C [2]

Storage
   4°C, 24 h [2]; 4°C, 8 days, 23% loss of activity [2]

6 CROSSREFERENCES TO STRUCTURE DATABANKS

PIR/MIPS code

Brookhaven code

7 LITERATURE REFERENCES

[1] Strack, D.: Z. Naturforsch.,35c,204–208 (1979)
[2] Nurmann, G., Strack, D.: Z. Pflanzenphysiol.,102,11–17 (1981)

1 NOMENCLATURE

EC number
2.4.1.121

Systematic name
UDPglucose:indole-3-acetate beta-D-glucosyltransferase

Recommended name
Indole-3-acetate beta-glucosyltransferase

Synonyms
Glucosyltransferase, uridine diphosphoglucose-indoleacetate
UDPG-indol-3-ylacetyl glucosyl transferase [3]
UDP-glucose:indol-3-ylacetate glucosyltransferase [1]
Indol-3-ylacetylglucose synthase [1]
UDP-glucose:indol-3-ylacetate glucosyl-transferase [2]
IAGlu synthase [2]
IAA-glucose synthase [3]

CAS Reg. No.
74082-56-7

---

2 REACTION AND SPECIFICITY

Catalysed reaction
UDPglucose + indole-3-acetate →
→ UDP + indole-3-acetyl-beta-1-D-glucose

Reaction type
Hexosyl group transfer

Natural substrates
UDPglucose + indole-3-acetate (first reaction in a series leading to the ester
conjugates of indole-3-acetic acid [2, 5], first enzyme-catalyzed reaction
leading from indole-3-acetic acid to the myo-inositol ester of indole-3-acetate
[3]) [2, 3, 5]

Substrate spectrum
1 UDPglucose + indole-3-acetate (i.e. indol-3-ylacetic acid [1], r [4], equilibrium away from ester formation and towards formation of indole-3-acetate
[4], specific for UDPglucose [1, 4]) [1–5]
2 UDPglucose + naphthalene-1-acetic acid [1]
3 More (not: 2,4-dichlorophenoxyacetic acid [1], oxindole-3-acetic acid [4],
7-hydroxyoxindole-3-acetic acid [4], low activity with phenylpropene
acids, such as p-coumaric acid [4]) [1, 4]

Product spectrum
  1 UDP + indole-3-acetyl-beta-1-D-glucose (i.e. 1-O-indol-3-
     ylacetyl-beta-D-glucose [1]) [1–5]
  2 UDP + ? [1]
  3 ?

Inhibitor(s)
  Phosphate buffer (100 mM, 50% inhibition) [3]; Inorganic diphosphate [4];
  Phosphatidyl ethanolamine [4]; Zeatin [4]; 2,4-Dichlorophenoxyacetic acid
  [4]; Indole-3-acid-myo-inositol [4]; Indole-3-acetic acid-glucan [4]; Gibberellic acid (weak) [4]; Abscisic acid (weak) [4]; Kinetin (weak) [4]; More (not:
  zeatin riboside) [4]

Cofactor(s)/prosthetic group(s)/activating agents
  Calmodulin (stimulates) [4]; Thiol compounds (stimulate) [4]

Metal compounds/salts
  Ca^{2+} (stimulates) [4]

Turnover number (min^{-1})

Specific activity (U/mg)
  0.798 [3]; 4.029 [5]; More [1]

K_m-value (mM)
  1.08 (indole-3-acetate) [1]; 2.5 (UDPglucose) [1]

pH-optimum
  7.1 [1]; 7.3–7.6 [3]

pH-range

Temperature optimum (°C)
  37 (assay at) [1, 3]

Temperature range (°C)

3 ENZYME STRUCTURE

Molecular weight
  46500 (Zea mays, gel filtration) [3]
  52000 (Zea mays, gel filtration) [4]

Subunits

Glycoprotein/Lipoprotein
  –

4 ISOLATION/PREPARATION

Source organism
Zea mays [1–5]

Source tissue
Kernels (immature [1]) [1–3, 5]; Liquid endosperm [4]; Embryo [4]

Localization in source

Purification
Zea mays (partial) [3, 5]

Crystallization
–

Cloned
–

Renatured
–

5 STABILITY

pH

Temperature (°C)

Oxidation

Organic solvent

General stability information
Loses activity during column chromatographic procedures, reactivated only fractionally by addition of column eluates [3]

Storage
–20°C, stable [3]; 4°C, stable [3]

6 CROSSREFERENCES TO STRUCTURE DATABANKS

PIR/MIPS code
PIR2:A54739 (maize)

Brookhaven code

7 LITERATURE REFERENCES

[1] Michalczuk, L., Bandurski, R.S.: Biochem. J.,207,273–281 (1982)
[2] Kowalczyk, S., Bandurski, R.S.: Plant Physiol.,.94,4–12 (1990)
[3] Leznicki, A.J., Bandurski, R.S.: Plant Physiol.,88,1474–1480 (1988)
[4] Leznicki, A.J., Bandurski, R.S.: Plant Physiol.,88,1481–1485 (1988)
[5] Kowalczyk, S., Bandurski, R.S.: Biochem. J.,279,509–514 (1991)

1 NOMENCLATURE

EC number
2.4.1.122

Systematic name
UDPgalactose:glycoprotein-N-acetyl-D-galactosamine 3-beta-D-galactosyltransferase

Recommended name
Glycoprotein-N-acetylgalactosamine 3-beta-galactosyltransferase

Synonyms
Galactosyltransferase, uridine diphosphogalactose-mucin beta-(1→3)-

CAS Reg. No.
97089-61-7

2 REACTION AND SPECIFICITY

Catalysed reaction
UDPgalactose + glycoprotein-N-acetyl-D-galactosamine →
→ UDP + glycoprotein D-galactosyl-1,3-N-acetyl-D-galactosamine

Reaction type
Hexosyl group transfer

Natural substrates
UDPgalactose + N-acetylgalactosaminide mucin (involved in O-glycan biosynthesis [5], participates in synthesis of the core portion of O-serine- and O-threonine-linked oligosaccharides in respiratory acidic mucins [1]) [1, 5]
UDPgalactose + glycophorin [2]

Substrate spectrum
1 UDPgalactose + N-acetylgalactosaminide mucin (e.g. asialo ovine submaxillary mucin [2, 4], asialo-agalacto-glycophorin [2], most active acceptor: macromolecular mucin glycopeptides with free N-acetylgalactosyl residues linked to polypeptide chain [1], high specificity for N-acetylgalactosaminyl residues linked to serine or threonine [1], glycosyl acceptor specificity study [3, 5], poor substrates are low-molecular weight acceptors, e.g. N-acetylgalactosamine [1, 4], N-acetylglucosamine [4], N-acetylgalactosaminitol, p-nitrophenyl-beta-N-acetylgalactosamide, glycopeptides with less than 5 amino acids containing terminal N-acetylgalactosaminyl residues [1], asialo-agalacto-(alpha$_1$-acid glycoprotein), ganglioside GM2 [4], ovomucoid, ovalbumin, fetuin devoid of sialic acid and galactose [1]) [1–5]

Product spectrum
    1 UDP + galactosyl-beta-1,3-N-acetylgalactosaminide mucin [1]

Inhibitor(s)

Cofactor(s)/prosthetic group(s)/activating agents
    Triton X-100 (activation and stabilization, unstable without) [1]

Metal compounds/salts
    Mn^{2+} (activation, 5–10 mM) [4]

Turnover number (min^{-1})

Specific activity (U/mg)
    More [4]; 0.81 [1]

K_m-value (mM)
    0.00035 (asialo Cowper's gland mucin) [1]; 0.02 (UDPgalactose) [1]; 0.025
    (UDPgalactose) [4]; 0.09 (asialo Cowper's gland mucin, calculated on the
    basis of terminal N-galactosaminyl residues) [1]; 5 (asialo ovine submaxil-
    lary mucin, in terms of N-acetylgalatosamine equivalents) [4]

pH-optimum
    More (pI: 6.40) [4]; 5–7 (broad) [4]; 6.9 (broad) [1]

pH-range

Temperature optimum (°C)
    37 (assay at) [1, 2, 4, 5]

Temperature range (°C)

3 ENZYME STRUCTURE

Molecular weight
    90000 (pig, gel filtration in the presence of Triton X-100) [1]

Subunits
    ? (x × 68000, chicken, SDS-PAGE) [4]
    Monomer (1 × 82000, pig, SDS/2-mercaptoethanol-PAGE) [1]

Glycoprotein/Lipoprotein
    —

4 ISOLATION/PREPARATION

Source organism
  Chicken [4]; Human [2]; Pig [1, 5]; Rat [3, 5]

Source tissue
  Colonic mucosa [5]; Embryo [4]; Erythrocytes [2]; Liver (rat [5]) [3–5]; Stomach (pig) [5]; Trachea mucosa [1]

Localization in source
  Membrane-bound (inner surface [1]) [1, 2, 4, 5]; Endoplasmic reticulum [1]; Golgi apparatus [1]; Microsomes [1, 4]; More (subcellular distribution) [1]

Purification
  Chicken (solubilized with 1% Triton X-100) [4]; Human (partial, solubilized with 3% Triton X-100) [2]; Pig (affinity chromatography on Sepharose 4B containing covalently bound mucin substrate) [1]; Rat (partial) [3]

Crystallization
  –

Cloned
  –

Renatured
  –

5 STABILITY

pH

Temperature (°C)

Oxidation

Organic solvent

General stability information
  Triton X-100 or Nonidet P-40, unstable without [1]

Storage
  3°C, at least 3 months, 0.1% Triton X-100 and 0.1 mg/ml albumin [1]

6 CROSSREFERENCES TO STRUCTURE DATABANKS

PIR/MIPS code

Brookhaven code

7 LITERATURE REFERENCES

[1] Mendicino, J., Sivakami, S., Davila, M., Chandrasekaran, E.V.: J. Biol. Chem.,257, 3987–3994 (1982)
[2] Hesford, F.J., Berger, E.G., Van den Eijnden, D.H.: Biochim. Biophys. Acta,659, 302–311 (1981)
[3] Brockhausen, I., Möller, G., Pollex-Kruger, A., Rutz, V., Paulsen, H., Matta, K.L.: Biochem. Cell Biol.,70,99–108 (1992)
[4] Furukawa, K., Roth, S.: Biochem. J.,227,573–582 (1985)
[5] Brockhausen, I., Möller, G., Merz, G., Adermann, K., Paulsen, H.: Biochemistry,29, 10206–10212 (1990)